Artificial Intelligence in Bioinformatics and Chemoinformatics

The authors aim to shed light on the practicality of using machine learning in finding complex chemoinformatics and bioinformatics applications as well as identifying AI in biological and chemical data points. The chapters are designed in such a way that they highlight the important role of AI in chemistry and bioinformatics particularly for the classification of diseases, selection of features and compounds, dimensionality reduction, and more. In addition, they assist in the organization and optimal use of data points generated from experiments performed using AI techniques. This volume discusses the development of automated tools and techniques to aid in research plans.

Features

- Covers AI applications in bioinformatics and chemoinformatics

- Demystifies the involvement of AI in generating biological and chemical data

- Provides an introduction to basic and advanced chemoinformatics computational tools

- Presents a chemical biology-based toolset for AI usage in drug design

- Discusses computational methods in cancer, genome mapping, and stem cell research

Artificial Intelligence in Bioinformatics and Chemoinformatics

Edited by

Yashwant V. Pathak
USF Health Taneja College of Pharmacy, University of South Florida,
USA, and Adjunct Professor, Faculty of Pharmacy,
Airlangga University, Surabaya, Indonesia

Surovi Saikia
Translation Research Laboratory, Department of Biotechnology,
Bharathiar University, Coimbatore, India

Sarvadaman Pathak
Chief Medical Officer, HIPAA Compliance Officer,
Universiti LLC–Tampa, USA

Jayvadan Patel
Formulation Scientist, Aavis Pharmaceuticals, USA and Professor Emeritus,
Faculty of Pharmacy, Sankalchand Patel University, India

Bhupendra Gopalbhai Prajapati
Professor of Pharmaceutics, Shree S.K. Patel College of Pharmaceutical
Education and Research, Ganpat University, Gujarat, India

CRC Press
Taylor & Francis Group
Boca Raton London New York

CRC Press is an imprint of the
Taylor & Francis Group, an **informa** business

First edition published 2024
by CRC Press
2385 NW Executive Center Drive, Suite 320, Boca Raton FL 33431

and by CRC Press
4 Park Square, Milton Park, Abingdon, Oxon, OX14 4RN

Library of Congress Cataloging-in-Publication Data
Names: Pathak, Yashwant V., editor. | Saikia, Surovi, editor. | Pathak, Sarvadaman, editor. |
Patel, Jayvadan, editor. | Prajapati, Bhupendra Gopalbhai, editor.
Title: Artificial intelligence in bioinformatics and chemoinformatics / edited by Yashwant V. Pathak (USF Health Taneja College of Pharmacy, University of South Florida, USA, and Adjunct Professor, Faculty of Pharmacy, Airlangga University, Surabaya, Indonesia), Surovi Saikia (Translation Research Laboratory, Department of Biotechnology, Bharathiar University, Coimbatore, India), Sarvadaman Pathak (Chief Medical Officer, HIPAA compliance officer, Universiti LLC, Tampa, USA), Jayvadan Patel (Formulation Scientist, Aavis Pharmaceuticals, USA and Professor Emeritus, Faculty of Pharmacy, Sankalchand Patel University, India), Bhupendra Gopalbhai Prajapati (Professor of Pharmaceutics, Shree S.K. Patel College of Pharmaceutical Education and Research, Ganpat University, Gujarat, India).
Description: First edition. | Boca Raton : CRC Press, 2024. | Includes bibliographical references and index. |
Summary: "The authors aim to shed light on the practicality of using machine learning in finding complex chemoinformatics and bioinformatics applications as well as identifying AI in biological and chemical data points. The chapters are designed in such a way that they highlight the important role of AI in chemistry and bioinformatics particularly for the classification of diseases, selection of features and compounds, dimensionality reduction and more. In addition, they assist in the organization and optimal use of data points generated from experiments performed using AI techniques. This volume discusses the development of automated tools and techniques to aid in research plans"– Provided by publisher.
Identifiers: LCCN 2023011979 (print) | LCCN 2023011980 (ebook) |
ISBN 9781032396576 (hardback) | ISBN 9781032405834 (paperback) | ISBN 9781003353768 (ebook) |
Subjects: LCSH: Artificial intelligence–Biological applications. | Cheminformatics. |
Bioinformatics. | Artificial intelligence. |
Classification: LCC QH324.25.A783 2024 (print) |
LCC QH324.25 (ebook) | DDC 570.285–dc23/eng/20230630
LC record available at https://lccn.loc.gov/2023011979
LC ebook record available at https://lccn.loc.gov/2023011980

ISBN: 978-1-032-39657-6 (HB)
ISBN: 978-1-032-40583-4 (PB)
ISBN: 978-1-003-35376-8 (EB)

DOI: 10.1201/9781003353768

Typeset in Times
by Newgen Publishing UK

Contents

Preface

The authors aim to shed light on the practicality of using machine learning in finding complex chemoinformatics and bioinformatics applications as well as identifying AI in biological and chemical data points. The chapters are designed in such a way that they highlight the important role of AI in chemistry and bioinformatics particularly for the classification of diseases, selection of features and compounds, dimensionality reduction, and more. In addition, they assist in the organization and optimal use of data points generated from experiments performed using AI techniques. This volume discusses the development of automated tools and techniques to aid in research plans.

Editors

Dr. Yashwant V. Pathak is Adjunct Professor at the Faculty of Pharmacy, Airlangga University, Surabaya, Indonesia. Dr. Pathak has over 9 years of versatile administrative experience in an academic setting as Dean (and over 17 years as faculty and as a researcher in higher education after his PhD) and has worked in student admissions, academic affairs, research, graduate programs, as Chair of the International Working group for USF Health (consisting of USF Medical, Nursing, Public Health, Pharmacy and Physiotherapy colleges), as Director of Nano Center, and now holds the position for Associate Dean for Faculty Affairs. Dr. Pathak is an internationally recognized scholar, researcher, and educator in the areas of healthcare education, nanotechnology, drug delivery systems, and nutraceuticals. Dr. Pathak has edited several books in the field of nanotechnology (Springer) and drug delivery systems (CRC Press), antibody-mediated drug delivery systems (John Wiley & Sons), and in the field of nutraceuticals (Taylor & Francis), artificial neural networks (Elsevier), and conflict resolution and cultural studies, with over 300 research publications, reviews, chapters, abstracts, and articles in educational research. He has authored many popular articles published in newspapers and magazines. He fluently speaks many languages and was trained and taught in English all through his career.

Dr. Surovi Saikia works as a Dr. DS Kothari Postdoctoral Fellow at the Department of Biotechnology, Bharathiar University, Coimbatore. She completed her PhD (Bioinformatics) from Dibrugarh University, Assam in 2020 and UG and PG from Tinsukia College and Dibrugarh University. She has 8 years of research experience with 18 publications in peer-reviewed national and international journals. Currently, she is working on algorithmic-based computational work.

Research Experience

- Dr. D.S Kothari Postdoctoral Fellow at Bharathiar University, Coimbatore, Tamil Nadu from 23/11/2021 to date.
- Senior Research Fellow-NFST Fellowship for PhD at CSIR-NEIST, Natural Product Chemistry Group, Jorhat, Assam from 01/06/2018 to 01/10/2020.
- Junior Research Fellow-NFST Fellowship for PhD at CSIR-NEIST, Natural Product Chemistry Group Jorhat, Assam from 01/06/2016 to 31/05/2018.
- Project Fellow–at CSIR-NEIST, Natural Product Chemistry Group, Jorhat, Assam from 08/04/2013 to 31/05/2016.
- DBT-studentship at CSIR-NEIST, Biotechnology Division from 02/07/2012 to 02/01/2013.

Dr. Sarvadaman Pathak 2016–2017 Global Clinical Scholars Research and Training Program. Harvard Medical School, Boston MA. A master's level program that provides clinicians and clinician-scientists advanced training in the methods and conduct of clinical research. The program also covers data analysis techniques using STATA, and access to datasets from various NIH trials. 2013–2014 Post-Doctoral Fellowship, in Traditional Chinese Medicine with a focus on Integration of Eastern and Western Medicine, Dalian Medical University, Dalian, P.R. China.

2007–2011 Doctor of Medicine (M.D.), Summa Cum Laude, Avalon University
School of Medicine, Bonaire, Netherlands Antilles (Caribbean Island)
2004–2007 Pre-Medicine/Pre-Pharmacy Track, University of Houston, Houston, TX

Professional Experience includes

2021–Present Chief Medical Officer, HIPAA compliance officer
Universiti LLC–A cloud-based healthcare provider focusing on behavioral and mental healthcare
  delivery using AR/VR and AI technology

Dr. Jayvadan Patel is a Formulation Scientist, Aavis Pharmaceuticals, USA, and Professor
Emeritus, Faculty of Pharmacy, Sankalchand Patel University, India. He has more than 26 years of
research, academic, and industry experience, and has published more than 270 research and review
papers in international and national journals. He has co-authored 20 books and contributed 116
book chapters in a book published by well-reputed publishers. His papers have been cited more than
5000 times, with more than 25 papers getting more than 50 citations each. He has a 35 h-index and
121 i10-index for his credit. He has guided 106 M. Pharm students and mentored 46 PhD scholars.
He is already decorated with a dozen awards and prizes, both national and international. He is
the recipient of the very prestigious "AICTE-Visvesvaraya Best Teachers Award-2020" by the All
India Council for Technical Education, Government of India and the APTI-young pharmacy teacher
award (2014) by the Association of Pharmaceutical Teachers of India. He is a reviewer of more than
75 and an editorial board member of 20 reputed scientific Journals. He has completed 12 industry
and government-sponsored research projects.

Dr. Bhupendra Gopalbhai Prajapati is a Professor in the Department of Pharmaceutics, Shree
S.K. Patel College of Pharmaceutical Education and Research, Ganpat University, North Gujarat,
India. He did his PhD at Hemchandracharya North Gujarat University, Patan. He did his PG and
UG from the M.S. University, Baroda. He has 20 years of experience in academics/industry (18+2).
He was awarded the Carrier Award for Young Teacher by AICTE, New Delhi in 2013 (5.3 lakh
rupees research prize). He was also awarded Distinguished Associate Professor in TechNExt India
2017 by CSI, Mumbai and the President Award of Staff Excellence in Research (2019 & 2021)
and Capacity Building (2020 & 2022) by Ganpat University. He claims on his name more than
120 publications. He fetched grants for Research Projects, Staff Development Programs, Seminars,
Conferences and Travel Grants from National and State Government agencies. He has also given his
guidance in industrial consultancy projects conducted at the institute. His one patent granted, five
patents published, and two applications were submitted at the Indian Patent Office in the field of
NDDS. He has delivered more than 50 expert talks and invited scientific sessions in several national
and international conferences and seminars. He is actively working in the field of lipid-based drug
delivery and nanotech formulations. He guided 8 PhD and 48 PG research scholars, while 6 PhD
and 4 PG research scholars are currently working under his guidance in the field of Nanoparticulate
Drug Delivery and Bioavailability Enhancement. In the last three years, seven prizes were secured
by his PG and PhD research scholars at national and international conferences. He is associated
with renowned journals as Section/Topic/Guest editor and reviewer. He is presently the editor of 12
international books. He contributed 50 book chapters and 28 were accepted for publication in the
year 2023. He is an active life member of IPA, APTI, and CRSIC. Recently his reviewer role was
recognized with the MDPI-Pharmaceutics 2022 Outstanding Reviewer Award, with a 500 CHF cash
prize.

Contributors

Aparna A
Translational Research Laboratory, Department of Biotechnology, Bharathiar University, Coimbatore–641 046, Tamil Nadu, India

Mohammed Unais AK
Translational Research Laboratory, Department of Biotechnology, Bharathiar University, Coimbatore–641 046, Tamil Nadu, India

Ashutosh Agarwal
Department of Pharmaceutical Quality Assurance, SSR College of Pharmacy, Sayli-Silvassa Road, Sayli, Silvassa-396 230, Union Territory of Dadra and Nagar Haveli & Daman Diu, India

Jay R. Anand
Department of Pathology and Laboratory Medicine, University of North Carolina, Chapel Hill, North Carolina, USA

Mayank Bapna
Principal, Neotech Institute of Pharmacy, Vadodara, Gujarat, India

Dharmendrasinh Bariya
A R College of Pharmacy and G H Patel Institute of Pharmacy, Vallabh Vudtabagar, Anand, India

Virupaksha A. Bastikar
Center for Computational Biology & Translational Research, Amity Institute of Biotechnology, Amity University Mumbai, Panvel Maharashtra

Alpana V. Bastikar
Computer Aided Drug Development Department, Naven Saxena Research & Technology, Kandla SEZ, Gandhidham, Gujarat

Bhargav Chandegra
School of Pharmacy, National Forensic Sciences University, Gandhinagar, Gujarat, India

Saloni Dalwadi
Anand Pharmacy College, Department of Pharmaceutics, Anand-388001, Gujarat, India

Aarohi Deshpande
MIT School of Bioengineering Sciences and Research, MIT-Art, Design and Technology University, Loni Kalbhor, Pune-412201, India

Rushabh Desarda
Department of Pharmaceutics, School of Pharmacy & Technology Management, SVKM'S NMIMS Deemed-to-be University, Shirpur, Maharashtra 425405, India

B. Duraiswamy
Department of Pharmacognosy, JSS College of Pharmacy, Ooty, Tamil Nadu-643001, India

Bhavinkumar Gayakvad
Graduate School of Pharmacy, Gujarat Technical University, Gandhinagar, Gujarat, 382028, India

Kevinkumar Garala
School of Pharmaceutical Sciences, Atmiya University, Rajkot–360005, Gujarat, India

Aarohi Gherkar
MIT School of Bioengineering Sciences and Research, MIT-Art, Design and Technology University, Loni Kalbhor, Pune-412201, India

Puja Ghosh
Department of Pharmacognosy, JSS College of Pharmacy, Ooty, Tamil Nadu-643001, India

Manas Joshi
Department of Comparative Development and Genetics, Max Planck Institute for Plant Breeding Research, Germany

Dignesh Khunt
Graduate School of Pharmacy, Gujarat Technical University, Gandhinagar, Gujarat, 382028, India

Muhasina KM
Department of Pharmacognosy, JSS College of Pharmacy, Ooty, Tamil Nadu-643001, India

Shreyash Kolhe
MIT School of Bioengineering Sciences and Research, MIT-Art, Design and Technology University, Loni Kalbhor, Pune-412201, India

Shama Mujawar
MIT School of Bioengineering Sciences and Research, MIT-Art, Design and Technology University, Loni Kalbhor, Pune-412201, India

Yashwant V. Pathak
USF Health Taneja College of Pharmacy, University of South Florida, USA and Faculty of Pharmacy, Airlangga University, Indonesia

V.Vijaya Padma
Translational Research Laboratory, Department of Biotechnology, Bharathiar University, Coimbatore–641 046, Tamil Nadu, India

Manishkumar S. Patel
Department of Tumor Biology, H. Lee Moffitt Cancer Center and Research Institute, Tampa, Florida, USA

Dhruv R. Parikh
Department of Pharmaceutical Sciences, Saurashtra University, Rajkot-360005, Gujarat, India

Sumita Bardhan
Shree S.K. Patel College of Pharmaceutical Education and Research, Ganpat University, Mehsana, Gujarat

Dhanabal S. Palaniswamy
Department of Pharmacognosy, JSS College of Pharmacy, Ooty, Tamil Nadu-643001, India

Priya Patel
Department of Pharmaceutical Sciences Saurashtra University, Rajkot–360005, Gujarat, India

Prexita Patel
Anand Pharmacy College, Department of Pharmacology, Anand-388001, Gujarat, India

Unnati Patel
The University of Alabama Huntsville, Alabama, USA

Jigna Prajapati
Acharya Motibhai Patel Institute of Computer Studies, Ganpat University, Gujarat, India

Prajesh Prajapati
School of Pharmacy, National Forensic Sciences University, Gandhinagar, Gujarat, India

Bhupendra Gopalbhai Prajapati
Shree S.K. Patel College of Pharmaceutical Education & Research, Faculty of Pharmacy, Ganpat University, India

Parixit Prajapati
Department of Pharmaceutical Chemistry, SSR College of Pharmacy, Sayli-Silvassa Road, Sayli, Silvassa-396 230, Union Territory of Dadra and Nagar Haveli & Daman Diu, India

Mihir Raval
Department of Pharmaceutical Sciences, Sardar Patel University, Vallabh Vidyanagar-388 120, Gujarat, India

Surovi Saikia
Translational Research Laboratory, Department of Biotechnology, Bharathiar University, Coimbatore–641 046, Tamil Nadu, India

Manish Shah
Department of Computer Science, University of North Carolina at Greensboro, Greensboro, North Carolina, USA

Jubie Selvaraj
Department of Pharmaceutical Chemistry, JSS College of Pharmacy, Ooty, Tamil Nadu-643001, India

Harshada Shewale
Department of Pharmaceutics, School of Pharmacy & Technology Management, SVKM'S NMIMS Deemed-to-be University, Shirpur, Maharashtra 425405, India

Akey Krishna Swaroop
Department of Pharmaceutical Chemistry, JSS College of Pharmacy, Ooty, Tamil Nadu-643001, India

Varun Talati
Center for Computational Biology & Translational Research, Amity Institute of Biotechnology, Amity University Mumbai, Panvel Maharashtra

Vaishali Thakkar
Anand Pharmacy College, Department of Pharmaceutics, Anand-388001, Gujarat, India

Atharva Tikhe
MIT School of Bioengineering Sciences and Research, MIT-Art, Design and Technology University, Loni Kalbhor, Pune-412201, India

Piyush Trivedi
Shivajirao Kadam Group of Institutions, Indore

Sudha Vengurlekar
Institute of Pharmacy, Oriental University, Indore

1 Bridging Bioinformatics and Chemoinformatics Approaches in Public Databases

Shama Mujawar, Manas Joshi, Aarohi Deshpande,
Aarohi Gherkar, Shreyash Kolhe,
Bhupendra Gopalbhai Prajapati

1 ARTIFICIAL INTELLIGENCE AND BIOINFORMATICS

1.1 INTRODUCTION TO BIOINFORMATICS AND ITS APPLICATIONS

The latter half of the 20th century saw the inception and culmination of various landmark projects that had a significant impact on the field of molecular biology. At the center of these landmark projects were two biomolecules (i.e., proteins and nucleic acids), and the projects focused on deciphering the information carried within them. The then widely accepted polypeptide hypothesis, which states that each gene codes for one polypeptide, bounded these two biomolecules together and established the direction of the flow of genetic information. On the successful completion of the first sequencing project, by obtaining the sequence of the insulin protein in 1955 (1,2), Frederick Sanger and colleagues highlighted that the protein molecules were chains of ordered amino acids. This sequence information further aided in determining the organization of protein molecules in 3D space. The work of Max Perutz and colleagues in deciphering the structure of haemoglobin (3) and John Kendrew and colleagues in deciphering the structure of myoglobin (4), through X-ray crystallography, were some of the landmark studies in the field of structural biology that further facilitated in solving structures of other protein molecules. In parallel to the advances in our understanding of the protein sequence and structure information, similar advances were being made in deciphering the nucleic acid molecules. The experiments from Avery, McLeod, and McCarty in 1944 (5) highlighted the role of nucleic acid molecules as the carriers of genetic information, which was later supported by the experiments from Hershey and Chase in 1952 (6). These studies were followed by the landmark findings from Watson, Crick, and Franklin in 1953 (7) that aided in the elucidation of the double helix structure of the DNA molecule. The protein sequencing endeavor was followed, albeit with some lag, by the first nucleotide sequencing project when Holley and colleagues were able to successfully sequence the alanine transfer RNA in 1965 (1,8). Later, Maxam and Gilbert (9), and Sanger and colleagues (10), independently, developed different DNA sequencing techniques. These techniques were a significant leap toward our understanding of the genetic makeup of species; however, one of their central drawbacks was the lack of speed in handling large sequencing volumes of sequence data. Next Generation Sequencing (NGS) techniques, which are based on similar principles as the Sanger sequencing technique, have now addressed these drawbacks by implementing a parallel sequencing approach and have enabled the undertaking of various large-scale sequencing projects like the species-specific whole genome sequencing and individual genome sequencing projects.

DOI: 10.1201/9781003353768-1

1

These seminal works contributed significantly not only to our understanding of biomolecules and their properties but also to the technological advancements that have enabled the rapid growth in the availability of biological data. Supported by these technical advances, biological data has grown exponentially in the recent past (Figure 1.1). However, this exponential increase in data availability raised two important challenges: interpreting the signal within the data and data storage. These challenges were gradually overcome by amalgamating molecular biology with computer sciences. In parallel to the developments in molecular biology, computer sciences also saw an accelerated development of powerful computers that enabled performing complex calculations in shorter timespans. However, it was mainly the ability to create customized software/tools that propelled the union of molecular biology and computer sciences. Such software/tools could be specifically tailored for performing the analysis of biological data. These factors culminated in the inception of a multidisciplinary field, bioinformatics, whose focus was on aggregating and analyzing biological data through the use of computers. Bioinformatics provided biologists with two important features to tackle the challenges posed by exponentially growing biological data: customized tools for the analysis of biological data and enhanced storage capacity. One of the earliest applications of such an "informatics"-based approach can be attributed to two individuals – Margaret Dayhoff and Robert S. Ledley. Margaret Dayhoff, also referred to as "the mother and father of bioinformatics" (11), was one of the first pioneers in the field of bioinformatics. In collaboration with Robert S. Ledley, she wrote COMPROTEIN, a computer program that aimed at elucidating the primary structure of a protein from sequencing data (12). This consequently resulted in the formation of the first sequence database, *Atlas of Protein Sequence and Structure* (13).

In the last couple of decades, with the advent of technological advances and enhanced computational capacities, the impact of bioinformatics on biological sciences has increased drastically. Consequently, bioinformatics has been applied across a wide array of domains like genomics, proteomics, metabolomics, etc., and has been implemented in solving various problems in the following domains.

1.1.1 Genomics – Bioinformatics is employed with a critical task to map the output fragmented reads that are obtained from sequencing experiments, to their respective sequences of origin. The presence of single nucleotide polymorphisms (SNPs) and insertions and/or deletion (indels) of genomic fragments make the task of a whole genome assembly particularly challenging. These challenges could be tackled on the sequencing level (i.e., by increasing the sequencing depth, and/or through bioinformatics tools like DeepVariant (14), GATK (15), VarScan (16) etc.), which help in inferring a genuine variation from sequencing error (17). In addition to these, through bioinformatics, various Genome-Wide Association Studies (GWAS) have enabled highlighting loci/variants that could be responsible for specific phenotypes.

1.1.2 Proteomics – Along similar lines as determining genome sequence data from sequencing experiments, bioinformatics is employed in inferring the peptide sequence data from the output fragmentation data of Mass Spectrometry (MS) experiments through either *de novo* peptide sequencing or by searching against a reference database (18). Functional characterization of the resultant protein sequence could then be performed by checking the protein identifier's Gene Ontology terms (19). In scenarios where the protein identifier is unavailable, online domain prediction tools like InterProScan (20) could be employed to highlight the potential functional domains within the protein and, consequently, functionally characterize the protein. In addition to inferences of the sequence and functional annotation of proteins, bioinformatics has also been tasked with identifying potential protein–protein interaction (PPI) networks. Various online databases like StringDB (21), BioGRID (22), IntAct (23), etc., contain information on PPI networks across different species. Here, edges in networks (connections between two proteins) are constructed either through data curation/experimental validation or bioinformatics-based predictions made based on various factors like co-occurrence of proteins, co-expression, protein homology, etc.

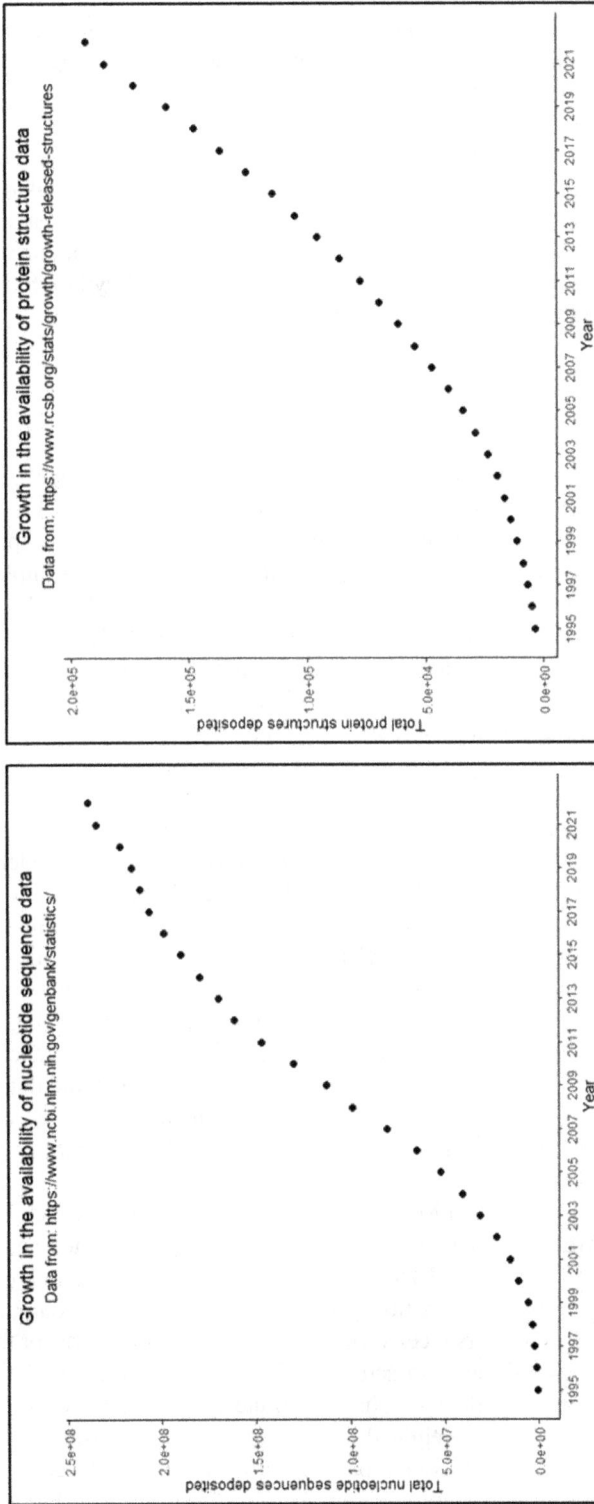

FIGURE 1.1 Increase in the number of deposited nucleotide sequences in GenBank and protein structures in RCSB over the last 28 years (1995–2022).

1.1.3 Metabolomics – Inferring relevant metabolites that could be responsible for specific phenotypes, including diseased conditions, has been an invaluable contribution from the domain of metabolomics to the field of disease biology. Through access to pathway information and statistical analyses, bioinformatics has been employed to highlight metabolites that could be pertinent to specific phenotypes.

An important driving force in the growing influence of bioinformatics has also been the evolution of computer programming languages. These languages provide a solid platform for importing biological data by supporting reading and writing functionalities across a wide array of data formats. Next, an open source package developing and accessing system enables processing and analyzing this data. The emergence of programming languages like Python has specifically made these tasks easier due to their high flexibility, simple vocabulary, and open source package management systems.

1.2 ARTIFICIAL INTELLIGENCE FOR BIOLOGY

Artificial Intelligence (AI), a domain that is focused on creating tools/programs that can perform functions in a manner such that they could mimic human intelligence, has grown into one of the most promising domains of this generation. Usually, AI consisted of a set of pre-fed instructions whose execution was conditional on stimuli. One of the earliest standout implementations of AI was the Deep Blue computer, which defeated Garry Kasparov (the then chess champion) in a game of chess in 1997. Deep Blue, a computer that was developed by IBM, was built based on the set of rules in chess. However, the game gets complicated with every move from the opponent, as that consequently opens up a vast array of possible outcomes. Here, the computer is tasked with keeping track of all the possible outcomes and making calculated moves that could lead to a victory. Albeit symbolic, the Deep Blue computer, gave a glimpse into the true potential of AI computers.

Given the complex nature of biological data, the application of AI required the algorithms not to rely on a set of predefined rules, but rather to create and constantly update models that could better define the system at hand (24). One of the central advantages of having a none-to-minimal set of predefined rules (human interference) is that the inferences could be truly unbiased and reflect the actual nature of the data. Machine learning (ML), an offshoot of AI, adopts a model-based approach wherein it "learns" from a set of "training" data. Depending on the nature of the training data, ML algorithms are broadly classified into two categories: supervised learning and unsupervised algorithms. In supervised learning, training data is incorporated with a set of labels that characterize the data and algorithms are tasked with establishing a set of rules that could define the assignment of labels to the data. In unsupervised learning, the training data is unlabeled and algorithms are tasked with identifying the potential labels and, similar to supervised learning, assigning data to labels (25).

Deep Learning (DL), an offshoot of ML, is an advanced AI algorithm, which implements similar functionality as that of ML, and is modelled after the highly interconnected neural network in the human brain. The structure of a typical DL network is made of multiple interconnected "layers" of neural networks that transmit information from the input layer to the output layer via hidden layers (Figure 1.2). The hidden layers receive an input from either the source or the previous neural layer, which is then processed through an activation function and the output is passed to the next neural layer (26). The individual elements that constitute these layers are known as perceptrons (27). An individual perceptron takes weighted inputs, processes them using an activation function, and produces a binary output, which is then passed to the next neural layer. Due to this complex multilayered structure, DL algorithms can identify non-linear relationships between different variables within the data thereby increasing the accuracy of the classification. Implementing the

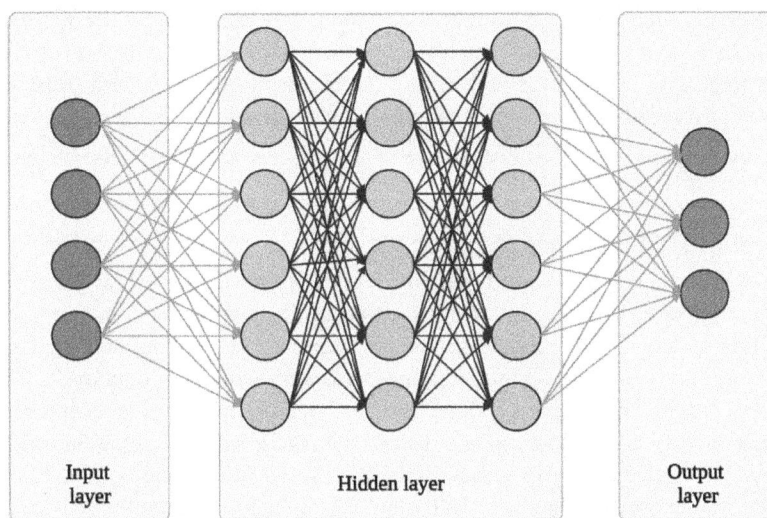

FIGURE 1.2   A typical neural network consisting of three layers, namely: input, hidden, and output layer. The flow of information is from input layer to the output layer, via the hidden layers, and is facilitated by perceptrons (shown here as circles) that use activation functions to process the weighted input data and produce a binary output. (Created with Biorender.com)

neural network approach, Convolutional Neural Networks (CNN) algorithms are based on a neural network approach and have been implemented in handling multi-dimensional data (26), specifically image data. In general, AI algorithms have been applied across a wide array of domains, like data mining, gene expression profiling, protein folding, etc.

1.3   APPLICATIONS OF AI IN SOLVING BIOLOGICAL PROBLEMS

The ability of AI algorithms in handling and interpreting complex biological data has had a significant impact on solving many longstanding questions in biology. The ability of these algorithms to process heterogenous data has enabled their implementation across various domains and solving the problems of those domains. Some of those are listed below.

1.3.1   The Protein-Folding Problem

Given the central role of proteins within the biochemical machinery of a cell, understanding their structural and sequence characteristics aids in understanding various biological processes. However, predicting protein folding (i.e., the 3D structure of a protein from its sequence) has been one of the challenging tasks in the field of structural biology. This problem could be addressed by homology modelling – constructing the 3D structure of a target protein by using a template structure whose sequence has high similarity with the target protein sequence. However, this approach suffers in scenarios where a homologous template is not available. AlphaFold (28) is an ML tool that employs a neural network approach in solving the structure of a protein from its sequence, which could also be employed in cases of the absence of a homologous protein structure. The structures of proteins resolved through AlphaFold are available at alphafold.ebi.ac.uk/.

1.3.2   AI in Disease Biology and Personalized Medicine

The application of ML in disease biology has had a significant impact on healthcare. This is specifically visible in the case of biomarker discovery, wherein AI algorithms can process multidimensional clinical data and can highlight specific molecular signatures that are specific to certain diseased conditions. These biomarkers could then be used to understand the disease progression. This information is important in terms of prescribing the treatment. In addition to this, deep learning algorithms have also aided in clinical decision-making by analyzing the medical image data like nuclear imaging, ultrasound, tomography, etc. These algorithms have been implemented in detecting early detection and metastasis of cancer.

1.3.3 The Problem of Identifying Genomic Variants from Sequence Reads

Whole genome sequencing experiments typically yield short read sequences, which are then required to be mapped to the reference genome. As discussed in the previous sections, one of the central challenges in this task is to differentiate between an error in sequencing and a genuine genomic variant as both are likely to leave a similar signature. DeepVariant (14) employs a CNN approach in addressing the problem of calling genomic variants. DeepVariant uses the sequencing read data around every candidate genomic variant, and, using an image pileup technique, internally calculates probabilities for each site to be a true variant. HELLO (29) uses a similar approach in highlighting genetic variants, but it bases its algorithm on the input sequencing data rather than the image pileup of the surrounding reads.

1.3.4 Predicting the Transcription Factor-Binding Sites

Transcription factors (TFs) perform an important task of gene regulation. Usually, TFs bind to the effector genes, in a sequence-specific manner, which ultimately results in their translation and, eventually, expression. The sites that TFs bind to to initiate the translation of genes are known as Transcription Factor-binding Sites (TFBS), and mutations occurring within these TFBSs could potentially switch off these regulatory interactions. These TFBSs are usually represented as consensus sequences that are obtained through various biochemical assays, with space for a certain degree of ambiguity per position. Hence, understanding the impact of genomic variants in TFBS and defining the overall binding model per TF becomes challenging. To address the problem of highlighting the sequence specificities of DNA- and RNA-binding proteins, DeeperBind (30) was one of the first methods that employed a deep learning approach with CNN. Following DeepBind, many other methods have been employed AI in understanding the binding models of DNA and RNA-binding proteins like DeepGRN (31), DeepRAM (32), etc. (33).

2 ARTIFICIAL INTELLIGENCE AND CHEMOINFORMATICS

2.1 INTRODUCTION TO CHEMOINFORMATICS

Chemistry, a field of science dealing with elements and matter, came to birth around 13 billion years ago when a pea-sized singularity violently expanded into the great supremacy we see today as the universe. It not only studies all the different combinations of matter that come together to form a large number of chemical structures (atoms, molecules, compounds etc.), but also plays a crucial role in everyday life from the very toothpaste we use to the fuel in our vehicles. This rich spectrum continues to expand into other aspects such as pharmaceuticals, nanotechnology, as well as drug discovery and development. With such a vast range of applications of chemistry comes a voluminous data that needs to be effectively managed. This is where the multidisciplinary field of chemoinformatics comes into the picture.

The next stepping stone of organizing such a saturated amount of data systematically is a challenge. This data includes structures, small molecules, formulas, and their properties. This led to the extension of a new branch altogether, which is widely known today as chemoinformatics. It focuses largely on designing a system to manage all the chemical data generated in order to manipulate and apply it to a bunch of other fields to make drug discovery simpler, more efficient, and effective (34,35). The following section briefly discusses the history of chemoinformatics and some of its applications.

It is possible to define chemoinformatics as a synthesis of chemistry and information technology that employs computational techniques to manage enormous volumes of chemical data that aid in establishing connections between chemicals and structural elements. It encompasses concepts such as chemical databases, quantitative structure-activity relationships (QSAR), compound property or spectral prediction, etc. Although these concepts have been around for over 40 years, the term cheminformatics was first coined by Frank Brown in 1998 (34,35). It was necessary to recognize this as a separate branch as scientists struggled to identify this work of in silico-based research and organization, which made it more difficult for them to correlate several disciplines together and find a common purpose.

Chemoinformatics not only allowed the organization of several million compounds into large databases that can be accessible to all, but also helped solve complex problems in chemistry such as determining the relationships between the structure of a compound and its biological activity or the influence of reaction conditions on chemical reactivity, which requires novel approaches to extract knowledge from data and model complex relationships (35,36). A few applications of chemoinformatics are discussed below.

2.1.1 Drug Discovery

The process of developing a lead chemical with potential target-binding pharmacokinetics that may be improved to satisfy FDA safe and effective standards is known as drug discovery. It involves identifying and verifying a target as well as a hit. Applications of chemoinformatics in drug discovery include compound selection, virtual library generation, virtual high-throughput screening, HTS data mining, and in silico ADMET (Absorption, Distribution, Metabolism, Excretion and Toxicity) early prediction (37).

Compound selection mainly focuses on choosing a relatively smaller set of compounds from a large pool. It usually deals with four main goals:

1. To be able to pick and get substances from other external sources that will enrich the diversity of the libraries already in existence
2. To be able to choose a subset from a corporate compound pool for screening that represents diversity
3. To be able to choose chemicals to create a combinatorial library with the greatest possible diversity
4. To be able to choose compounds with new scaffolds that are related to known ligands but different from them from the existing compound collections (37)

The sole distinction between virtual high-throughput screening and experimental high-throughput screening is that the former emphasizes compound selection more than the latter. It tries to minimize the number of applicants who must undergo testing in an experimental setting.

In chemoinformatics, ML techniques are commonly utilized to forecast the biological and chemical features of novel chemicals that can be used in practical settings. The Statue of Unity, which is 182 meters tall, can be used as a concrete illustration of this. According to predictions oxidation will cause the brown structure to turn green after around 100 years, resembling the Statue of Liberty. Several ML approaches such as recognition patterns unveil relationships between chemical structures and their respective biological activities. For lead identification and optimization, a variety of techniques and methods are employed. Some of these include virtual screening, molecular databases, data mining, high-throughput screening (HTS), QSAR, protein ligand models, structure-based models, microarray analysis, property calculation, and ADMET properties (38,39).

The main purpose of ADMET analysis is to understand drug metabolism – a process by which the body gets rid of the drug mainly by modifying it chemically (also referred to as biotransformation). A drug is significantly reduced after being metabolized by the liver (Cytochrome P450 enzymes in the liver metabolize through oxidation, hydrolysis, and hydroxylation). The ADMET properties allow developers to investigate all the properties of a potential drug to help reduce potential risks during clinical trials. They determine whether or not a compound is suitable to proceed to the clinical stage. Developers can understand the safety and efficacy of a drug candidate necessary for regulatory approval. For example, ethanol is readily absorbed from the gastrointestinal track, but poorly absorbed through skin.

HTS is a laboratory automation assay technique used to swiftly screen a large number of chemical compounds for the desired activity. A method for predicting the preferred orientation of a molecule to a protein when attached to another molecule to create a stable complex is called docking (40). This method involves studying the interactions between two macromolecules.

2.1.2 Computer-Assisted Synthesis Design

Computer-Assisted Synthesis Design (CASD) is another area in chemoinformatics that requires the application of AI techniques (41). A chemical database is designed with various types of machine comprehensible chemical representations in order to store their data so as to retrieve and manipulate it later. An example of one such chemical format is Simplified Molecular Input Line Entry Specification (SMILES). It is a linear chemical notation format that describes the connection table and the stereochemistry of a molecule. Another aspect of chemoinformatics is the structural representation, which is often associated with Molecular Modeling, Structure Searching, Computer-Assisted Structure Elucidation (CASE), etc. These are concerned with analyzing the structure as a whole, checking similarities and diversities, and also identifying key characteristics such as bond angles, rings, and aromaticity. Structure descriptors (topological, geometrical, hybrid or 3D, electronic) are also important when considering the physical, biological, and chemical properties of the compound of interest (41). An example of a program that is based on CASD is Logic and Heuristics Applied to Synthetic Analysis (LHASA). It uses a heuristic approach and was developed by Corey in the late 1960s and is still used today. It consists of a special chemical language called CHeMistry TRaNslator (CHMTRN) that helps in searching for disconnections. It is an easy-to-learn programming language whose syntax is similar to English, therefore making it more user-friendly for chemists.

2.2 CHEMOINFORMATICS: FROM BIG DATA TO ARTIFICIAL INTELLIGENCE

As discussed in section 2.1, chemoinformatics mainly deals with chemical data that is usually stored in large databases and used to extrapolate knowledge. This data can get extremely complex with multiple structures and properties that are difficult to interpret manually. Big data therefore refers to large and complicated data sets that are bigger by reference to the past and to the methods and devices available to deal with them.

According to the book *Data Mining for Business Analytics*, there are a number of obstacles to overcome when analyzing big data, including the volume of data, the velocity at which it is generated and changed, the variety of the data being generated, and the veracity of the data, which refers to the fact that it is being produced by organic distributed processes and is not subject to the same controls or quality checks as data collected.

The big data generated in chemistry is known as "BIGCHEM" and includes large databases that have been made available by experimental techniques such as high-throughput screening, parallel synthesis, etc. (42). In order to draw insight from this data, it necessary to mine it so as to get rid of all the noise (sometimes due to experimental errors during assay, lack of standardization while annotating biological data or errors while extracting the data values, units, chemical type, etc.) that is present in the original data. Various ML methods are used to analyze this data, which ultimately helps in visualizing it in silico. It can then be used to predict outcomes in the future. This section of the chapter will focus on the role of AI in chemoinformatics.

2.2.1 Artificial Intelligence

Artificial intelligence lends a helpful hand to BIGCHEM. When considered practically, many experts predict that AI will eventually replace the work that people currently do in numerous industries, leading to widespread unemployment. However, AI can actually be extremely reassuring when it comes to big data by minimizing the amount of work that is spent processing chemical data. Neural networks in relation to big data are an illustration of AI and ML techniques that are widely employed in chemoinformatics. It is a time-tested, adaptable data-driven approach for categorization or prediction. The architecture of the mathematical model employed for pattern recognition is largely responsible for its great predicted accuracy. A number of neurons in the input layer, a hidden layer or layers, and the output layer hold it together (38,39). Each link between neurons in a typical architecture has a weight. Prior to being tested on instances that have not yet been observed, the weights are adjusted during the training phase as the network learns how to connect input and output data. This model is based on and very similar to the neural networks of the human brain and mimics the way a human learns. In chemoinformatics, these can be used to analyze and predict the bioactivity, toxicological, pharmacological, and physicochemical properties, such as hERG blockade, aquatic toxicity, drug clearance, pKa, melting point, and solubility.

2.2.2 Deep Neural Network

Deep Neural Network (DNN), a subtype of deep learning, is also a very useful ML algorithm applied to in silico ADMET analysis. It makes use of several layers of a neural network, where each layer is dependent on the previous one. It can learn multiple levels of representations at different levels of abstraction (43). This powerful algorithm not only allows one to make predictions based on the SMILES chemical format, but also has the ability to create new, more expressive representations as compared to the traditional chemical fingerprinting ones that have been around for over 28 years (43).

A scaffold represents the composition of the smallest core ring sub-structure common in a set of compounds (44). These can get extremely complicated according to various metrics used to determine their synthetic as well as structural complexity. Hits that are too complex can cause in challenging chemistry, and thus, slowing down the SAR development process to a great extent.

Due to the different in vivo and in vitro biological effects of enantiomers as well as diastereomers, it becomes more difficult to interpret and analyze the number of chiral centers in a molecule. Some examples of algorithms that are used for scaffold analysis are ScaffoldTree and HierS. Databases can be analyzed relatively faster using ScaffoldTree, HierS, and similar techniques with linear computing time dependence and chemically logical clustering (45). Scaffold hopping is another technique that is used widely in the drug discovery process, which mainly aims at jumping in a chemical space to find new bioactive structures originating from a known active compound by the modification of the central core (44).

2.3 APPLICATION OF CHEMOINFORMATICS AND AI IN DRUG DISCOVERY

It is common knowledge that the process of developing new drugs involves screening vast chemical libraries for promising hits, which are then carefully examined to produce leads. These leads are then optimized to mitigate their negative side effects and increase their beneficial effects, which are then confirmed by clinical trials. This procedure is designed to ensure that new medications are quick to market, safe, and efficient for patients. Each drug's market introduction typically costs $1 billion and takes 15 to 20 years to complete. In addition to being expensive and time-consuming, the process is also extraordinarily difficult and fraught with uncertainty. A lead chemical that appeared to be a viable medicine in the early trials may turn out to be ineffective in the later stages of research, resulting in a significant loss of resources. However, using AI and chemoinformatics techniques can significantly reduce this uncertainty by foreshadowing which compounds would be selected and worked on. Modern deep learning neural network techniques and potent AI algorithms like support vector machines have been shown to be more promising than traditional ML techniques. Some of the uses of chemoinformatics in drug development and AI will be briefly discussed.

Virtual screening plays a very crucial role in the initial stages of the drug discovery process by allowing researchers to analyze molecules computationally or in silico to predict the protein–ligand interactions. This helps them eliminate a large number of molecules therefore reducing the laboratory experiments and also significantly speeding up the process. Compound screening can be done using a variety of open source, free online cheminformatics tools. One such software is ZINCPharmer, which applies pharmacophore search screening techniques to the entire ZINC database that is available for purchase. ZINCPharmer allows users to import LigandScout and MOE pharmacophore definitions and their discovered pharmacophore attributes straight from the structure database (46). A pharmacophore is a specific area of a molecule or drug that has the necessary geometric configuration of atoms and functional groups as well as a shape that can bind to a target and trigger a biological reaction. A well-defined pharmacophore can act as a lead molecule in the drug discovery process.

Due to its speed, ligand-based virtual screening (LBVS) continues to be one of the most popular and commonly utilized cheminformatics applications for assisting in drug discovery. It uses information of ligands or small molecules interacting with the target to identify new potent compounds. Ligand information can include binding affinities, chemical structures, physicochemical properties, etc. Some examples of ML methods are QSAR, similarity search, and pharmacophore mapping. LBVS is preferred when there is little to no information available on the structure of the target. Structure-based virtual screening (SBVS), on the other hand, is used when there is sufficient structural information of the target protein available (specifically the crystal structure). It focuses on simulating the interactions between a ligand and its protein. Molecular dynamic simulations, docking, homology modeling, and pharmacophore modeling are some of the examples of ML methods used for this type of virtual screening.

Virtual screening can further be categorized into pharmacophore-based virtual screening (PBVS) and docking-based virtual screening (DBVS). Pharmacophore modeling, a structure-based method, is based on the concept of "stripping" functional groups of their original chemical makeup in order to categorize them into a relatively small number of pharmacophore types based on their predominant physicochemical characteristics (47). It is often used to evaluate the probability of molecules binding to the protein-binding sites and also to identify potential binders based on interactions. LigandScout, a common software used for both structure and ligand-based pharmacophore modeling, involves new high-performance alignment algorithms for excellent prediction quality with unprecedented screening speed (47). It allows the automatic construction and visualization of 3D pharmacophores from structural data of macromolecule/ligand complexes. Hydrogen bond donors and acceptors are represented by directed vectors in the LigandScout algorithm's chemical characteristics, along with positive and negative ionizable regions, spheres for lipophilic regions, and hydrogen bond donors and acceptors. Additionally, the LigandScout model incorporates spatial data on regions that are inaccessible to any prospective ligand, increasing selectivity and reflecting potential steric restrictions (48). The resultant pharmacophore model is automatically updated to include excluded volume spheres that are sterically forbidden places.

Docking-based virtual screening on the contrary focuses on computational fitting of molecules into a receptor active site using advanced algorithms, followed by the scoring and ranking of these molecules to identify potential leads. Typically, the method begins with a 3D structure of the target protein, after which chemicals from databases are docked into the target protein using docking software (47). The ligand that interacts with the receptor in the best possible way is selected for further optimization once the docking results have been scored. After this, the ligands are generally subject to various binding assays. A popular docking software is AutoDock, which allows prediction of binding of small molecules (substrates or leads) to the receptor of the known 3D structure. Another example of a protein–ligand docking software is GOLD (27) – Genetic Optimization for Ligand Docking – which is based on a genetic algorithm to dock flexible ligands into protein-binding sites.

In DBVS, the 3D structure of the target protein is crucial and must be known either experimentally (X-ray crystallography techniques, NRM studies, etc.) or using computational techniques (homology modeling) (49). The compound database source (ZINC, for instance) typically correlates to the compound collection available from chemical suppliers. In-silico filters (based on the Lipinski's rule of five) are applied to reduce the number of compounds with low oral bioavailability. To further examine molecules, a filter that eliminates substances with poor chemical stability, reactivity, or toxicity can be used. After that, the ligands are rated according to their binding affinities (assumed to be the sum of independent terms). Based on them, the results are scored in accordance with the relative binding affinities of various ligands, whereas the ligands are ranked by the quality of the solutions (49). The next step is to conduct a post-processing stage analysis of the compounds to make sure that their interactions with the target protein and conformations are appropriate. The compounds are then sent to be assayed.

3   BRIDGING BIOINFORMATICS AND CHEMOINFORMATICS APPROACHES

3.1   BIOINFORMATICS APPROACHES

The Human Genome Project (HGP) could be perceived as one of the landmark endeavours in the field of biology that specifically gave a significant impetus to the growth of the bioinformatics domain. The overwhelming accumulation of data from sequencing experiments raised two important challenges: data storage and analysis. Bioinformatics, which had already started gaining importance in the latter half of the 20th century, emerged as one of the key domains that could aid in overcoming those two challenges. Today, the tasks of submission and retrieval of data from various biological experiments have been facilitated by dedicated databases. Most of these databases are free to access and are maintained and curated by experts. In addition to this, some databases like Ensembl (rest. ensembl.org/), UniProt (www.uniprot.org/help/programmatic_access), etc., also enable programmatic access to their data via REST-API endpoints. By enabling open access to the available data and hosting data originating from different parts of the world, such databases are at the forefront of addressing biological problems, specifically problems that demand immediate action. To exemplify, since the early stages of the SARS-CoV-2 outbreak, it had become imperative to sequence samples from infected patients and make these sequences openly available for analysis to better characterize the virus and contain its spread. Under such circumstances, data shared from individual contributors on databases, specifically GISAID (50) (gisaid.org/), has been a pioneering force in tackling the pandemic. We discuss various such databases throughout this chapter.

However, besides playing a key role in the storage and access of biological datasets, bioinformatics has played an even more important role in interpreting the signal contained within these biomolecules. Such bioinformatics-based approaches have seen applications across a wide array of fields/domains. Some of these approaches/applications are discussed in detail here.

3.1.1   Human Health

The Human Genome Project (51) was estimated to cost around $3 billion. However, due to the advances in sequencing technologies, the overall cost of sequencing an entire human genome has been reduced significantly in the last 20 years (Figure 1.3). The magnitude of reduction in sequencing price is also outpacing the hypothetical Moore's law (52) distribution, which exhibits the phenomenon of the

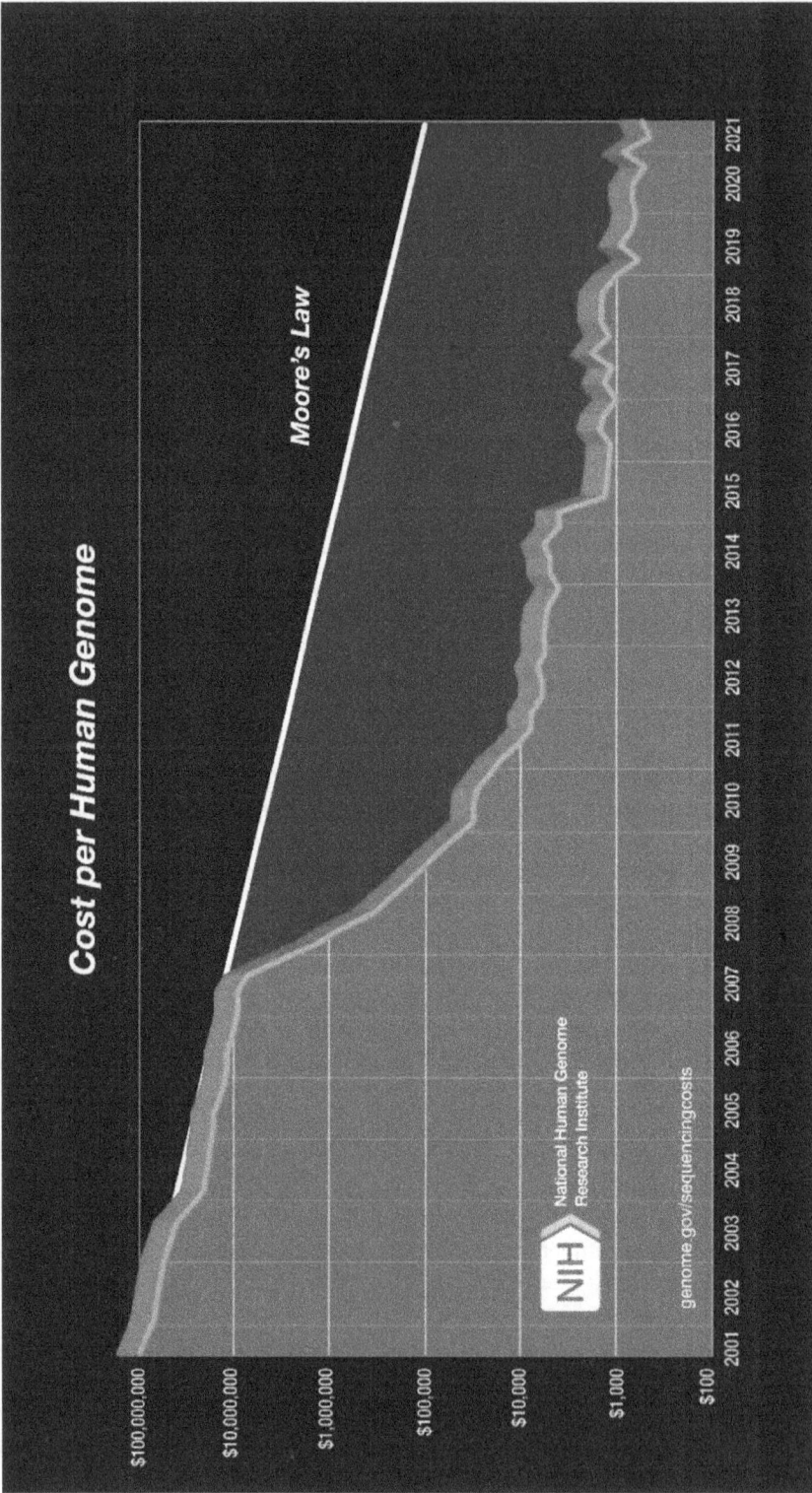

FIGURE 1.3　The significant reduction in the cost of sequencing the whole human genome in the last 20 years. Figure downloaded from www.genome.gov/sequencingcosts.

doubling of computational power every year. Bioinformatics-based genome assembly pipelines have provided a significant boost in enabling such ease in reconstructing the entire genome of species. To exemplify, CANU (53) is a genome assembly tool that uses sequencing reads from PacBio or Oxford Nanopore as the starting points to generate the whole genome sequence. As an additional package, HiCANU relies on the internal algorithm of CANU but uses HiFi sequencing reads generated from PacBio to generate genome assemblies. NECAT (54) is a recently introduced genome assembly tool that uses long reads obtained from Nanopore sequencing technology from reconstructing whole genome sequences to lower error rates. BUSCO (55) serves as a quality assessment tool for genome assemblies by checking the "completeness" of the assembled genomes.

Today, genome sequencing has developed its niche in the commercial market. Various ventures have developed sophisticated pipelines that could perform whole genome sequencing of keen participants from all over the world for a feasible price range, and document results in a way such that they could also be understood by individuals who have no prior background in life sciences. These results range from inferring ancestry, predicting various traits, and importantly, susceptibility to various diseases. The power of diagnosing genetic diseases and vulnerability to disorders has grown significantly in the recent past by being able to monitor the genome and its consequential downstream products, like RNA and proteins. In parallel to genome sequencing, the techniques involved in genome editing have also seen a significant boost in the recent past concerning their precision in targeting specific genomic locations. Bioinformatics has been used extensively here to identify the precise genome regions as targets of interest. Specifically, the introduction of the CRISPR (Clustered Regularly Interspaced Short Palindromic Repeats) system and application of CRISPR-Cas9, an immunological protein mainly involved in the defence of bacteria against viruses, has revolutionized genome engineering. The field of personalized medicine has benefited significantly due to these advances in genome sequencing and editing. Prescribing patient-specific therapies and tackling diseases through genome editing are set to reinvent the manner of medical treatments. As described before, bioinformatics has been at the forefront of documenting and storing genome sequences. The 1000 Genomes project (56) (www.internationalgenome.org/) is a database dedicated to cataloguing sequence and variant information obtained through various human genome sequencing experiments from different geographical regions and populations. Such population-specific variant annotation helps in understanding the population-specific segregating frequencies of potentially beneficial and deleterious alleles (Figure 1.3).

3.1.2 Evolutionary Biology

Genomes could be perceived as the carriers of footprints that are created by the forces of natural selection, and genetic drift, over long evolutionary time scales. The observable phenotype of every species, which can be directly inferred from its genotype, can partially be explained by the evolutionary forces in action that penalize those genomic changes that could potentially be harmful to fitness and reward certain genomic changes that could be beneficial for fitness. Hence, unsurprisingly, genome analysis has been at the centre of evolutionary biology for a long time. Here, genomes can help in deducing the species-specific evolutionary trajectories. With an increase in the ease of accessing genomic data, via constantly maintained sequence databases, undertaking such evolutionary studies has become easier. Taking advantage of this, various bioinformatics tools have been developed to make inferences on species-specific evolutionary patterns. ANGSD (57) is a freely available tool that directly uses the raw data from sequencing experiments, processes it internally, and estimates various evolutionary statistics (like Fst, Linkage disequilibrium, etc.) that aid in estimating the evolutionary path of species. Relate (58) enables detecting the ancestral population structures and gene flow for a given species. Such inferences could be particularly interesting in terms of understanding the evolution and separation of different species and their settlement in specific environmental niches. At the centre of such evolutionary inferences are genomic changes with respect to a reference genome, and using these to predict the similarity and differences between two species or populations. ANNOVAR (59) aids in highlighting such genomic changes by annotating the genetic variants on different criteria like protein-coding changes mutations, variants falling within conserved regions, etc. Population genetics, a field focusing on inferring the action of selection acting within species, relies on the information of such variants within populations of species. Tools like vcftools (60), bcftools (61), plink (62), etc., are used extensively for performing various operational tasks and inferring evolutionary statistics.

3.2 CHEMOINFORMATICS APPROACHES

Adopting an informatics-like approach in biology was comparatively straightforward as compared to chemistry. Biomolecules, mainly nucleic acids and proteins, are made up of specific constituent elements whose variable and repetitive arrangements give rise to different structures and properties. Nucleic acids are made up of four different bases and proteins are made up of 20 different amino acids. Characterizing these biomolecules through sequencing is comparatively uncomplicated when these constituent elements are known. However, complex chemical structures are made of constituent atoms whose arrangements could vary by comparing different functional groups. Hence, one of the earliest steps in adopting an informatics-like approach in chemistry (i.e., chemoinformatics) was to digitize such complex chemical structures. Today, various small molecules databases like ZINC (www.zinc.docking.org/) and PubChem (www.pubchem.ncbi.nlm.nih.gov/) enable access to various small molecules, with some of the molecules also having drug-like properties. Just like bioinformatics, chemoinformatics has grown into a field that has not only played a pivotal role in creating and maintaining vast datasets but also made a significant impact in various domains. Chemoinformatics has specifically played an important role in the field of drug discovery by transforming the steps involved in the treatment of diseased conditions. At the heart of a typical drug discovery pipeline is a hypothesis, which is usually based on a literature study, that highlights a certain target molecule whose activation or inhibition could have a beneficial effect in healing the diseased state. On identifying the target molecule, the search ensues for identifying small molecules that could potentially have a drug-like property. This search could be based on data mining, wherein potential candidates that have been reported for their inhibition or exhibition of the target molecule are reported (63). Otherwise, this search is conducted via screening a large number of small molecules for the desired activity against the target molecule. Traditionally the process of synthesizing a wide array of such small molecules was a laborious task. However, in the recent past, Combinatorial Chemistry (CC) has been extensively employed for the synthesis of such potential small molecules through constituent building block elements. Today, information on properties and structures of various such small molecules could be accessed through dedicated databases. Such databases are discussed extensively throughout this chapter. The desired drug molecule needs to be effective against the target molecule, but at the same time should not result in any other undesired effect. This process of shortlisting a potential drug molecule, given the target molecule, is known as screening and could occur in one of many ways. High-throughput screening (HTS) usually employs assay technologies to directly analyze a large number of molecules against the target. Virtual screening is an information-based approach that adopts a docking approach wherein a large number of small molecules could be "docked" against the target molecule and their mode of action can be ranked based on quantifying their binding affinity scores against the target molecule. On identifying a core set of drug molecules, the molecules are now passed through an intensive Structure Activity Relationship (SAR) phase, where they are characterized in detail on their magnitude of action against the target molecule (64).

3.3 BIOINFORMATICS AND CHEMOINFORMATICS AT AN INTERFACE WITH SYSTEMS BIOLOGY

As elucidated in the previous sections, chemoinformatics has made a significant impact in the domain of drug discovery in the recent past. On the other hand, a comparatively older field, bioinformatics

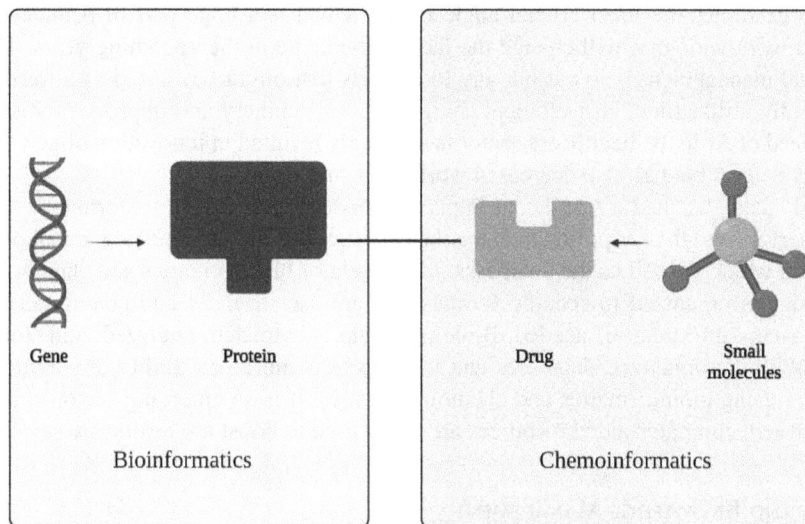

FIGURE 1.4 Drug discovery as the interface between bioinformatics and chemoinformatics. Adapted from (36). (Created with Biorender.com)

has been at the forefront in the inception and revolution of various "omics," including genomics and proteomics disciplines. A conjunction of these two fields within the domain of drug discovery is shown in Figure 1.2.

A central challenge within the drug discovery domain is characterizing the entire interaction profiles of drug molecules once they are introduced within a biological system. Systems biology has been employed in the recent past to understand the underlying molecular mechanism of these inter-action profiles (65). Here, systems biology employs data from the "omics" disciplines to construct a representative network of the biological system at hand. Chemical genomics, or chemogenomics, has emerged as a promising field that could aid in obtaining an in-depth understanding of the cel-lular, specifically genomic, response to the introduction of chemical compounds. Chemogenomics has rapidly facilitated the identification of potential drug candidates and testing their validation against a target molecule (64). Similarly, chemoproteomics is a discipline that integrates medi-cinal chemistry with proteomics to obtain a proteome-specific interaction profile of drug usage. Specifically, chemoproteomics attempts at quantifying the impact of the drug on the biological system by observing the proteome expression and post-translational modification data (66). Using these two approaches, the gene or protein that has been impacted due to the introduction of the drug molecule can be highlighted. Consequently, by obtaining further information on the affected gene or protein, and by integrating systems biology, information on specific processes/pathways that can be potentially disrupted can also be deduced (Figure 1.4).

4 PUBLICLY AVAILABLE RESOURCES IN BIOINFORMATICS AND CHEMINFORMATICS

In this section, we discuss different aspects of resources available to the general public and other communities in the field of bioinformatics and chemoinformatics. Databases can be of various types including biological databases, chemical databases, composite databases, RNA databases, pheno-type databases, etc. Each one of these databases is specialized to contain comprehensive, verified, easy-to-understand, and accurate information, which not only helps the research communities with project work but also drives them to gain so much more relevant information about the same. This chapter also examines the most crucial angle of AI, which is a huge part of healthcare systems incorporated worldwide that will change the face of medicine in the upcoming years. The goal of AI in medical management is to cut out any loose ends that might be caused as a form of human error, efficiently utilize time, cost-efficient equipment of machinery and improve productivity. The increasing need of AI in the healthcare sector has not only resulted in innovation of new technology to cure many defects but has also decreased work labor in hospitals.

The word "data" has had a significant impact in modern days. All the information to be transmitted, the servers and the script in which the data is extracted and manipulated are all organized in a general term which we call as the databases. In the field of bioinformatics and chemoinformatics, this knowledge is organized in specific formats that are user friendly and contain all the related information (meta-information) needed. Biological data is extracted, analyzed, and stored in specific ways. With the organized databases and tools there is more probability of solving emerging challenges by using bioinformatics and chemoinformatics. It is an emerging sector in information technology where computer-aided resources are being used to boost the healthcare sector.

4.1 DATA AND KNOWLEDGE MANAGEMENT

Data management is an important aspect in determining and analyzing biological data. In bioinformatics, this knowledge management is called *e2e*, which is a representation of the data in XML format, also known as eXpressML (67). It uses operations like query written language using databases like PDB or SwissPROT or any publication data. Data such as expression of genes, the information mostly based on the mechanism and the pathway used in the expression makes the whole data to be robust and quantitative. It is required to make the data clean and arrange it using the algorithms (Figure 1.5).

In bioinformatics, knowledge management can be defined as a systematic process that creates, captures, shares, and analyzes information with the aim of improving analytical performance.

There are a number of technologies involved in knowledge management that will collect, differentiate or integrate, store, and share information to the end user.

- **Intranet:** Secure internal networks, to provide an ideal environment for sharing information accessed using a standard browser. Information Retrieval Engines – search engines are an absolute necessity and are an integral part of knowledge management.
- **Groupware:** Facilitate information sharing via email, online discussions, databases, and related tools. Its collaborative features can result in the creation of stores of untapped knowledge.
- **Database Management Systems**: Computer databases are common repositories of information. KMS can be constructed to incorporate the information that is stored in an organization and is accessible by all.
- **Data Warehousing and Data Mining**: Data warehouses are centralized repositories of information. Data mining refers to specialized tools that allow the organization to convert increasingly complex sets of data into useful information.
- **Push Technologies:** Delivering appropriate information to individuals based on specific criteria (68).
- **Collaboration**: Expert modeling and decision-making analysis that will lead to more collaboration, information expertise, and insight sharing among knowledge workers.
- **Visualization and Navigation Systems**: Relationship between knowledge elements and holders of knowledge.

Knowledge management (KM) processes consist of knowledge creation, storage, distribution, and application to generate knowledge and data in a systematic way as shown in Figure 1.6.

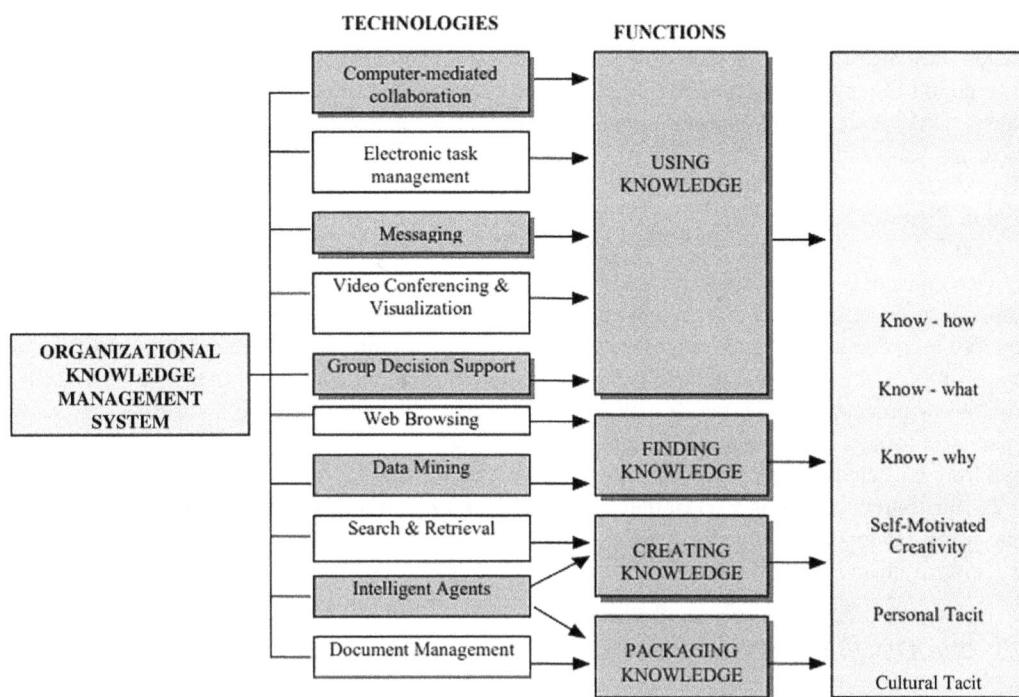

FIGURE 1.5   Knowledge Management System Organized Techniques.

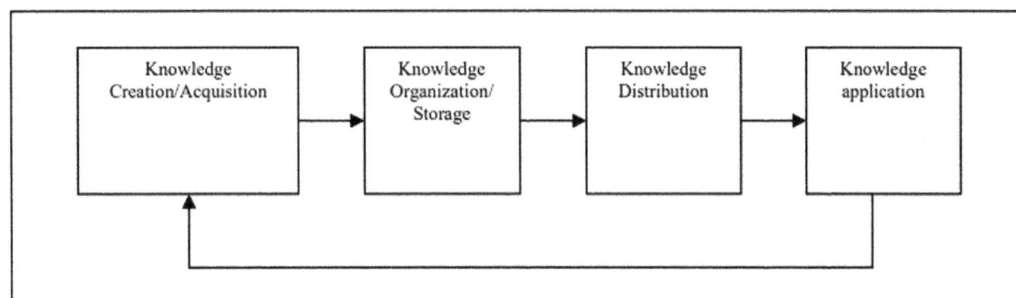

FIGURE 1.6   Cycle of Knowledge Management Process.

4.2   DATA EXTRACTION AND TRANSFORMATION

The main aim of data retrieval and transformation is accumulating data from various sources to be used and represented as one unit. Various tools come in handy while performing these data management tasks. GeoBoost2, Bert, Oracle clinical, CLOBNET, GenomicsDB, SNPator, etc., are some of the commonly used tools in data extraction. These tools can also give highly specific information from structured or unstructured lines of data on a database. Most of these tools are NLP (Natural Language Processing) driven. CLOBNET was used for predicting primary therapy outcomes of patients with high-grade serous ovarian cancer (HGSOC) on the basis of EHR data (69).

Data transformation is the process of converting data structures from various sources into a common format or a particular standard. A simple example of data transformation is converting Word files (*.docx*) into PDF files (*.pdf*) or vice versa. The common tasks performed during every data transformation include data annotation, data validation, and data aggregation. According to the FAIR cookbook, ELT-Tools data annotation is the process of data labeling, whereas data validation validates the data in terms of content and format to check if it has a set standard. Data aggregation is representation of the data in a compiled format. Basic tools used are Informatica, Talend, OMOP, etc. Data transformation in bioinformatics and cheminformatics is done by various querying tools such as PGA, Query Tabular, DISQOVER, OpenRefine-metadata-extension, TransMART, etc. Chemotion is an important data repository performing various tasks and is often used in chemoinformatics (70).

Next generation sequencing (NGS) technology has resulted in massive amounts of proteomics and genomics data. ETL (extraction, transformation, loading) is an important step in designing data analytics applications. ETL requires proper understanding of features of data. Data format plays a key role in understanding data, representation of data, space required to store data, data I/O during processing of data, intermediate results of processing, in-memory analysis of data, and overall time required to process data. Different data mining and ML algorithms require input data in specific types and formats. It will help researchers and developers in choosing appropriate data format to be used for a particular tool or algorithm (71).

To illustrate that there is a set of data that has the sense that it can be linked to each other like a lineage we can modify or transform that data in a view like an hierarchical tree so that data is clearer to demonstrate and for further studies.

4.3 Big Data Analysis

In the whole genome of the organism, there are various things to consider with respect to genome size, its GC content, and genes, which can be manipulated as a show cause of mutagenesis. Similarly to chemoinformatics the rate of the reaction, the ligand-binding sites studies, molecule structure, 2D or 3D alignment, and many other factors make the data large and noisy to analyze.

Big data can be analyzed by ML algorithms, that is, supervised and unsupervised algorithms (Figure 1.7).

There are R programming languages that can be used in ML (72). Machine learning will simply make the particular build up model to get trained as the model will get newer and newer data from various sources.

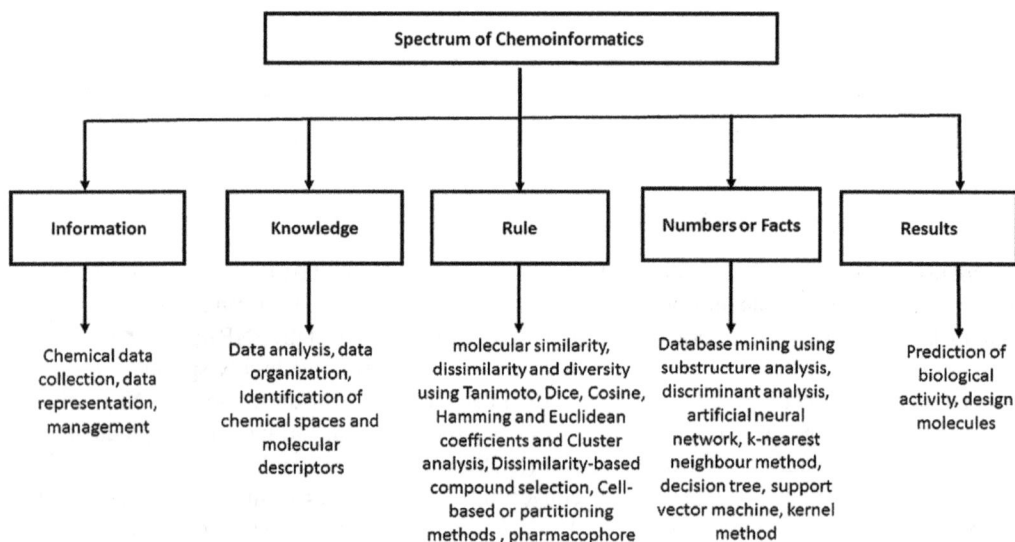

FIGURE 1.7   The Spectrum of Chemoinformatics.

There are two types of approaches to data analysis: deep analysis, which shows how there are integrated working between the regulatory systems, and broad analysis, which maintains and shows the principles at different expression levels.

The largest deep datasets are the Cancer Genome Atlas, which measures somatic mutations, copy number variations, mRNA expressions, protein expression, and miRNA expression. Unsupervised ML uses ML algorithms to analyze and cluster unlabeled datasets. These algorithms discover hidden patterns or data grouping without any need for human intervention. An unsupervised algorithm called PCA is used to identify hidden features and the variability in the given dataset (73).

Since evolutionary divergence from a common ancestor, genes can be found in the different organisms, with identical functions and/or protein motifs. The way to do this is by sequence analysis. Hence, to analyze and identify big data, it is the major part to work on in the modern world (74).

Publicly Available Resources

Bioinformatics is a scientific subdiscipline that involves using computer technology to collect, store, analyze, and disseminate biological data and information, such as DNA and amino acid sequences or annotations about those sequences whereas chemoinformatics is related to the use of physical chemistry theory. There are numerous publicly available resources that contain vast databases related to biological and chemical data for a single protein or a related drug. These databases as a whole are linked to each other via a unique ID or an accession number, which forms a cluster of data called metadata.

5 BIOCHEMOINFORMATICS: INTEGRATING BIOINFORMATICS AND CHEMOINFORMATICS

With the increase in the availability of biomolecular data, several parallel technological advancements have enabled the elucidation of the information contained within these molecules. These technological advances consequently gave rise to various "omics" disciplines that were focused on thoroughly characterizing specific biomolecules. For example, genomics, proteomics, metabolomics, and transcriptomics are disciplines that are focused on characterizing entire sets of genes, proteins, metabolites, and transcripts (mRNA), respectively, within a biological system.

5.1 GENOMICS

Genomics is one of the first "omics" disciplines that gained importance with the increase in the amount of gene sequences that were propelled by advances in the sequencing technologies. Specifically, the inception and completion of the species-specific whole genome sequencing projects, including the Human Genome Project (51), enabled characterizing the genetic makeup of species. On the other hand, sequencing has also enabled to deduce the structure and make-up of species within an environmental niche through metagenomics, which is usually based on the diversity analysis of a small sample of genes (75), of which the majority are ribosomal genes due to their comparatively high conservation across phylogenetic clades. In the recent past, metagenomics has made a significant contribution to microbial ecology by broadening our understanding of the microbial makeup of various environmental niches. Over the last two decades, the availability of reference and whole genome sequences have increased exponentially. RefSeq (76) (www.ncbi.nlm.nih.gov/refseq), a sequence database contained within the NCBI repository, catalogues annotated reference genomic, transcript, and protein sequences. This database has seen rapid increase in the number of deposited sequences and contains a total of 321,282,996 reference accessions from various species (www.ncbi.nlm.nih.gov/refseq/statistics/, DOA–11th July, 2022). Given the reference sequences, highlighting naturally occurring variants, which could either be single position (single nucleotide polymorphisms, SNPs) or multiple position (insertions and/or deletions, indels) variants, have garnered attention due to their potential roles in altered phenotypes. dbSNP (77) (www.ncbi.nlm.nih.gov/snp/) is a database that catalogues genomic variants. To further explore the clinical impacts of a given genetic variant, dbSNP also aggregates annotation information on clinically relevant variants, which is available on the ClinVar database (78) (www.ncbi.nlm.nih.gov/clinvar/). ClinVar is an archive containing information on clinically relevant variants within the human genome that are annotated through various community projects.

On occurrence, natural variants could be subjected to the actions of natural selection, depending on their impact on the overall "fitness" of species. Population genetics attempts to elucidate the action of selection on such variants by observing the changes in their frequencies over time within populations. Similar to population genetics, comparative genomics could be employed to infer the action of selection on genomic elements by comparing their levels of conservation across genetically related species via whole genome alignments. Such approaches are aimed at inferring the proportional of functional elements within a given genome. Functional genomics attempts to assign genetic sequences to their function by correlating genotype and phenotype. To exemplify, accumulating data from various biochemical assays like ChIP-seq, ATAC-seq, etc., databases like ReMap (79) (remap.univ-amu.fr/), JASPAR (80) (jaspar.genereg.net/), etc., highlight the biochemically active regions within the genome that exhibit some binding affinity for transcriptional regulators. However, specifically in the context of disease biology, Genome Wide Association Studies (GWAS) have garnered importance in the recent past in highlighting potential loci within the genome that could be responsible for change in phenotype. From a pool of observed genomic data, GWAS attempts to associate genomic variants to their consequential phenotype through various statistical inferences.

5.2 PROTEOMICS

Consequential to the focus on genomics, protein characterization has also gained impetus given its central role in various biological processes. The term "proteome" refers to the entire set of proteins within a given biological system. Similar to genomic sequences, characterizing the sequence information, and additionally structural information, is an important starting point for proteomics studies. Proteomics data can be obtained from either shotgun proteomics or targeted proteomics (81), which both heavily employ mass spectrometry techniques for elucidating the peptide content within a given sample. Shotgun proteomics is an exploratory approach used to highlight the underlying protein makeup for a given sample. On the other hand, targeted proteomics are aimed at identifying specific proteins within a given sample via monitoring the signature peptides, also referred to as proteotypic peptides, that are specific to the protein(s) of interest (82). Similar to targeted proteomics, proteogenomics technique aims to identify the peptide content within a given sample by comparing it to a given protein sequence database. This protein sequence database consists of the desired protein sequence(s) of interest, and is obtained from genomic and transcriptomic sequence information (83). PeptideAtlas (84) (www.peptideatlas.org/) and PRIDE (85) (www.ebi.ac.uk/pride/) are two of the central databases containing data from proteomics experiments that is uploaded by individual contributors.

UniProt Knowledgebase (86) (UniProtKB) is a comprehensive dataset containing sequence and other diverse protein information obtained from various species (www.uniprot.org/uniprotkb). The plethora of proteins hosted on UniProtKB can be broadly classified into two categories on the basis of the manner of protein annotation, Swiss-Prot identifiers and TrEMBL identifiers, where the former are manually annotated and the latter are automatically/computationally annotated protein sequences. A majority of the sequences hosted on UniProtKB are obtained by translating the coding sequences deposited in the International Nucleotide Sequence Database Collaboration (87) (INSDC, www.insdc.org/). In addition to the sequence information, since the *2022_02* Uniprot release, protein entry pages now also contain information on the predicted 3D of the protein. These predicted structures are made available via AlphaFoldDB (28), which is an extensive implementation of AlphaFold2, a recently proposed tool that employs ML for predicting protein structure from sequence information.

The RCSB PDB (Research Collaboratory for Structural Bioinformatics Protein Data Bank) is one of the central protein structure databases (88) (www.rcsb.org/) that aggregates information on experimentally inferred structures that is uploaded by individual contributors. Besides the structure and sequence information on proteins, various association studies have enabled the construction of protein networks to highlight interaction between proteins via various evidences. String database (21) (string-db.org/) is a protein–protein interaction database, where the networks are developed on the basis of various criteria like protein co-expression, protein homology, gene co-occurrence, etc. Similar to String database, IntAct (23) (www.ebi.ac.uk/intact) and BioGrid (22) (thebiogrid.org/) are other examples of interaction databases that contain information on the protein–protein interaction networks; these networks are inferred from various sources of information, including data curation, individual submissions, etc.

5.3   COMPLEX CHEMICAL STRUCTURES

Representing chemical structure is a challenging task mainly due to their complex structures. One of the earliest attempts on molecular representations was the Wiswesser Line Notation (WLN) (89). WLN codified chemical structures, including the information on the nature of chemical bonds, into a line notation. Along similar lines, connection tables were also used as a molecular representation technique, where all the atoms and bonds within a molecule were represented as separate tables (90). Due to the tabular representation, connection tables were comparatively easier to understand as compared to the line notations. Following these seminal molecular representations, various other recently proposed representation formats have been proposed, some of which are listed below:

- SMILES (Simplified Molecular Input Line Entry Specification) –Similar to WLN, SMILES is a line notation format that condenses the atom and bond information from molecules into a single line (91). SMILES format is readable in most of the molecular editors available today, and could be used to reconstruct the two–of 3D molecular structure.
- SDF (Structure Data File) – The SDF format is a tabular representation of a given chemical structure. In principle this format is very similar to the connection table; however, besides the information on the atom and their respective bonds, the SDF format contains other meta-information, such as chemical formula of the molecule (if known), 3D coordinates of every atom, and a summary of the total number of atoms and bonds contained within the molecule.
- InChI (International Chemical Identifier) – Introduced by the IUPAC (International Union of Pure and Applied Chemistry), InChI takes an in-depth approach to chemical representation, thus enabling a unique identifier per molecule (92). InChI is primed to have a wide array of applications as it could be applied in the organic domain, and also to an extent on the inorganic domain of chemistry.

Due to the advances in structure elucidation techniques and increase in storage capacities, the availability of structural information has seen a significant increase. Today, various small molecule databases enable access to structure information of a wide array of molecules. Some of these databases are listed here:

- PubChem (pubchem.ncbi.nlm.nih.gov/) – This is an open dataset hosted by the National Institute of Health (NIH), which contains information on various small molecules and, in some cases, also on large molecules. For every molecule entry, PubChem provides information on structure, chemical and physical properties, toxicity, etc.

- ChEMBL (www.ebi.ac.uk/chembl/) – This is a manually curated database of bioactive molecules hosted by the European Bioinformatics Institute (EBI). Here, every molecule has information relating to its potential drug activity.
- ChemSpider (www.chemspider.com/) – Similar to PubChem, ChemSpider contains information on a plethora of small molecules that is curated and maintained by the developers.

Developing a novel therapeutic molecule is a daunting task; here development is termed as a process that encompasses identifying the molecule, processing it through rigorous clinical trials, and marketing. It has been documented that, on average, the process of developing such a novel therapeutic molecule could take up to 12 years or more and cost $2.6 billion (93).

6 ARTIFICIAL INTELLIGENCE IN HEALTHCARE

In this review article we discuss different aspects of resources available to the general public and other communities in the field of bioinformatics and chemoinformatics. Databases can be of various types including biological databases, chemical databases, composite databases, RNA databases, phenotype databases etc. Each one of these databases is specialized to contain comprehensive, verified, easy to understand, and accurate information that not only helps the research communities with project work but also drives them to gain so much more relevant information about the same.

Artificial intelligence is a huge part of healthcare systems worldwide, and will change the face of medicine in the upcoming years. The goal of AI in medical management is to cut out any loose ends that might be caused as a form of human errors, efficiently utilize time, cost-efficient equipment of machinery and improve productivity. The increasing need of AI in the healthcare sector has not only resulted in innovation of new technology to cure many defects but has also decreased labor work in hospitals.

Artificial intelligence is a field in computer science that aids in developing machinery that is trained to make quick, logical decisions based on mathematical algorithms merged into them. Artificial intelligence can pick up small details, communicate, display, provide visualization, and make human decisions. Several aspects like Natural Processing Language, deep learning, ML, big data analysis, implants, and non-invasive devices are key parts of AI.

Developing countries experience a more powerful impact of AI in healthcare due to its competence in being used wherever medical practitioners are short-handed. There have been numerous incidents where AI devices have been proven to be more efficient in maintaining healthcare than humans. In this section, we look at some of the chief aspects of AI and how it has overturned the medical community.

Artificial intelligence has been on the rise in recent years, with the help of which healthcare systems and treatment plans have been established globally. AI technology is superseding human intelligence in several ways from cybersecurity to contributing towards the appropriate medication facilities. Every other industry in alignment with the healthcare sector has been positively influenced by the grossly increasing adaptation of AI all over the world which is estimated to reach a figure of 194.14 billion (2022–2030). Analyzing the forecast, AI would become a prioritized sector trustworthy enough to be invested additionally into to get lesser complexities into healthcare projects. In this section we discuss the advancements in AI optimizing its efficiency in the coming years with a Compound Annual Growth Rate (CAGR) of 38.1%.

6.1 IMPLICATIONS OF AI ML

The increasing magnitude of datasets has been the core point for technology to begin. Big data has created an urgent need for ML algorithms and various pattern distribution techniques to simultaneously handle numerous health records. The use of scatterplots, correlation coefficients, supervised and unsupervised ML techniques aids the medical community in achieving accurate diagnosis of diseases that can not be visualized as efficiently by humans.

Machine learning is learning from experience with respect to a class of tasks for which appropriate performance measures should be defined that improves over time. It consists of two major categories: supervised and unsupervised algorithms. Supervised ML is when input is categorical and output is a numeric value (e.g., calculating temperature).

Deep learning is a part of data science that subsumes the methodology of how humans interpret things and builds useful technology formulated from the same. A simple example is face detection used to unlock a smartphone. Here, the deep learning system identifies and records face patterns. In terms of medicine, wearable devices track patient details for conveying heart conditions, overall health, blood pressure, sugar levels, etc. It is a multilayer network that learns from huge amounts of structured and unstructured data. Higher and complex layers of representation can be solved with deep learning (94). It is more efficient for analyzing images like CT scan or MRI, X-rays, etc., since it is less dependent on human interaction. Mathematical formulae involved are more complicated as it is a fully functioning computational system used for detection of various patterns in medicine.

In 2019 unsupervised clustering algorithms from ML were applied in diagnostics and to monitor changes in Alzheimer's patient behavior. Clustering algorithm is a ML technique that groups similar patterns together and keeps it from outliers, which helps determine the probable cause of data being unfinished. This helped quite a lot in revealing the patterns of changes that took place in patients that were missed by the medical team. Various studies were conducted for studying the differing neurological patterns in a group of patients. Algorithms such as K-means, multilayer clustering, and hierarchical agglomerative clustering are used to visualize the neurological and behavioral deficits. Artificial intelligence has given us the advantage of being able to earlier diagnose such genetically inherited disorders that still lack a cure.

Latent profile analysis is an algorithm massively applied to Alzheimer's study as it gives a more scattered analysis. However, it was being used in younger age groups for probable cause of depression and anxiety which are the peak for mental health issues.

As of 2022, the leading cause of death is coronary artery disease (CAD) and ischemic heart disease. CAD often goes unnoticed until a heart attack episode as it advances over the years until it cuts off an artery completely. Both of these heart complications can be diagnosed effectively using physical checkups and tests run by doctors that would remain insufficient without the implementation of multiple imaging tests such as computerized tomography (CT scan), EKG, angiography, MRI, etc. These imaging tests use artificial intelligence. Medical imaging and accurate analysis are the key to surgeries for analyzing any abnormalities in the patient; without which diagnosis of any disorder would ultimately hamper patient care. Advancement of such ML techniques and scanning of biological data through these can produce reliable and quick results. Deep learning (DL) has also been used for predicting if a person has CAD or not. Recently, DL algorithms have been effectively able to detect CAD using electrocardiograph (ECG) signals, which was an example of AI-built technology wherein ML & DL techniques compared various findings, symptoms, tests and reported ECG as an accomplished model for diagnosing CAD (95). The results of these scans and tests are one of the major deciding factors for medical practitioners to plan the entire course of treatment. After accurate diagnosis of an abnormality is found accurate diagnosis would help in reducing chronic healthcare.

6.2 AI IN BIOINFORMATICS

Needleman Wunsch algorithm is a sequence alignment algorithm that performs basic global alignment on protein and amino acid sequences. In bioinformatics databases there are proteins that do not have the protein structure, but just have their amino acid residue sequence. Through needleman wunsch algorithm generation of different global alignment proteins derived from the single template can be done.

The new proteins which will be received as a result can be homologous proteins, and by this we can get that specific gene's protein. Several applications can be implied using the very same AI system. A model can be trained with AI in such a way that the user has to enter a template on the basis of which the model will identify similar sequence homologies, analyze scores, and hits by itself. Furthermore, protein modeling can be done on the basis of this acquired data. This is a simple representation of how a model developed with AI can work if incorporated with various algorithms.

6.3   Transforming Healthcare with AI

Over the last decade, drastic changes in data handling technologies have been reported. This was assembled when Boolean algebra, data mining, and extraction tools were fused together to build systems supporting the healthcare sector. The foremost application of introducing AI in the healthcare industry was done for medical diagnostics. The functioning of multiple data mining tools was a prerequisite and a crucial part to achieve the same.

Early detection of genetically inherited disorders (such as Alzheimer's mentioned previously) and other abnormalities in the body became much easier with technological advances. Incorporating mathematical formulas, distribution pattern, mapping of images, scans for ruling out, and furthermore communing out symptoms of a patient for effective treatment and nursing facilities was soon accomplished. Advancement of machinery combined with human knowledge led to the construction of AI.

Natural language processing (NLP) is basic analysis of unstructured knowledge or clinical notes built as a part of AI. It plays a central role in understanding human behavior and speech and text recognition that can be applied to improve patient care. NLP is most widely used by the healthcare sector. It usually captures and utilizes incomplete clinical data that is converted to sensible data understandable by doctors. NLP helps to reduce overall healthcare costs and has improved healthcare to a large extent (96). Recent studies have shown that NLP systems are being used for early diagnosis of aggressive diseases that do not show any symptoms until after a life-threatening event. Kawasaki disease is an example, in which misdiagnosis of any kind or ruling out any symptoms can put a patient's life in danger. NLP algorithms and formulae have been shown to have a sensitivity of 94% and specificity of 77.5%, which surpasses other clinical tests.

Furthermore, this is the point where fuzzy C-means might sound more appealing. Hypothesis of non-specific data reduced down to a range of possible values is the basic ideal of fuzzy means in data mining. The unreliability of data stimulates the introduction of fuzzy means, which can handle high levels of vagueness in data.

- Calculations of big data are one major application.
- As the number of cases dependent on fuzzy logic (FK) are increasing, it is used as a factor for selecting process of treatment for disease.
- Rate of risks and considerable output.

6.4   AI in Drug Design and Development

AI-enabled systems are able to identify new drug targets, their toxic traits, and the action of a particular drug following the pattern of distribution in the body. These algorithms were the basis for establishing a "drug discovery platform." It permitted companies to reinstate drug policies after reinvestigating existing drugs and active metabolites. The need to handle large datasets containing biological data of a chemical, non-uniformity of these datasets, lack of sorting them has developed the need for quantitative structure-activity relationship (QSAR). QSAR is a computational-based model that identifies inhibitors, analyzes the ADME profile of a compound, and basically facilitates the relationship between the physical properties and biological activity of a compound. This tool is best used in the prediction of complex 3D structures of proteins that may give a clear idea of the components present in the protein. Drawbacks of QSAR led to development of deep learning studies, which are used for testing and recording safety, efficacy, adverse effects, and rate of successful workups of patients. Successful clinical trials are conducted globally now as we are able to invest more and get more precise results out of these AI machines. Installing AI in the pharmaceutical industry has become one of the topmost applications and transformations in the timeline (97).

Virtual healthcare has been around since the late 19th century. One such product of virtual healthcare is "Molly," a virtual nursing assistant created by Sense.ly. Sense.ly is an AI-powered virtual caregiver that can virtually monitor patient's progress, follow-ups from doctors, and track overall health. It asks patients to update their status through a smartphone app. It has also been tested to give off mental health counseling and empathy sensing which makes it a robot capable of showing human emotions (98). Several hospitals and investors believed that Molly would take up a huge part in medical innovations and research technology. If made affordable and accessible to the general public, it would make people live longer. It is now providing assistance to "massive healthcare providers including National Health Services of the UK" according to https://techcrunch.com/.

Furthermore, according to www.gavi.org/, new treatment plans for COVID-19 from AI are under development. This technology has already been customized for discovering new antibiotics. COVID-19 patients could be diagnosed using algorithms applied on CT scans of lungs of patients. A high-performance deep learning algorithm was set up after training of 131 CT volumes collected from Jan. 24 to Feb. 6, 2020. This helped in directly clustering out COVID-positive patients from COVID-negative patients within seconds of results (99). As the CT scans are not the only prime factor for starting treatment, AI systems started to cross-match CT scans of lungs to the other patient data such as blood grouping, age, medical history, etc. This made it possible to match any symptoms or possible contact with the virus.

A major study conducted by researchers at MIT was the diagnosis of Covid using cough sounds. An AI speech processing model was developed that powered the audial biomarker feature for scanning COVID-19 patients from cough recordings. They were balanced in a large audio dataset. With this a patient saliency map was given to each patient that non-invasively monitored patient progress over time and with real patient data. The results showed a COVID-19 sensitivity of 98.5% with a specificity of 94.2% when cross-checked with patients who took an official COVID test. Asymptomatic patients showed 100% sensitivity and 83.2% specificity (100).

As can be seen, NLP, ANN, ML, robotics, big data analysis, and deep learning have made AI successful in building a strong reputation in healthcare and other fields.

6.5 FUTURE PROSPECTS AND CHALLENGES

In the near future scientists believe it is likely that AI systems will take over human practice. In the past decade AI has gone from not existing to being installed in every industry possible (101). New systems and tools are always opening up to us as opportunities for uncovering answers to medical problems. Mass testing and tracking of clinical trials and data would not be possible without computers. As per the "rule-based expert system" pent down by ForeSee medical (software company in San Diego) the fact that ML will soon replace the current systems with innovations that are better than the present times cannot go unnoticed (102).

Moreover, latent profile analysis contributes a huge amount in identifying eating disorders in several age groups which were then worked on for attaining required results. Innovations and advancements are and never were a hurdle for AI to develop, we just needed to have the knowledge to use it appropriately. Without these studies and algorithms we would never have achieved the level we have now. It has enabled us to diagnose many diseases and even manage the pandemic.

However, there are some challenges that need addressed to discover more of what AI has to offer to us. One of the foremost challenges is the interpretation of data shown by systems. Meticulously calculated algorithms displayed in ML techniques are not easy to interpret by humans. It uses specialized tools which are not as handy for every AI application. It is unknown for a layman to see how deep learning models can efficiently produce results. This lack of knowledge and understanding is the main limitation of AI (103). Another challenge may be the lack of personnel to be able to operate AI devices accurately. AI systems are built to perform a particular task with proficiency after reading a large amount of data, but they cannot complete the whole procedure without any human intervention. Lack of conviction on the data produced by AI can be an additional problem as not all machines in existence are entirely certain about their results. More tests and appropriate time and amount of training must be given to cut off any outliers and preserve the result prediction theory. Data scientists and software engineers have a clear understanding of what AI is capable of and the targets to achieve the R&D goals. Moreover, AI can be developed with overfitting and underfitting algorithms applied as and when necessary (104). This implies that AI technology and human correlation go hand in hand, and it is the foundation of what bioinformatics has to achieve in the future.

Luckily, we haven't stopped making advancements. Incision-less surgeries, 4D imaging, mechanical organs, and stem cell therapy are all achievable due to this technology (105). It is also reported to be assisting surgeons in complex surgeries. One successful experiment leads to the hope that others will work too. All of this is the reason AI is being called the future of software development. This is what keeps humans going in the very direction of inventing new AI-enabled systems every day.

7   CONCLUSION

Artificial intelligence has undoubtedly changed the way we do things. From the collection of small data in computers to retrieving big data from large systems, leading to new databases that can be used by all the communities now. Artificial intelligence can be used to sort the database in the field of bioinformatics and chemoinformatics. AI focuses on the data that has been collected from the previous test data so that it can train itself to be more "intelligent," accurate, and precise.

In bioinformatics, AI can be used to cluster a broad set of related data and represent them with the help of high-profile images, categories, and represent it in a much simpler way. Bioinformatics databases are a large repository of complex DNA sequences, protein structures, sequence formats, dendrograms, phylogenetic trees, etc., which allows this data to be visualized, analyzed, and accessed easily using ML tools like decision trees, ANN, K-Nearest neighbor algorithm, etc. It collects information from numerous databases making big data and performs BLAST search on a target sequence resulting in all the similar sequences matching our target; which also means, big data was successfully analyzed by BLAST algorithm to form a much smaller data from which homology can be determined.

The whole genome of the organism can be visualized in the form of links or connections between the set of genes (ANN). The complete sequence homology and similarity can be fully determined by making training models having the core principle of bioinformatics studies. Artificial neural networks have been specially employed for dealing with large network sets by expressing relations and network connections between two data points. Its multilayer perception has risen above to prove to us its dominance in the majority of bioinformatics tools. The systems biology domain is currently running on ML algorithms and using signal transduction networks, representing enormous amounts of dissimilar to similar molecules via ANN, graphical models, genome-wide association, enzyme prediction and function, and most importantly modeling genetic networks. Building such genetic networks is the core to understanding the markers of a disease or to rule out certain proteins. AI will continue in this field with the development of other AI inventions helping us to better understand the relationship between two components. Bioinformatics is one of the major philanthropists to the present-day AI modernization.

In chemoinformatics, big data, also known as BigChem, mostly involves large databases with tons of data about the properties, characteristics, and structures of chemical compounds. In order for this data to be meaningful and comprehensive, it must be analyzed thoroughly. This is where AI comes into play, as several deep learning techniques such as CNN and RNN are used, allowing AI to establish new heights in this field using databases such as DrugBank or chEMBL to generate data having the same chemical compound in any drug. This tells the user about similarities in the content between two drugs. Cancer-related drug research is now using AI as there are various sets of pathways and metabolic processes that are difficult to interpret and follow, and AI makes the task easier as the pathways and metabolic process of the drugs get assembled into a drug delivery process. Data visualization techniques are called on to solve random and abstract patterns and make sense of of the data and suggest neural pathways for chemical compounds. Chemoinformatics can visualize a chemical at its molecular level to get information about the bonds and energies present, which in turn can be used to dock it with another molecule via different AI applications. Analyzing and visualizing compounds is becoming challenging and demanding day after day for which advanced ML models and algorithms can be applied to handle such big data. Therapeutic and pharmacological information of biological molecules can be easily retrieved using smart ML technology. Other AI applications under development including identifying the active site on a molecule and relating its structure with any other similar molecules to better understand the molecule better and modify it for other purposes. Visualizing complex compounds is growing in popularity because of advanced ML algorithms that are applied to various tools used in chemoinformatics. Although more studies are needed on the accuracy of support vector machine (SVM), it is also important to note that new technologies such as SVM are being tried for decreasing the number of false outcomes, making data more robust, increasing efficiency of decision trees and so on in the field of chemoinformatics.

In light of AI being used everywhere in the modern world today, it is no surprise how rapidly it is being used to make raw data more precise, reliable, and purposeful. Bioinformatics and chemoinformatics are such growing areas of the industry that more research and efforts into it will actually result in ML and informatics going hand in hand. The visualization of biological data will be less hampered in the process of exponentially big data falling on systems. AI ML is the basis of what bioinformatics and chemoinformatics will accomplish. Without the development of AI ML, we will still lack precision, accuracy, time, and management of big raw data will be failing. Exploration of data will be much easier even for a heterogenous data set. Computationally challenging databases can be improved to develop highly standardized, comprehensive, and easy-to-access databases with the developing of completely new tools.

REFERENCES

1. Sanger F. Sequences, sequences, and sequences. Annu Rev Biochem [Internet]. 1988 Jun [cited 2022 Sep 2];57(1):1–29. Available from: pubmed.ncbi.nlm.nih.gov/2460023/
2. Sanger F. Chemistry of insulin. Science (80-) [Internet]. 1959 May 15 [cited 2022 Sep 2];129(3359):1340–4. Available from: www.science.org/doi/10.1126/science.129.3359.1340
3. Perutz MF. Hemoglobin structure and respiratory transport. Sci Am [Internet]. 1978 [cited 2022 Sep 2];239(6):92–125. Available from: pubmed.ncbi.nlm.nih.gov/734439/
4. Kendrew JC, Bodo G, Dintzis HM, Parrish RG, Wyckoff H, Phillips DC. A Three-Dimensional Model of the Myoglobin Molecule Obtained by X-Ray Analysis. Nat 1958 1814610 [Internet]. 1958 Mar 1 [cited 2022 Sep 2];181(4610):662–6. Available from: www.nature.com/articles/181662a0
5. Avery OT, Macleod CM, McCarty M. STUDIES ON THE CHEMICAL NATURE OF THE SUBSTANCE INDUCING TRANSFORMATION OF PNEUMOCOCCAL TYPES: INDUCTION OF TRANSFORMATION BY A DESOXYRIBONUCLEIC ACID FRACTION ISOLATED FROM PNEUMOCOCCUS TYPE III. J Exp Med [Internet]. 1944 Feb 2 [cited 2022 Sep 2];79(2):137. Available from:/pmc/articles/PMC2135445/?report=abstract

6. HERSHEY AD, CHASE M. INDEPENDENT FUNCTIONS OF VIRAL PROTEIN AND NUCLEIC ACID IN GROWTH OF BACTERIOPHAGE. J Gen Physiol [Internet]. 1952 Sep 9 [cited 2022 Sep 2];36(1):39. Available from:/pmc/articles/PMC2147348/?report=abstract

7. Watson JD, Crick FHC. Molecular Structure of Nucleic Acids: A Structure for Deoxyribose Nucleic Acid. Nat 1953 1714356 [Internet]. 1953 Apr 25 [cited 2022 Sep 2];171(4356):737–8. Available from: www.nature.com/articles/171737a0

8. Holley RW, Apgar J, Everett GA, Madison JT, Marquisee M, Merrill SH, et al. Structure of a Ribonucleic Acid. Science (80-) [Internet]. 1965 Mar 19 [cited 2022 Sep 2];147(3664):1462–5. Available from: www.science.org/doi/10.1126/science.147.3664.1462

9. Maxam AM, Gilbert W. A new method for sequencing DNA. Proc Natl Acad Sci [Internet]. 1977 Feb 1 [cited 2022 Sep 2];74(2):560–4. Available from: www.pnas.org/doi/abs/10.1073/pnas.74.2.560

10. Sanger F, Nicklen S, Coulson AR. DNA sequencing with chain-terminating inhibitors. Proc Natl Acad Sci [Internet]. 1977 Dec 1 [cited 2022 Sep 2];74(12):5463–7. Available from: www.pnas.org/doi/abs/10.1073/pnas.74.12.5463

11. Moody G. Digital code of life: how bioinformatics is revolutionizing science, medicine, and business. 2004 [cited 2022 Sep 2];389. Available from: www.wiley.com/en-us/Digital+Code+of+Life%3A+How+Bioinformatics+is+Revolutionizing+Science%2C+Medicine%2C+and+Business-p-9780471327882

12. Gauthier J, Vincent AT, Charette SJ, Derome N. A brief history of bioinformatics. Brief Bioinform [Internet]. 2019 Nov 27 [cited 2022 Sep 2];20(6):1981–96. Available from: academic.oup.com/bib/article/20/6/1981/5066445

13. Strasser BJ. Collecting, Comparing, and Computing Sequences: The Making of Margaret O. Dayhoff's Atlas of Protein Sequence and Structure, 1954–1965. J Hist Biol 2009 434 [Internet]. 2009 Dec 24 [cited 2022 Sep 2];43(4):623–60. Available from: link.springer.com/article/10.1007/s10739-009-9221-0

14. Poplin R, Chang PC, Alexander D, Schwartz S, Colthurst T, Ku A, et al. A universal SNP and small-indel variant caller using deep neural networks. Nat Biotechnol 2018 3610 [Internet]. 2018 Sep 24 [cited 2022 Sep 2];36(10):983–7. Available from: www.nature.com/articles/nbt.4235

15. McKenna A, Hanna M, Banks E, Sivachenko A, Cibulskis K, Kernytsky A, et al. The Genome Analysis Toolkit: A MapReduce framework for analyzing next-generation DNA sequencing data. Genome Res [Internet]. 2010 Sep 1 [cited 2022 Sep 2];20(9):1297–303. Available from: genome.cshlp.org/content/20/9/1297.full

16. Koboldt DC, Chen K, Wylie T, Larson DE, McLellan MD, Mardis ER, et al. VarScan: variant detection in massively parallel sequencing of individual and pooled samples. Bioinformatics [Internet]. 2009 [cited 2022 Sep 2];25(17):2283–5. Available from: pubmed.ncbi.nlm.nih.gov/19542151/

17. Ning L, Liu G, Li G, Hou Y, Tong Y, He J. Current challenges in the bioinformatics of single cell genomics. Front Oncol. 2014;4 JAN:7.

18. Chen C, Hou J, Tanner JJ, Cheng J. Bioinformatics Methods for Mass Spectrometry-Based Proteomics Data Analysis. Int J Mol Sci [Internet]. 2020 Apr 2 [cited 2022 Sep 2];21(8). Available from:/pmc/articles/PMC7216093/

19. Mi H, Muruganujan A, Ebert D, Huang X, Thomas PD. PANTHER version 14: more genomes, a new PANTHER GO-slim and improvements in enrichment analysis tools. Nucleic Acids Res [Internet]. 2019 Jan 8 [cited 2022 Sep 2];47(D1):D419–26. Available from: academic.oup.com/nar/article/47/D1/D419/5165346

20. Jones P, Binns D, Chang HY, Fraser M, Li W, McAnulla C, et al. InterProScan 5: genome-scale protein function classification. Bioinformatics [Internet]. 2014 May 1 [cited 2022 Sep 2];30(9):1236–40. Available from: academic.oup.com/bioinformatics/article/30/9/1236/237988

21. Szklarczyk D, Gable AL, Nastou KC, Lyon D, Kirsch R, Pyysalo S, et al. The STRING database in 2021: customizable protein-protein networks, and functional characterization of user-uploaded gene/measurement sets. Nucleic Acids Res [Internet]. 2021 Jan 8 [cited 2022 Sep 2];49(D1):D605–12. Available from: pubmed.ncbi.nlm.nih.gov/33237311/

22. Stark C, Breitkreutz BJ, Reguly T, Boucher L, Breitkreutz A, Tyers M. BioGRID: a general repository for interaction datasets. Nucleic Acids Res [Internet]. 2006 Jan 1 [cited 2022 Sep 2];34(Database issue):D535. Available from:/pmc/articles/PMC1347471/

23. del Toro N, Shrivastava A, Ragueneau E, Meldal B, Combe C, Barrera E, et al. The IntAct database: efficient access to fine-grained molecular interaction data. Nucleic Acids Res [Internet]. 2022 Jan

7 [cited 2022 Sep 2];50(D1):D648–53. Available from: academic.oup.com/nar/article/50/D1/D648/6425548

24. Hassoun S, Rosa E. Non-Parametric Variational Inference with Graph Convolutional Networks for Gaussian Processes; A Tool for Predicting the Dark Side of Enzymes View project Metabolomics View project. Artic Integr Comp Biol [Internet]. 2021 [cited 2022 Sep 2]; Available from: doi.org/10.1093/icb/icab188

25. Tarca AL, Carey VJ, Chen X wen, Romero R, Drăghici S. Machine Learning and Its Applications to Biology. PLOS Comput Biol [Internet]. 2007 [cited 2022 Sep 2];3(6):e116. Available from: journals.plos.org/ploscompbiol/article?id=10.1371/journal.pcbi.0030116

26. Angermueller C, Pärnamaa T, Parts L, Stegle O. Deep learning for computational biology. Mol Syst Biol [Internet]. 2016 Jul 1 [cited 2022 Sep 2];12(7):878. Available from: onlinelibrary.wiley.com/doi/full/10.15252/msb.20156651

27. Rosenblatt F. The perceptron: A probabilistic model for information storage and organization in the brain. Psychol Rev [Internet]. 1958 Nov [cited 2022 Sep 2];65(6):386–408. Available from:/record/1959-09865-001

28. Jumper J, Evans R, Pritzel A, Green T, Figurnov M, Ronneberger O, et al. Highly accurate protein structure prediction with AlphaFold. Nat 2021 5967873 [Internet]. 2021 Jul 15 [cited 2022 Sep 2];596(7873):583–9. Available from: www.nature.com/articles/s41586-021-03819-2

29. Ramachandran A, Lumetta SS, Klee EW, Chen D. HELLO: improved neural network architectures and methodologies for small variant calling. BMC Bioinformatics [Internet]. 2021 Dec 1 [cited 2022 Sep 2];22(1):1–31. Available from: bmcbioinformatics.biomedcentral.com/articles/10.1186/s12859-021-04311-4

30. Hassanzadeh HR, Wang MD. DeeperBind: Enhancing prediction of sequence specificities of DNA binding proteins. Proc–2016 IEEE Int Conf Bioinforma Biomed BIBM 2016. 2017 Jan 17;178–83.

31. Chen C, Hou J, Shi X, Yang H, Birchler JA, Cheng J. DeepGRN: prediction of transcription factor binding site across cell-types using attention-based deep neural networks. BMC Bioinformatics [Internet]. 2021 Dec 1 [cited 2022 Sep 2];22(1):1–18. Available from: bmcbioinformatics.biomedcentral.com/articles/10.1186/s12859-020-03952-1

32. Trabelsi A, Chaabane M, Ben-Hur A. Comprehensive evaluation of deep learning architectures for prediction of DNA/RNA sequence binding specificities. Bioinformatics [Internet]. 2019 Jul 15 [cited 2022 Sep 2];35(14):i269–77. Available from: academic.oup.com/bioinformatics/article/35/14/i269/5529112

33. Zeng Y, Gong M, Lin M, Gao D, Zhang Y. A Review about Transcription Factor Binding Sites Prediction Based on Deep Learning. IEEE Access. 2020;

34. Engel T. Basic overview of chemoinformatics. J Chem Inf Model [Internet]. 2006 [cited 2022 Sep 12];46(6):2267–77. Available from: pubs.acs.org/doi/full/10.1021/ci600234z

35. Bhalerao SA, Verma DR, L D'souza R, Teli NC, Didwana VS. Chemoinformatics: The Application of Informatics Methods to Solve Chemical Problems. RJPBCS. 2013;4(3).

36. Firdaus Begam B, Satheesh Kumar J. A Study on Cheminformatics and its Applications on Modern Drug Discovery. Procedia Eng. 2012 Jan 1;38:1264–75.

37. Xu J, Hagler A. Chemoinformatics and Drug Discovery. Mol A J Synth Chem Nat Prod Chem [Internet]. 2002 [cited 2022 Sep 12];7(8):566. Available from:/pmc/articles/PMC6146447/

38. Hessler G, Baringhaus KH. Artificial Intelligence in Drug Design. Mol 2018, Vol 23, Page 2520 [Internet]. 2018 Oct 2 [cited 2022 Sep 12];23(10):2520. Available from: www.mdpi.com/1420-3049/23/10/2520/htm

39. Mitchell B.O. JBO. Machine learning methods in chemoinformatics. Wiley Interdiscip Rev Comput Mol Sci [Internet]. 2014 Sep 1 [cited 2022 Sep 12];4(5):468–81. Available from: onlinelibrary.wiley.com/doi/full/10.1002/wcms.1183

40. Ripphausen P, Nisius B, Bajorath J. State-of-the-art in ligand-based virtual screening. Drug Discov Today [Internet]. 2011 [cited 2022 Sep 12];16(9–10):372–6. Available from: pubmed.ncbi.nlm.nih.gov/21349346/

41. Polanski J. 4.14 Chemoinformatics.

42. Tetko I V., Engkvist O. From Big Data to Artificial Intelligence: chemoinformatics meets new challenges. J Cheminform [Internet]. 2020 Dec 1 [cited 2022 Sep 12];12(1):1–3. Available from: jcheminf.biomedcentral.com/articles/10.1186/s13321-020-00475-y

43. Bhhatarai B, Walters WP, Hop CECA, Lanza G, Ekins S. Opportunities and challenges using artificial intelligence in ADME/Tox. Nat Mater 2019 185 [Internet]. 2019 Apr 18 [cited 2022 Sep 12];18(5):418–22. Available from: www.nature.com/articles/s41563-019-0332-5

44. Ertl P, Schuffenhauer A, Renner S. The scaffold tree: an efficient navigation in the scaffold universe. Methods Mol Biol [Internet]. 2011 [cited 2022 Sep 12];672:245–60. Available from: pubmed.ncbi.nlm.nih.gov/20838972/

45. Duffy BC, Zhu L, Decornez H, Kitchen DB. Early phase drug discovery: Cheminformatics and computational techniques in identifying lead series. Bioorg Med Chem. 2012 Sep 15 ;20(18):5324–42.

46. Koes DR, Camacho CJ. ZINCPharmer: pharmacophore search of the ZINC database. Nucleic Acids Res [Internet]. 2012 Jul [cited 2022 Sep 12];40(Web Server issue). Available from: pubmed.ncbi.nlm.nih.gov/22553363/

47. Chen Z, Li HL, Zhang QJ, Bao XG, Yu KQ, Luo XM, et al. Pharmacophore-based virtual screening versus docking-based virtual screening: a benchmark comparison against eight targets. Acta Pharmacol Sin 2009 3012 [Internet]. 2009 Nov 23 [cited 2022 Sep 12];30(12):1694–708. Available from: www.nature.com/articles/aps2009159

48. Wolber G, Langer T. 0: 3-D pharmacophores derived from protein-bound ligands and their use as virtual screening filters. J Chem Inf Model [Internet]. 2005 [cited 2022 Sep 12];45(1):160–9. Available from: pubs.acs.org/doi/full/10.1021/ci049885e

49. Tuccinardi T. Docking-based virtual screening: recent developments. Comb Chem High Throughput Screen [Internet]. 2009 Mar 4 [cited 2022 Sep 12];12(3):303–14. Available from: pubmed.ncbi.nlm.nih.gov/19275536/

50. Shu Y, McCauley J. GISAID: Global initiative on sharing all influenza data–from vision to reality. Eurosurveillance. 2017 Mar 30;22(13):30494.

51. International Human Genome Sequencing Consortium. Initial sequencing and analysis of the human genome. Nature [Internet]. 2001 Feb 15 [cited 2021 Jun 17];409(6822):860–921. Available from: www.nature.com

52. Steehler JK. Understanding Moore's Law: Four Decades of Innovation (David C. Brock, ed.). 2007 [cited 2022 Sep 1]; Available from: www.JCE.DivCHED.org

53. Koren S, Walenz BP, Berlin K, Miller JR, Bergman NH, Phillippy AM. Canu: scalable and accurate long-read assembly via adaptive k-mer weighting and repeat separation. Genome Res [Internet]. 2017 May 1 [cited 2022 Sep 5];27(5):722–36. Available from: genome.cshlp.org/content/27/5/722.full

54. Chen Y, Nie F, Xie SQ, Zheng YF, Dai Q, Bray T, et al. Efficient assembly of nanopore reads via highly accurate and intact error correction. Nat Commun 2021 121 [Internet]. 2021 Jan 4 [cited 2022 Sep 5];12(1):1–10. Available from: www.nature.com/articles/s41467-020-20236-7

55. Simão FA, Waterhouse RM, Ioannidis P, Kriventseva E V., Zdobnov EM. BUSCO: assessing genome assembly and annotation completeness with single-copy orthologs. Bioinformatics [Internet]. 2015 Oct 1 [cited 2022 Sep 5];31(19):3210–2. Available from: academic.oup.com/bioinformatics/article/31/19/3210/211866

56. Auton A, Abecasis GR, Altshuler DM, Durbin RM, Bentley DR, Chakravarti A, et al. A global reference for human genetic variation. Nature. 2015;526(7571):68–74.

57. Korneliussen TS, Albrechtsen A, Nielsen R. ANGSD: Analysis of Next Generation Sequencing Data. BMC Bioinformatics [Internet]. 2014 Nov 25 [cited 2022 Sep 5];15(1):1–13. Available from: bmcbioinformatics.biomedcentral.com/articles/10.1186/s12859-014-0356-4

58. Speidel L, Cassidy L, Davies RW, Hellenthal G, Skoglund P, Myers SR. Inferring Population Histories for Ancient Genomes Using Genome-Wide Genealogies. Mol Biol Evol [Internet]. 2021 Aug 23 [cited 2022 Sep 5];38(9):3497–511. Available from: academic.oup.com/mbe/article/38/9/3497/6299394

59. Wang K, Li M, Hakonarson H. ANNOVAR: functional annotation of genetic variants from high-throughput sequencing data. Nucleic Acids Res [Internet]. 2010 Sep 1 [cited 2022 Sep 5];38(16):e164–e164. Available from: academic.oup.com/nar/article/38/16/e164/1749458

60. Danecek P, Auton A, Abecasis G, Albers CA, Banks E, DePristo MA, et al. The variant call format and VCFtools. Bioinformatics [Internet]. 2011 Aug 1 [cited 2022 Sep 5];27(15):2156–8. Available from: academic.oup.com/bioinformatics/article/27/15/2156/402296

61. Li H, Barrett J. A statistical framework for SNP calling, mutation discovery, association mapping and population genetical parameter estimation from sequencing data. Bioinformatics [Internet]. 2011 Nov 1

[cited 2022 Sep 5];27(21):2987–93. Available from: academic.oup.com/bioinformatics/article/27/21/2987/217423

62. Purcell S, Neale B, Todd-Brown K, Thomas L, Ferreira MAR, Bender D, et al. PLINK: a tool set for whole-genome association and population-based linkage analyses. Am J Hum Genet [Internet]. 2007 [cited 2022 Sep 5];81(3):559–75. Available from: pubmed.ncbi.nlm.nih.gov/17701901/

63. Yang Y, Adelstein SJ, Kassis AI. Target discovery from data mining approaches. Drug Discov Today [Internet]. 2009 Feb [cited 2022 Sep 5];14(3–4):147–54. Available from: pubmed.ncbi.nlm.nih.gov/19135549/

64. Hughes JP, Rees SS, Kalindjian SB, Philpott KL. Principles of early drug discovery. Br J Pharmacol [Internet]. 2011 Mar [cited 2022 Sep 5];162(6):1239. Available from:/pmc/articles/PMC3058157/

65. Duran-Frigola M, Rossell D, Aloy P. A chemo-centric view of human health and disease. Nat Commun 2014 51 [Internet]. 2014 Dec 1 [cited 2022 Sep 5];5(1):1–11. Available from: www.nature.com/articles/ncomms6676

66. Bantscheff M, Drewes G. Chemoproteomic approaches to drug target identification and drug profiling. Bioorg Med Chem. 2012 Mar 15;20(6):1973–8.

67. Adak S, Batra VS, Bhardwaj DN, Kamesam P V., Kankar P, Kurnekar MP, et al. A system for knowledge management in bioinformatics. Int Conf Inf Knowl Manag Proc [Internet]. 2002 [cited 2022 Sep 12];638–41. Available from: www.ncbi.nlm.nih.gov/geo/

68. Clarke P, Cooper M. Knowledge management and collaboration. 2000 [cited 2022 Sep 12];30–1. Available from: sunsite.informatik.rwth-aachen.de

69. Isoviita V-M, Salminen L, Azar J, Lehtonen R, Roering P, Carpén O, et al. Open Source Infrastructure for Health Care Data Integration and Machine Learning Analyses. JCO Clin Cancer Informatics. 2019 Dec 27;(3):1–16.

70. Tremouilhac P, Lin CL, Huang PC, Huang YC, Nguyen A, Jung N, et al. The Repository Chemotion: Infrastructure for Sustainable Research in Chemistry**. Angew Chemie Int Ed [Internet]. 2020 Dec 7 [cited 2022 Sep 12];59(50):22771–8. Available from: onlinelibrary.wiley.com/doi/full/10.1002/anie.202007702

71. Ahmed S, Ali MU, Ferzund J, Sarwar MA, Rehman A, Mehmood A. Modern Data Formats for Big Bioinformatics Data Analytics. Int J Adv Comput Sci Appl [Internet]. 2017 [cited 2022 Sep 12];8(4). Available from: www.ijacsa.thesai.org

72. Gentleman R. R Programming for Bioinformatics. R Program Bioinforma [Internet]. 2008 Jul 14 [cited 2022 Sep 12]; Available from: www.taylorfrancis.com/books/mono/10.1201/9781420063684/programming-bioinformatics-robert-gentleman

73. Greene CS, Tan J, Ung M, Moore JH, Cheng C. Big Data Bioinformatics. J Cell Physiol [Internet]. 2014 Dec 1 [cited 2022 Sep 12];229(12):1896–900. Available from: onlinelibrary.wiley.com/doi/full/10.1002/jcp.24662

74. De Brevern AG, Meyniel JP, Fairhead C, Neuvéglise C, Malpertuy A. Trends in IT Innovation to Build a Next Generation Bioinformatics Solution to Manage and Analyse Biological Big Data Produced by NGS Technologies. Biomed Res Int [Internet]. 2015 [cited 2022 Sep 12];2015. Available from: pubmed.ncbi.nlm.nih.gov/26125026/

75. Thomas T, Gilbert J, Meyer F. Metagenomics–a guide from sampling to data analysis. Microb Inform Exp [Internet]. 2012 Dec [cited 2022 Sep 2];2(1):3. Available from:/pmc/articles/PMC3351745/

76. O'Leary NA, Wright MW, Brister JR, Ciufo S, Haddad D, McVeigh R, et al. Reference sequence (RefSeq) database at NCBI: current status, taxonomic expansion, and functional annotation. Nucleic Acids Res [Internet]. 2016 [cited 2022 Sep 2];44(D1):D733–45. Available from: pubmed.ncbi.nlm.nih.gov/26553804/

77. Sherry ST, Ward M, Sirotkin K. dbSNP—Database for Single Nucleotide Polymorphisms and Other Classes of Minor Genetic Variation. Genome Res [Internet]. 1999 Aug 1 [cited 2022 Sep 2];9(8):677–9. Available from: genome.cshlp.org/content/9/8/677.full

78. Landrum MJ, Lee JM, Benson M, Brown GR, Chao C, Chitipiralla S, et al. ClinVar: improving access to variant interpretations and supporting evidence. Nucleic Acids Res [Internet]. 2018 Jan 4 [cited 2022 May 9];46(D1):D1062–7. Available from: academic.oup.com/nar/article/46/D1/D1062/4641904

79. Hammal F, de Langen P, Bergon A, Lopez F, Ballester B. ReMap 2022: a database of Human, Mouse, Drosophila and Arabidopsis regulatory regions from an integrative analysis of DNA-binding

sequencing experiments. Nucleic Acids Res [Internet]. 2021 Nov 9 [cited 2021 Nov 17]; Available from: academic.oup.com/nar/advance-article/doi/10.1093/nar/gkab996/6423925

80. Castro-Mondragon JA, Riudavets-Puig R, Rauluseviciute I, Berhanu Lemma R, Turchi L, Blanc-Mathieu R, et al. JASPAR 2022: the 9th release of the open-access database of transcription factor binding profiles. Nucleic Acids Res [Internet]. 2022 Jan 7 [cited 2022 Sep 2];50(D1):D165–73. Available from: academic.oup.com/nar/article/50/D1/D165/6446529

81. Evanko D, Doerr A. Targeted proteomics. Nat Methods 2010 71 [Internet]. 2009 Dec 21 [cited 2022 Sep 2];7(1):34–34. Available from: www.nature.com/articles/nmeth.f.284

82. Faria SS, Morris CFM, Silva AR, Fonseca MP, Forget P, Castro MS, et al. A Timely shift from shotgun to targeted proteomics and how it can be groundbreaking for cancer research. Front Oncol. 2017 Feb 20;7(FEB):13.

83. Nesvizhskii AI. Proteogenomics: concepts, applications and computational strategies. Nat Methods [Internet]. 2014 Nov 1 [cited 2022 Sep 2];11(11):1114–25. Available from: pubmed.ncbi.nlm.nih.gov/25357241/

84. Deutsch EW. The PeptideAtlas Project. Methods Mol Biol [Internet]. 2010 [cited 2022 Sep 2];604:285. Available from:/pmc/articles/PMC3076596/

85. Perez-Riverol , Csordas A, Bai J, Bernal-Llinares M, Hewapathirana S, Kundu DJ, et al. The PRIDE database and related tools and resources in 2019: improving support for quantification data. Nucleic Acids Res [Internet]. 2019 Jan 8 [cited 2022 Sep 2];47(D1):D442–50. Available from: academic.oup.com/nar/article/47/D1/D442/5160986

86. Bateman A, Martin MJ, O'Donovan C, Magrane M, Alpi E, Antunes R, et al. UniProt: the universal protein knowledgebase. Nucleic Acids Res [Internet]. 2017 Jan 4 [cited 2022 Sep 2];45(D1):D158–69. Available from: academic.oup.com/nar/article/45/D1/D158/2605721

87. Karsch-Mizrachi I, Takagi T, Cochrane G, Collaboration on behalf of the INSD. The international nucleotide sequence database collaboration. Nucleic Acids Res [Internet]. 2018 Jan 1 [cited 2022 Sep 2];46(Database issue):D48. Available from:/pmc/articles/PMC5753279/

88. Berman HM, Westbrook J, Feng Z, Gilliland G, Bhat TN, Weissig H, et al. The Protein Data Bank. Nucleic Acids Res [Internet]. 2000 Jan 1 [cited 2022 Sep 2];28(10):235–42. Available from: academic.oup.com/nar/article/28/1/235/2384399

89. Wiswesser WJ. Historic Development of Chemical Notations. J Chem Inf Comput Sci [Internet]. 1985 Aug 1 [cited 2022 Sep 2];25(3):258–63. Available from: pubs.acs.org/doi/abs/10.1021/ci00047a023

90. Gasteiger J. Chemoinformatics: a new field with a long tradition. Anal Bioanal Chem [Internet]. 2005 Sep 22 [cited 2022 Sep 2];384(1):57–64. Available from: europepmc.org/article/med/16177914

91. Weininger D. SMILES, a Chemical Language and Information System: 1: Introduction to Methodology and Encoding Rules. J Chem Inf Comput Sci [Internet]. 1988 Feb 1 [cited 2022 Sep 2];28(1):31–6. Available from: pubs.acs.org/doi/abs/10.1021/ci00057a005

92. Heller SR, McNaught A, Pletnev I, Stein S, Tchekhovskoi D. InChI, the IUPAC International Chemical Identifier. J Cheminform [Internet]. 2015 May 30 [cited 2022 Sep 2];7(1):1–34. Available from: jcheminf.biomedcentral.com/articles/10.1186/s13321-015-0068-4

93. Mohs RC, Greig NH. Drug discovery and development: Role of basic biological research. Alzheimer's Dement Transl Res Clin Interv [Internet]. 2017 Nov 1 [cited 2022 Sep 2];3(4):651–7. Available from: onlinelibrary.wiley.com/doi/full/10.1016/j.trci.2017.10.005

94. Cheng CH, Wong KL, Chin JW, Chan TT, So RHY. Deep Learning Methods for Remote Heart Rate Measurement: A Review and Future Research Agenda. Sensors 2021, Vol 21, Page 6296 [Internet]. 2021 Sep 20 [cited 2022 Sep 12];21(18):6296. Available from: www.mdpi.com/1424-8220/21/18/6296/htm

95. Alizadehsani R, Khosravi A, Roshanzamir M, Abdar M, Sarrafzadegan N, Shafie D, et al. Coronary artery disease detection using artificial intelligence techniques: A survey of trends, geographical differences and diagnostic features 1991–2020. Comput Biol Med. 2021 Jan 1;128:104095.

96. Iroju OG, Olaleke JO. Information Technology and Computer Science. Inf Technol Comput Sci [Internet]. 2015 [cited 2022 Sep 12];08:44–50. Available from: www.mecs-press.org/

97. Paul D, Sanap G, Shenoy S, Kalyane D, Kalia K, Tekade RK. Artificial intelligence in drug discovery and development. Drug Discov Today [Internet]. 2021 Jan 1 [cited 2022 Sep 12];26(1):80. Available from:/pmc/articles/PMC7577280/

98. Kopalle PK, Gangwar M, Kaplan A, Ramachandran D, Reinartz W, Rindfleisch A. Examining artificial intelligence (AI) technologies in marketing via a global lens: Current trends and future research opportunities. Int J Res Mark. 2022 Jun 1;39(2):522–40.

99. Wang X, Deng X, Fu Q, Zhou Q, Feng J, Ma H, et al. A Weakly-Supervised Framework for COVID-19 Classification and Lesion Localization from Chest CT. IEEE Trans Med Imaging. 2020 Aug 1;39(8):2615–25.

100. Laguarta J, Hueto F, Subirana B. COVID-19 Artificial Intelligence Diagnosis Using only Cough Recordings. IEEE Open J Eng Med Biol. 2020;1:275–81.

101. Bartoletti I. AI in healthcare: Ethical and privacy challenges. Lect Notes Comput Sci (including Subser Lect Notes Artif Intell Lect Notes Bioinformatics) [Internet]. 2019 [cited 2022 Sep 12];11526 LNAI:7–10. Available from: link.springer.com/chapter/10.1007/978-3-030-21642-9_2

102. Char DS, Shah NH, Magnus D. Implementing Machine Learning in Health Care — Addressing Ethical Challenges. N Engl J Med [Internet]. 2018 Mar 3 [cited 2022 Sep 12];378(11):981. Available from:/pmc/articles/PMC5962261/

103. Ahmad MA, Eckert C, Teredesai A. Interpretable Machine Learning in Healthcare. Proc 2018 ACM Int Conf Bioinformatics, Comput Biol Heal Informatics [Internet]. 2018 Aug 15 [cited 2022 Sep 12];21:559–60. Available from: doi.org/10.1145/3233547.3233667

104. Makhlysheva A, Bakkevoll PA, Nordsletta AT, Linstad L. Artificial intelligence and machine learning in health care [Internet]. 2018 Sep [cited 2022 Sep 12]. Available from: ehealthresearch.no/files/documents/Faktaark/Fact-sheet-2018-09-Artificial-intelligence-and-machine-learning-in-healthcare.pdf

105. Chhikara BS, Varma RS. Nanochemistry and Nanocatalysis Science: Research advances and future perspectives. J Mater Nanosci [Internet]. 2019 Apr 1 [cited 2022 Sep 12];6(1):1–6. Available from: pubs.iscience.in/journal/index.php/jmns/article/view/877

2 An Introduction to Basic and Advanced Chemoinformatics Computational Tools

Surovi Saikia, Aparna A, Mohammed Unais AK,
Yashwant V. Pathak, V.Vijaya Padma

1 INTRODUCTION

Huge societal and technological progress can be achieved by bringing new functional molecules using Artificial Intelligence (AI), which depends primarily on how a molecule is written to be understood by computers. Strings remain a common method for molecular representation graphs. SMILES (Simplified Molecular Input Line Entry System), the most popular one, has empowered cheminformatics since its inception, but it is not adequate in the context of AI and Machine Learning (ML) in chemistry [1]. Methods in chemoinformatics studies include a number of overlapping computer methodologies in different fields of chemistry [2]. Big data is everywhere and is gaining popularity with two challenges in relation to incompleteness of data and inconsistency in quality and reliability of data [3] as the final inferences depend on these two factors. These factors also point towards solving the challenges associated with inhomogeneous biological and chemical data, which are difficult to resolve and specifically true for biological data that are affected by many factors such as protocols, timing, and place of performing the experiment. Inorganic chemistry has a rabbit hole with regard to the modeling and prediction of chemical reactions. The performance of string-based and non-string representations in terms of ML has been quantified, and a basic understanding of chemical language by both human and AI scientists has also been studied [4].

In chemical sciences, outcome prediction of an organic transformation is really challenging as it depends on various chemical and physical properties of reagents such as solubility, acidity, basicity, electrophilicity, nucleophilicity, and polarity of the solvent as well as other experimental conditions such as atmosphere, temperature, concentration, and pressure [5]. Quantum mechanics-assisted techniques of computer simulations with approximations that are up-to-date mark such as DFT (density functional theory) calculations are assisting experimental procedures [6]. Drastic impact has been observed in the designing of experiments for a specific problem by these computational methods including advancement in the *in silico* modeling approach of QC (quantum chemistry). Besides QC, AI-guided chemoinformatic methods are at the leading front of chemical property prediction as they offer a rapid solution for rapid screening of molecules [7]. In the past decade, application of ML to chemistry has enabled the property prediction of organic molecules such as lipophilicity, solubility, electrophilicity, etc. [8], as well as related outcomes such as reaction barriers, selectivity, yields, energies of bond dissociation, and enantioselectivity [9].

This chapter provides details on the various aspects of the cheminformatics field and the ways in which it is critical for drug discovery and the introduction of various new computational tools.

DOI: 10.1201/9781003353768-2

2 CHEMINFORMATICS APPLICATION IN DRUG DISCOVERY

The drug discovery process consists of seven steps: disease selection, target assumption, lead compound recognition (inspection), lead optimization, pre-clinical trial, clinical trial, and pharmacogenomic improvement. Traditionally, such processes are carried out sequentially and if any of the steps become slow, the entire pipeline slows down. Bottlenecks are generated by these slow steps, which include time, manufacturing, and testing costs of chemicals. To overcome such hurdles HTS (high-throughput screening) and automated drug screening were introduced, which made it easier to test thousands of compounds per day for a drug [10]. The introduction of combinatorial chemistry along with HTS made drug discovery faster and more fascinating. The CC (combinatorial chemistry) approach was a breakthrough but did not yield many drug candidates; to bypass this construction of chemically diverse libraries was introduced with the help of computational studies, later termed as cheminformatics.

2.1 Selectivity and ADMET Properties

A chemically diverse library provides a handful of target compounds that do not possess any drug characteristics, which was solved by developing technologies to scrutinize drug like compound from a wide variety of chemically synthesized compounds. Later, drug-like screening and filtering technologies were introduced, which was a partial solution, as some drug candidates did not show much promise in terms of biological nature, which finally led to optimizing the compound according to Absorption, Distribution, Metabolism, Excretion, and Toxicity (ADMET) parameters. The collaborative refinement of all these interrelated factors is a serious obstacle in pharmaceutical research. *In-silico* approach for ADMET prediction is a key player in pharma research and development, assessing pharmacological activity ahead of schedule in the process of drug development, which has so far been vital for directing hit-to-lead and lead-optimization initiatives. Advancement in cheminformatics provided a new platform to identify ADMET targets and extract knowledge from it for speeding up the drug discovery process [11].

There are wide varieties of ADMET prediction tools including QikProp, DataWarrior, MetaTox, MetaSite, and StarDrop. Due to less efficacy and safety concerns the reliability of pharmacokinetic (PK) properties declined. To overcome this, fully integrated ADMET tools that can exclude unwanted compounds were introduced methods such as k-nearest neighbor (k-NN), support vector machines (SVM), random forest (RF), and artificial neural networks (ANNs). k-NN is a non-linear method for pattern recognition with >85% accuracy; SVM is a linear approach used in regression and classification analyses with 81.5% performance; RF is an ensemble method for classification and regression tasks with 80% accuracy, which can also predict the coefficient of blood-brain portioning (log BB), which is useful in predicting the compounds that can clearly cross BBB (blood brain barrier); and ANNs are the most widely used ADMET tool, which in regression and classification analysis mimic the human brain, all with a correlation coefficient of >0.86. A novel ensemble approach AECF (adaptive ensemble classification framework) strategically entails the steps of data balancing, model generation, model integration, and ensemble optimization to tackle high-dimensionality and unbalance issues all at once [12].

Several free web-based tools were developed to access ADMET predictions, such as ADMETlab (admet.scbdd.com), which is the most thorough platforms ADMET analysis based on rule of five (RO5) and 31 predictive models. FAF-Drugs4 is a free accessible web tool (fafdrugs4.mti.univ-paris-diderot.fr/) that has many features like drug-likeliness relied on PK properties. MetStabOn is a free online platform for metabolic stability prediction with different training sets for human (5234 compounds), rat (2829), and mouse (1136) [13]. SwissADME is a recently established web-based tool for ADMET analysis (www.swissadme.ch/index.php), which is used to predict 10 different physiochemical properties and is one of the most practical tools used in ADMET prediction. Several other web-based tools like vNN (vnnadmet.bhsai.org/) and Hit Dexter 2.0 (hitdexter2.zbh.uni-hamburg.de/) are also available for ADMET prediction and analysis.

2.2 OPTIMIZATION OF COMPOUNDS

Once a compound is identified through ADMET studies it can be optimized as per the need to enhance its potential as a drug candidate using an *in silico* approach using cheminformatic tools. These tools can structurally modify compounds in such a way that it can boost the physiochemical and pharmacokinetics properties. Optimization of natural leads into drugs or drug candidates enhances not only drug efficacy but also enhances the ADMET profiles and chemical responsiveness. Direct chemical manipulation of functional groups, SAR (structure-activity relationship) directed optimization, and pharmacophore-oriented molecular design using natural templates are examples of optimization strategies. To assist in product and process development, principles of both fundamental medicinal chemistry and cutting-edge computer-aided drug design methods can be used [14].

Direct chemical modification includes isosteric replacement and addition and alteration of the ring systems based on chemical strategy, which conditions that chemically related structured compounds have similar bioactivity. SAR and SAR-directed optimization use chemical and biological information of compounds for optimization. Over 30% of anti-cancer drugs have been derived by these two methods. Pharmacophore-oriented optimization methods focus on the pairing or blending of pharmacophores from distinct chemical classes of anti-cancer agents in order to create new varieties. Eliminating redundant chiral hubs and scaffold hopping are two instances of this approach, but this approach can change the core structure at the end [14].

There are many online accessible platforms for drug optimization like ADMETlab (admet.scbdd. com/), which predicts drug-likeliness, ADMET profiling, and subsequent prioritization of chemical structures; DruLiTo (www.niper.gov.in/pi_dev_tools/DruLitoWeb/DruLito_index.html) based on SVM (support vector machine) and QSAR (quantitative structure activity relationship) for physiochemical properties analysis and drug-likeliness; Drugmint (crdd.osdd.net/oscadd/drugmint/) based on SVM algorithm for drug-likeliness; QED Score and Optimization; SwissADME (www.swissadme.ch/) for physicochemical properties, ADME, rule-based drug-likeness, and optimization [31]; SwissBioisostere (www.swissbioisostere.ch/) specifically for optimization based on Hussain-Rea algorithm; and DataWarrior (openmolecules.org/datawarrior/) for physicochemical properties, rule-based drug-likeness, toxicity prediction, prioritization, and optimization (through the generation of Structure–Activity Landscape Index) based on Stereo-enhanced Morgan-algorithm. All these is put together with a great effort to optimize the hit compound as per our requirement to increase efficacy and decrease toxicity.

2.3 PREDICTIVITY AGAINST STRUCTURAL DIVERSITY

The majority of ADMET mockups are created on the basis of minor sets of chemical compounds ranging from tens to hundreds, which are considered as less significant ones. As we go deeper into the SAR studies related problems which give much more complex problem than that of predictivity versus diversity, which states that when the diversity of the chemical compound increase lesser is the chance of finding the appropriate SAR models. Similarly, if the SAR model exist the SAR information will increase parallelly with the diversity of the chemical compound [15]. The likeliness of constructing a single, effectual QSAR model tends to decrease as the specificity of a collection of compounds intensifies, which is especially true for linear QSAR models. An increase in structural diversity broadens the applicability domain even though structurally different molecules could really demonstrate exercise vs. a single target, highlighting the importance of polyspecificity in drug discovery which is truly the case in the other ADMET *in silico* studies. This is due to the fact each bioactivity has a distinct mechanism, and there exist numerous mechanisms through which toxicity is defined. In order to address this problem, data mining tools are used to combine these mechanisms [16].

2.4 DATA MINING

Data mining refers to the method for recognizing patterns or trends in data. The modern era of drug discovery is fully focused on data from various sources like HTS, CC, and other new tools involved in drug discovery. Data from these branches are growing very fast at an exponential rate and digging out information from the raw data is an important goal in this process. The difficult task of extracting underlying, unfamiliar, and incredibly valuable data that contains information is termed as knowledge discovery. Data mining involves six common classes of tasks:

  i)   anomaly detection; for unusual data records or data errors,
 ii)   association rule learning; which helps in finding out the relationships between variables
iii)   clustering; to find out similar groups and structures in the data
 iv)   classification; to generalize know structure
  v)   regression; for modeling data with least errors
 vi)   summarization; for visualization and report generation

Descriptor computations, structural similarity matrices, and classification algorithms represent the most typical data mining strategies for use in cheminformatics. The success of data mining relies on the descriptor selection. Correct descriptor selection helps to understand the computer program to solve, which mainly relies on four criteria: 1) bioactivity related, 2) informative, 3) independent, and 4) easy to extract. Thorough studies have shown that 2D descriptors is well functioned than 3D descriptors. The core technology of data mining is pattern recognition, and regression analysis and classification are two main techniques used for pattern recognition. In addition to this there are some common patterns such as Markush structure, fingerprint, 3D pharmacophore, decision tree classification, hierarchical clustering, non-hierarchical clustering, and SOM (self-organization map). For comparing patterns, similarity or distant metrics are used, and classification algorithms also play a key role in data mining as it helps to develop different algorithms that can be used to retrieve data easily and effectively from the source. Source data are extracted mainly by two algorithm types: hierarchical and partitional. The former is used in rearranging objects in a tree structure, while the latter is used for non-parametric data, but cannot be used on a large amount of data. One way to tackle this is decision trees, which were introduced like RP (Recursive Partitioning), which can run 20,00,000 compounds in less than an hour. The main drawback of this algorithm is there are too many solutions [17].

3 CHEMINFORMATICS IN DATABASE CREATION

Databases are methods for computer systems to store information in a way that can be tracked down. There exist two different approaches for database creation, DBS (database backup system) and DBMS (database management system) for handling and storing huge amount of data using different software. Based on origin, nature, or design databases are classified as hierarchical, network, relational, and object based. Cheminformatics seeks to enhance chemical decision making by stashing and summarizing information in a sustainable manner, as well as by providing multiple tools and protocols enabling the use of applications and data across multiple systems and mining the many chemical property spaces in a time- and space-efficient manner [18].

3.1 MOLECULAR DESCRIPTORS

Molecular descriptors are computational depictions of a molecule obtained through the application of a well-defined algorithm to a known molecular portrayal or a well-defined experimental procedure, which includes diversity analysis, library design, and virtual screening. Many modern chemoinformatic and bioinformatic approaches depend on molecular descriptors to encode a wide range of molecular information. They grasp specific molecular features (e.g., geometry, shape, pharmacophores, or atomic properties) and have a direct impact on the outcome, performance, and applicability of computational models. Many cheminformatics and bioinformatic applications depend on the molecular descriptors, which include protein–ligand interaction prediction, molecular similarity analysis, ligand-based virtual screening, drug design, and environmental/toxicological hazard estimation [19].

Thousands of molecular descriptors have been proposed over the years, ranging from simple bulk properties to complex 3D interpretations and molecular fingerprints of thousands of bits. There is a step-by-step approach which leads to the development, calculation, pretreatment, and use of MD, which starts with data set preparation comprising molecular representations (in 0D, 1D, 2D, 3D, and 4D models, in addition to 5D-QSAR and 6D-QSAR) and structure curation followed by molecular descriptor calculation and processing, which comprises descriptor calculations that are subdivided into classical MD (molecular dynamics) and binary fingerprints; dimensionality reduction, which focuses on removal of missing-valued descriptors, low variance filter, and high-correlation filter and scaling of descriptor. The final step is computational modeling, which consists of similarity search and QSAR modeling [20].

3.2 SIMILARITY INDEX

The similarity index helps in prediction of the molecular behavior of structurally related compounds as it is very helpful in correlating two different molecules with the use of computational tools mainly through binary fingerprints. Molecular similarity is solely focused on ligand-based virtual screening and molecular informatics. It is mainly used to identify compounds with the same activity as that of a given compound (similarity analysis) and in SAR derivations to justify read-across applications and is also widely used in diversity selection [21]. Binary fingerprint is a widely used and reliable approach for similarity measurements. A fingerprint is a fixed-length bit string in which the presence of molecular fragments is recorded (as one or more bits set to 1) by a hashing algorithm, fingerprints of two molecules can be used to compare their similarity using distance measure. Out of several binary fingerprints fragment-based Daylight and Tripos UNITY 2D fingerprints are widely accepted and used [22].

The activity of a chemical compound is described and recorded by topological, physicochemical, and electronic descriptors. Comparing the numerical values from these descriptors the similarity of compounds can be easily predicted. In addition to this, the quantum mechanical wave function is also used in similarity prediction. It is classified into seven different types which are: 1) based on constitution, which represents the chemical compounds as graph; 2) configuration and conformation, which focuses on the 3D structure of compounds; 3) physiochemical properties; 5) quantum chemistry approach; 6) reactions; and 7) distance between real valued descriptors. The development of computational methods such as the SRD (sum of ranking differences) and factorial ANOVA made similarity predictions easier and more reliable between multiple compounds at the same time [23].

3.2.1 Construction of Chemical Space

The chemical space is an orchestra of all possible molecules and encompasses the entire world of both known and unknown molecules, along with the transitions necessary to generate these molecules and their interactions with one another (i.e., a network of chemical reactions connecting the molecules in the space). Chemical space offers more uplifting depictions in order to find the right compound and has been extensively used in drug discovery. The GDB (Human Genome Database) database, which was constructed using the results of the GENG (Game ENGine for Rust Programming Language) program, helped to produce the actual list of all available molecules of a certain size. Most GDB molecules are formed at intermediate polar atom-to-carbon ratios with clogP values between −2 and 2. Lipinski's criteria for oral bioavailability, as well as lead-likeness and fragment-likeness criteria, are met by these molecules, owing to the fact that these criteria primarily limit molecular size.

Different algorithms are used to create chemical space, to compare compounds and retrieve useful information; SPROUT is one such algorithm that helps growing molecules be modified to protein-binding sites. SYNOPSIS is also such an algorithm that works on the same principle of SPROUT, EVOLUATOR for interactive molecule selection, Other genetic algorithms such as Skelgen, TOPAS, Flux, ADAPT, GANDI, and MEGA are also used to travel in chemical space to find more suitable one. SPACESHIP is a genetic algorithm used to study mutation and selection between two molecules [24]. Chemical space depends on the dimensions and map, and helps to measure the distance between two compounds; these are very helpful in determining the fitness values, which is helpful in drug discovery. Such maps are constructed by PCA analysis, ChemGPS system, and SOM-maps, which classifies according to bio activity. The advancements in small molecule drug discovery have driven chemistry into unknown chemical spaces.

3.3 Specificity

The concept of polyspecificity is complementary to polypharmacology and refers to the activity of structurally different towards a given target. This term is rather uncommon in chemoinformatics literature. but it is prevalent as a description for antibody specificity [25]. Polypharmacology and polyspecificity are two complementary terms mathematically as evident from different publications and mathematical expressions for these terms [26]. The CLARITY program developed at Chemotargets serves this purpose for polypharmacology targets [27].

4 ADVANCES IN CHEMOINFORMATICS COMPUTATIONAL TOOLS

4.1 SMILES

SMILES (Simplified Molecule Input Line Entry System) is a chemical notation that encodes molecular structure as a single line of text allowing users to represent chemical structures in computer language. They are easily learned and the most popular flexible line notations used in the past three decades. Created in 1986 at the US EPA (US Environmental Research Laboratory) by David Weininger and further developed by Daylight Chemical Information Systems they capture connectivity and stereochemical configuration [30]. Differences in implementation are observed as there is no formal specification in the SMILES format with several ambiguities. In this language, it is important to state the physical model of the molecule based on the valence model of chemistry. SMILES grammars explain complex structure and properties like chirality, ions, shapes, and stereochemistry, which has little resemble to underlay quantum mechanical reality of electrons, protons and neutrons. The molecules are defined as chains of atoms written as letters in strings, and molecules with branches are defined in parentheses, with ring closure indicated by two matching numbers.

OpenSMILES, a specification for SMILES in 2007, was initiated by Craig James as a community approach as a comprehensive chemical language that represents atom and bond symbols specified using ASCII characters. SMILES are compact and understandable unique strings used as universal identifiers for specific chemical structures described as a graph concept with nodes and edges as bonds. It is hosted under the Blue Obelisk project, with the intent to contribute and comment from entire computational chemistry community. There exist many independent SMILES software packages in Java, C, C++, Python, FORTRAN, and LISP. Various weaknesses in the representation of SMILES were found with this concept, as multiple strings represented the same molecule. Secondly, SMILES has no mechanism to confirm with respect to syntax and physical properties whether the molecular strings are valid and no particular interpretation as a molecular graph exist and showed a negative impact on the validity of computer designed molecules based on evolutionary or deep learning methods due to lack of semantic and syntactic robustness [30]. The solution for this weakness paved the way for the design of special ML models to enforce robustness such as DeepSMILES and SELFIES [31].

4.2   DeepSMILES

Until now, most of the existing models like SMILES uses single atom-based representations; other more advanced representations show molecular property prediction like SELFIES and DeepSMILES (Self-supervised pre-training and heterogeneity-aware deep Multiple Instance LEarning) is syntax transformed from SMILES string and suited for ML [32]. The syntax problems listed below are addressed by DeepSMILES:

- They are single-digit ring closure instead of two unmatched rings.
- Considering the postfix notation, the paired parentheses are avoided by DeepSMILES.

The problem of unbalanced parentheses has been avoided by DeepSMILES syntax by only using the close parentheses, where the number of parentheses indicates the branch length. They also use single symbols at ring closing location instead of ring closure symbols. Recently DeepSMILES were used to analyze the whole slide images of Hematoxyln and Eosin stained tumor tissues that proved to have a 40% increase in performance [33]. These notations are used independently to interpret the output of a generative model, which converts DeepSMILES to SMILES strings. In order to switch between the DeepSMILES and SMILES, pure Python modules are used only where minimal changes are need within standard cheminformatics libraries.

4.3   InChI

Over the last three decades, SMILES has been the platform in chemoinformatics, but weaknesses in the format resulted in the development of new molecular string representations leading to InChI (International Chemical Identifier), which is enforced by post-processing canonicalization via tools such as RDkit [34]. SMILES have never been a unique molecular graphs representation and it is difficult to construct large-scale databases with unique structure mapping. To tackle this issue, IUPAC developed open-source software in 2013 to encode molecular structures to standardize the search within databases. The strings of InChI contains six main layers with multiple sublayers where each layer represents different information about the molecule (e.g., atomic connections, charges, stereochemistry, chemical formula, etc.). InChI syntax has many advantages such as canonical representations and straightforward linking in databases and helps to standardize the output from different cheminformatics toolkits [32]. InChIKey is the complemented hashed, fixed length counterpart compact, which is also an open source, non-proprietary, chemical identifier with principle features like:

- Uniqueness in identification
- Structure-based approach

- Can be applied to classic organic chemistry and to a large extent to inorganic chemistry
- Hierarchical and layered and can generate the same InChI structures under different styles and levels of granularity (standard InChI for inter-operability)

InChIs are more expressive than SMILES, as they can specify the mobile and immobile hydrogen atoms. A single InChI is only used for resonance structure instead of multiple in SMILES. Apart from these, other disadvantages for InChI like hierarchical structure and syntax makes it non-human readable. This format follows a lot of arithmetic and grammatical rules for sampling a new molecule, which is difficult for generative modeling. It consistently disconnects the bond to metal atoms leading to loss of important bonding and stereochemical information. It is found that compared to SMILES in ML-based applications, InChI has poor performance due to the aforementioned reasons [35].

4.4 SELFIES

SMILES was introduced in the late 1980s and has had several limitations over the years. Thus, a new language was introduced in 2020 known as SELFIES (SELF-referencing Embedded Strings), a 100% robust, human-readable and string-based representation of molecules that describes molecular graphs representing every valid molecule. SELFIES are always unique as they are independent and can be directly applied to arbitrary machine learning models without any adaption of models. As compared to SMILES, outputs of SELFIES are more valid and diverse and use the generative model. Today SELFIES are used broadly in chemistry and have found their way into AI and ML. They can be used for representation of ordinary molecules like isotopes and charged and radical species. This format can also represent stereochemistry and chirality as an objective approach with SMILES, which are also termed to be a surjective mapping from strings to molecular graphs [4].

5 ADDITION TO PubChem INFORMATION

PubChem is an open and public repository for information on chemical substances and their biological role. It was launched in September 2004 as an initiative of the US NIH (National Institute of Health) under the research program. The Molecular Libraries Roadmap (MLP) initiative, aimed at discovering chemical probes through high-throughput screening of small molecules that regulate the gene activity, has grown drastically and become a key chemical information resource that serves various scientific domains like cheminformatics, medicinal chemistry, chemical biology, biochemistry, and drug discovery. It contains one of the largest corpuses of publicly available chemical information and has three related databases for substance, compound, and BioAssay, which are SID (SubstanceID), CID (CompoundID), and AID (AssayID), respectively. It also organizes various data into these three interlinked databases, where the Substance database stores depositor-contributed information (www.ncbi.nlm.nih.gov/pcsubstance). The substance-configured unique chemical structures are stored in a Compound database (www.ncbi.nlm.nih.gov/pccompound). Further, these various assays on these compounds are stored in BioAssay (www.ncbi.nlm.nih.gov/pcassay). In total, PubChem gives comprehensive information about chemicals for drug discovery. Multiple programmatic access routes exploit PubChem to build automated virtual screening pipelines. According to the update in 2022, the data in PubChem is cumulatively provided by 890 sources (https://pubchem.ncbi.nlm.nih.gov/sources/), which includes MLP and non-MLPs, such as government agencies, pharmaceutical companies, chemical vendors, publishers, and chemical biology resources. These data contain information about small molecules including miRNAs, siRNAs, peptides, lipids, macromolecules, carbohydrates, and many others. The non-MLPs are mainly literature-derived data that is manually extracted from thousands of scientific articles through

BindingBD and ChEMBL. They also consist of data from Hazardous Substances Data Bank, DrugBank, etc., mainly chemical records for pharmacology, drug target information, and toxicology. PubChem also receives data from regulatory agencies such as the US FDA (Food and Drug Administration) (www.fda.gov/ForIndustry/DataStandards/StructuredProductLabeling/) on FDA-approved chemicals that include pharmacological classifications and UNIIs (Unique Ingredient Identifiers). Organizations like SureChEMBL (www.surechembl.org/) and IBM (www.almaden. ibm.com/) contribute data related to patented information. Likewise, bioactivity data of molecules (binding assay, ADMET assay, and functional assay) is given by ChEMBL and the experimentally determined binding affinities of protein–ligand complex of the same molecules are provided by BindingBD.

5.1   Expandability in Search Terms

5.1.1   Integration with Entrez for Textual Search

PubChem's three databases and other NCBI databases like PubMed, Genome, Nucleotide, Taxonomy, Biosystems, Gene Expression Omnibus (GEO), etc., use Entrez as a search and retrieval system. The convenience of searching in Entrez is it can be initiated from the PubChem homepage or searched from the NCBI homepage. The search criteria also help to gather data from various PubChem tools, documents, services, etc. When the search is initiated in Entrez, if the specific database is not selected, it searches all the databases against the query and lists the number of records found in each. If the Entrez search finds multiple records for the query, they are finally displayed as document summary report (DocSum). Another feature developed in the search particularly for the Unix terminal window is "Entrez direct" (Edirect), which provides access to NCBI's suite of interconnected databases from the Unix terminal window such as publication, gene, structure, sequence, expression, variation etc., which are entered as command-line arguments. As previously mentioned, the DocSum report contains specific information regarding the data with links. The Entrez system uses powerful options like indices, filters, cross-links, history, etc.

Entrez History

Entrez automatically keeps tracks of a user's searches, caches them, and allows to combine with Boolean logic (AND, OR, NOT, etc.), which helps to keep track of searches and return to them later. It also helps users to avoid receiving and sending large lists of identifiers and used as an input to various Pubchem tools for further analysis or manipulation.

Entrez Filters and Indices

These can include the particular aspects of records where indices contain numeric data, text, etc. The indexed terms in any Entrez database can be found in the drop-down menu in the Advanced Search Builder page with an "Advanced" link next to the "Go" option. Filters in Entrez are Boolean bits for all the records in the database, and indicate true or false (i.e., whether or not has a particular property). It relates to the majority of the Entrez filters in the PubChem databases and adds filters in the entered query to subset the Entrez searches.

5.1.2   PubChem Search Using Non-Textual Query

Primarily Entrez is a text-based search system and cannot be used for searching for chemical structures that use a data type specific to PubChem. The search tool "Chemical structure" enables one to query, and various other related search types like identity, molecular formula, sub- or super-structure, and 2D/3D similarity searches can be run as subsets in the compound database. As the PubChem homepage only accepts textual keywords alone it is difficult to input non-textual queries; however, PubChem Sketcher allows to use both textual and non-textual searches.

Molecular Formula Search
This search allows searching for molecular formulas that contain a certain type and number of elements and returns the default molecules that match the stoichiometry.

Identity Search
It is known as locating particular chemical structure seems to be identical to query chemical structure. Among several predefined options in the PubChem chemical structure, the identity search allows the user to choose highly similar ones. In order to obtain this the search should be expanded by adding a "plus" near the "option" section heading.

Similarity Search
Molecular similarity, also called chemical structure similarity or chemical similarity, predicts the properties of chemical compounds and is used to design chemicals with desired properties. These computational methods help to scrutinize structurally similar molecules with similar physiological and biological properties. This search plays an important role in cheminformatics as there is no direct and easy method for definition of molecular similarity but mathematical definition that agrees. Thus, the molecular similarity is the criteria for similarity searches of the query.

Two-dimensional (2D) and three-dimensional (3D) similarity search methods are broad classifications for molecular similarity. In particular, 2D similarity methods are called molecular fingerprints where structural keys are the most common information of structure, considering the molecule into binary string (0 & 1). According to the definition of the fingerprint the fragments can be defined in different ways for molecular fingerprints. PubChem subgraph fingerprints are the only fingerprint used by PubChem. Thus, by analyzing the molecular fingerprints of two molecules structural similarities can then be predicted. These molecular similarities are quantified using similarity coefficient or similarity score. PubChem subgraph fingerprints use the *"Tanimoto coefficient"* to analyze this similarity score [37].

Similar to 2D similarity search, the 3D conformer tab in the PubChem chemical structure search uses 3D structure (conformations) of molecules to search. Grant and coworkers suggested the PubChem's 3D similarity method known to be Atom-centered Gaussian–shape comparison method which was implemented in ROCS (Rapid overlay of Chemical Structures) (ROCS). Three metrics quantify 3D molecular similarity: shape, color, and combo Tanimoto. There are variety of query formats to search the chemical structure such as SMARTS, InChI, SMILES, CID, SDF, and molecular formula. Similarly, the chemical structure can be manually drawn using the PubChem Chemical Structure Sketcher.

6   DATABASE OF 3D STRUCTURES

2D methods are very useful and computationally inexpensive, but they lack the ability to relate the biological and functional similarity of diverse molecules. Here comes the role of the 3D information where the compatible functional groups would likely bind to the appropriate biological moiety having a Lock and Key property. These predominantly help in docking, modeling, and structure-based drug design [38].

6.1.1   PubChem3D

PubChem3D is an open repository that provides information about the biological activities of small molecules. PubChem3D is a platform that helps to search, visualize, subset, export, and analyze chemical structures and their biological data. PubChem provides the 3D conformer description that records in the PubChem compound database with conditions like; which is having ≤50 non-hydrogen atoms and not too large, ≤ 15 rotatable bonds and too flexible, with single covalent unit and consist of only supported elements like N, O, F,P, Cl, Br, S etc., has fewer than six undefined atoms and those atoms recognized by MMFF94s force field [39].

6.1.2 ChemSpider

ChemSpider has a collection of over 67 million structures from different sources. It is an open access chemical structure database. They can link the open and closed access chemistry journals, environment data, Chemical Entities of Biological Interest (ChEBI), Wikipedia, The Kyoto Encyclopedia of Genes and Genomes (KEGG), chemical vendors, and other patent databases. There are many additional features for physiochemical properties and chemical structures, such as structure identifiers like IUPAC, InChI, SMILES, and Index names [40]. Public access for the database has been provided since 2009 by the Royal Society of Chemistry (RSC). ChemSpider integrates the data into RSC publications and uses validated names to search in PubMed, Google Scholar, RSC journals, books, and databases [41].

6.1.3 3D-e-Chem-VM

3D-e-Chem-VM (http://3d-e-chem.github.io/3D-e-Chem-VM/a) is a free, open source, and virtual machine that analyses protein–ligand interaction data, and integrates cheminformatics and bioinformatics. Using graphical programming, 3D-e-Chem-VM can combine and analyze small molecules and protein structural information. It uses a novel chemical and biological data analytics tool to efficiently exploit the structural and pharamacological ligand-protein interactions from protein-wide databases (e.g., PDB, ChEMBLdb) and customized information systems [42].

6.1.4 LigandBox

LigandBox (http://ligandbox.protein.osaka-u.ac.jp/ligandbox/) contains 4 million compounds collected from 37 chemical vendors and approved drugs and biochemical compounds from PDB, KEGG_COMPOUND, and KEGG_DRUG databases, which are essential for virtual screening by molecular docking. *myPresto* is the molecular simulation program that is used to generate the 3D conformations. The database provides two services for compound searches: chemical/property ID search and chemical structure search. The structure search is mainly done using the maximum common substructure and the descriptor search uses the k combo program. Thus, this helps users find similar compounds in minutes. The database promotes a wide range of research in the field of biochemistry, bioinformatics, chemical biology, life science, and medical science to find active chemical compounds using virtual screening [43].

6.2 DEALING WITH RANDOMNESS OF COMPOUNDS

The SMARTS-based pattern extraction algorithm using decision trees and random forest was employed for data mining over a 60-cell line dataset, which was identical to PASS methods. RF (random forest) models due to favorable features like independence on feature selection and overcoming the over-fitting problem. The MACCS key was used to predict the anti-cancer activity of the compounds [44]. Ensemble feature selection is used to select a suitable set of descriptors from a sample of random compounds so that it optimizes the performance of multiple models [45]. Finding "activity cliffs" within a random dataset that are structurally similar compounds but with a huge difference in activity is a challenge for traditional ML-based QSAR methods [46].

6.3 RSS FEEDS

RSS is the first format developed for syndication and is commonly used by websites and blogs. These RSS feeds add value to PubChem data by generating RSS documents intended to be viewed by a RSS reader such as Google or Sage where each entry in the feed is an individual hit that matches the query from the database. The user also has the option to bookmark the view for later use [47].

7   CHEMOINFORMATICS WITH SEMANTIC WEB TECHNOLOGIES

Due to the influence of Semantic Web Technologies (interconnected web pages) bioinformatics-related subjects are evolving stronger and more well-defined as researchers are able to merge biochemical, chemical, biological, and medical information. Many factors are involved in this development such as importance of storing the source and well curated metadata, concept of chemical identifiers and their role in chemical and biology, and most importantly a grasp of the contribution made by Semantic Web has the capability of bringing these stages of research lifecycle to profusion. Concepts such as open source, open access, and open collaboration are a part of this relationship [48].

7.1   INTEGRATION AND MANAGEMENT OF DATA

Huge quantities of experimental data are at reach now due to combinatorial, high-throughput methodologies, new visualization and spectroscopic methods, with increasing computational techniques. To make possible the best use of the ever-increasing data, specifically the amount of fund that goes into its collection, access, curation, preservation, and discovery remains to be the core issues of Semantic Web vision [49]. The science of informatics has become itself a discipline in its own right for handling and extraction of this data. Quality and reliability are the two most critical aspects of data for drug discovery, particularly access to quality data, due to which a shift has been observed in chemoinformatics from techniques to management, organization, and integration of large amounts of data [50]. A systematic and standardized approach is required for implementing Semantic Web in life sciences [51]. When Semantic Web was compared with chemoinformatics and bioinformatics it was found that the differences in quantity of data depend on the amount of public funding spent in the bioinformatics area [52]. Ownership complexity, potential for income generation in addition to native complexity of chemical space are parts of the basic problem of data management.

There is a preference among scientists to store data in flat files, which is not appropriate data reuse, curation, or preservation [53], which otherwise is preferred to be stored in LIMS (laboratory information management systems) and relational databases (for dynamic data) for large-scale data preservation [54]. Recently, cloud storage has added a new dimension to the management of large data sets. Pharmaceutical R & D data are complex with large volume and so it is not always possible to bring together both data and information from multiple sources [55]. Here, semantics is required to derive and interpret information, with Chem2Bio2DRF projects as on one of the best examples [56]. Chemoinformatics application is dependent on the process of data exchange and the development of CHEMINF (Chemical Information Ontology) will help towards the description of chemical entities [57]. SADI (Semantic Automated Discovery and Integration) and CHESS (Chemical Entity Semantic Specification) were two new features implemented by the CHEMINF group for annotating chemical descriptors and entities [58]. Thus, advances in drug discovery research can be achieved by collection and integration of data from multiple sources.

7.2   METADATA

Metadata now plays an important role in gaining interoperation activity between the gap of chemoinformatics and bioinformatics which exited around 2005 lacking integration. Metadata is critical in understanding the vision of Semantic Web and aids in the performance of machines to perform the important steps required for integration such as data discovery, data interrelation, and chemoinformatics initiation techniques that act on that data. Metadata is defined as "data about data" and it is the heart of the CMCS (multi-scale chemical sciences) project, which attach importance in discovering data and preserving its provenance were regarded as important 10 years later. The tools provided by Semantic Web Technologies enforce the metadata concepts across the various communities that work with semantic metadata [59].

The reliability of metadata depends on the data capture as early in the research procedure. This capture at source requires both manual and automatic recording providing no additional burden on researchers in manual one while automatic one should also capture context with relation to data [60]. CombeChem is an example of this with regard to automatic data capture there exist a caution with relation to data standards of international level by the instruments, as quality data is essential for Semantic Web Technologies [53].

7.3   LEXIS IN CHEMOINFORMATICS

It is imperative to find a common vocabulary to understand or communicate in chemoinformatics and Semantic Web plays provides this common vocabulary. With advances in the chemical Semantic Web common vocabulary and basic ontologies have been developed [61]. Many metadata constructs provide formal data descriptions with controlled vocabularies that are used by many chemoinformatics tools such as CML (Chemical Markup Language) [62].

Ontologies in chemistry are not as well developed as in life sciences but progress in this field is quite encouraging. The first ontology was Semantic Eye by Casher and Rzepa [63]; ChemCloud is another attempt, but it still requires more ontologies to represent existing information in databases [64]. ChEBI (Chemical Entities of Biological Interest) is the recent established ontology with frequent updates [65]. Ontology sets of ChemAxion aim to provide a base for basic chemistry by providing a place to store the origin of records with *ChemAxionMeta* [66].

An understated common goal from the union of chemoinformatics and Semantic Web is the storage of records. In this regard, the oreChem project reflects the importance of the relationships that exist between pedigree of data and trust of those data [67]. To simply this checking on pedigree and validity, repositories requires maximum data about the data they house, Semantic Web provides a solution to this by providing tools for data capturing and storage.

7.4   ACCESS AND DISCOVERY LEVELS

Methods used to find the semantics in the contents of documents were in use before the advent of Semantic Web. Text-mining techniques are applied to biomedical literature to identify characteristic data to extract information for chemical reactions and raw data if well presented should be able to be retrieved by text-mining techniques such as CCDBT (Collaborative Chemistry Database Tool), which is a repository of raw chemical data generated by computational chemistry packages [68]. Metadata from raw data are extracted using a sequence of parsers and to occupy a database for future query based on the metadata model. The use of systems that are compatible with Semantic Web technologies are essential in labs to facilitate the discovery process [53], but the researcher's point of view is quite different from that of a computer, and has to be considered before a computer system is made to work for humans. Semantic Web technologies can be used for storage and access to molecular properties and structures [69]. The influence of the Semantic Web is still yet to be felt on structure search of patent search which was recently reviewed [70]. The need to control access to protect intellectual property is also essential for secure Semantic Web [71].

7.4   LINKING DATA

The concept of linked data is fundamental to Semantic Web. The concerns that exist for web design include four principles for inserting linked data on the web: to identify object with URIs (Universal Resource Identifier), to use http URIs, serving information as per the URIs and to connect data into web [72]. The InChI and InChiKey are the two most important tools for linking raw and processed data of molecule. InChi identifiers are used by the eCrystals archive [73] for linking data of X-ray structures produced by the UK NCS (National Crystallography Service) [74]; they also preserve these raw and processed data, thus making it available for scrutiny and refinement. URIs should be dereferenceable so that human or software can access the linked data [72] and W3C page shows that growing number and range of compliant datasets, which are a part of the emerging web for linked data [75]. An infrastructure to include pharmaceutical database, chemical and biochemical is adopted by ChemCloud to link the data [76].

8   FUTURE PROSPECTS IN CHEMOINFORMATICS

In recent times, many online tools in the public domain have helped the chemoinformatics community in drug discovery research. The ever-increasing amount of data has led to the birth of many new servers designed to manage this large amount of data in biology and chemistry. Contributions to these types of servers were also made by the requirement for analysis of structure such as SAR (structure–activity relationships) and SmART (structure multiple–activity relationships). However, these servers require timely updates for their maintenance as they may become obsolete without this maintenance. Table 2.1 shows recent developments in the chemoinformatics tools, databases, directories, and web resources used in the drug discovery process.

Web servers are currently reviewed extensively as mentioned in the literature [77-81]. CCSAM [73] (Center for Computational Science for the Americas) and D3 [82] (Distributed Drug Discovery) are two very good initiatives taken up to support the drug discovery efforts. The former is directed towards interdisciplinary projects to aid academic, innovative, technology research, and education projects, while the latter trains students all over the world in computational analysis, synthesis, and biological screening with special attention to neglected diseases. Recent studies have shown that deep learning plays a critical role in drug discovery, but performance has not been found to be significantly better as compared to simple methods when streamlined specifically towards drugs and targets. Thus, deep learning has the potential to be explored for biological pathways and target identification. mtk-QSBER [83] (multi-tasking quantitative structure-biological effect relationships), which simultaneously accounts and combines different types of biological and chemical data covering ADMET, biological activity, and toxicity, has been applied for *de-novo* design known as multi-scale *de-novo* drug design [84].

Expert opinions suggest that empirical rules such as Lipinski's RoF have been widely used, but its effect is reduced significantly due to its hard threshold, which can be softened to closely mirror similar property sets. To clear the way for phenotype-based drug discovery, high-content screening is an important technology [85]. Special focus on organ, cellular, and organism phenotypes along with enzyme and receptor details is a critical aspect of drug design. Even molecules that have activities in cell-based and enzyme-based assays and some may be active in cell-based but may be inactive in enzyme assays or vice versa are challenging to deal but they provide some invaluable information about a drug's mechanism of action.

9   CONCLUSIONS

Recent data explosions in the availability of publicly available chemical data provide us an opportunity to use these data and apply computational techniques to extract the underlying structure-activity

TABLE 2.1
Recent Development in Chemoinformatics Tools, Databases, Directories, and Web Resources Used in the Drug Discovery

Database	Interpretation
DrugBank 5.0	Database of investigational drugs
Pharos	User interface
DrugCentral	An online compendium
PubChem	Access and addition of new chemical data
Tox21	Evaluate safety of compounds
ChEMBL	Large scale database with bioactivity data

Methods	Examples in Use
Structure Search	Virtual screening; Data mining
Collection and storing chemical data	Chemical database organizing; Analysis and organization of high-throughput screening
Descriptor calculation	Development of quantitative and qualitative structure-property (activity) relationships; Chemical data communication
Analysis of diversity	Design and optimization of compounds; chemical library comparisons; collection of compound dataset; virtual and similarity screening
Molecular modeling	Fragment and *de-novo* structure design; binding affinity calculation; 3D structure prediction of molecular targets; docking
Chemical information visualization	Interpretation of virtual and experimental high-throughput data; compound selection and classification; pattern recognition
Synthesis design with computers	Synthetic feasibility assessment; *de-novo* design
Biological activity prediction and *in vivo* compound characteristics	Data shaving; virtual screening; identification of lead and hit molecules
Prediction and calculation of compound properties	Compound filtering; ADME-Tox prediction; metabolism prediction; hit and lead optimization

Web Resources	Brief Description
D-Tools	Public database for diversity analysis of compounds
ChemBench	Public web portal for distribution and development of QSAR models
VLS3D	Directory of 3,300 online tools and databases compiled for more than 10 years

relationship. In this chapter, we aimed to explain the basic and advanced techniques on the forefront of chemoinformatics. Chemoinformatics is utilized for drug discovery, database design, and advanced chemoinformatics computational tools, and PubChem database can be enhanced using chemoinformatics and application of Semantic Web Technologies in chemoinformatics along with the future prospects in this field.

CONFLICT OF INTEREST

The authors declare none, financial or otherwise.

ACKNOWLEDGMENTS

We thank the Vice Chancellor, Bharathiar University, Coimbatore-641046, Tamil Nadu for providing the necessary facilities. We also thank UGC-New Delhi for Dr. D S Kothari Fellowship ((No.F-2/2006 (BSR)/BL/20-21/0396)), DST-INSPIRE (DST/INSPIRE/2019/IF190185), RUSA (BU/RUSA2.0/BCTRC/2020/BCTRC-CT03).

REFERENCES

1. Zubatiuk T, Isayev O. Development of multimodal machine learning potentials: toward a physics-aware artificial intelligence. Acc Chem Res. 2021; 54: 1575–1585.
2. von Lilienfeld OA, Müller KR, Tkatchenko A. Exploring chemical compound space with quantum-based machine learning. Nat Rev Chem. 2020; 4: 347–358.
3. Terayama K, Sumita M, Tamura R, Tsuda K. Black-box optimization for automated discovery. Acc Chem Res. 2021; 54, 1334–1346.
4. Krenn M, Ai Q, Barthel S, Carson N et al. SELFIES and the future of molecular string representations. Patterns (N Y). 2022; 14; 3(10):100588.
5. Sun D, Gao W, Hu H, Zhou S. Why 90% of clinical drug development fails and how to improve it? Acta Pharm Sin B12. 2022; 3049–3062.
6. Geerlings P, De Proft F, Langenaeker W. Conceptual density functional theory. Chem Rev.2003; 103:1793–1874.
7. Varnek A, Baskin I. Machine learning methods for property prediction in chemoinformatics: quo vadis? J Chem Inf Model. 2012; 52:1413–1437.
8. Saini V, Sharma A, Nivatia D. A machine learning approach for predicting the nucleophilicity of organic molecules. Phys Chem.2022; 24:1821–1829.
9. Jorner K, Brinck T, Norrby P-O, Buttar D. Machine learning meets mechanistic modelling for accurate prediction of experimental activation energies. Chem Sci.2021; 12:1163–1175.
10. Hecht P. High-throughput screening: beating the odds with informatics-driven chemistry. Curr Drug Discov. 2002; 21–24.
11. Ferreira LL, Andricopulo AD. ADMET modeling approaches in drug discovery. Drug Discov Today. 2019; 24(5), 1157–1165.
12. Shen M, Xiao Y, Golbraikh A, Gombar VK, Tropsha A. Development and validation of k-nearest-neighbor QSPR models of metabolic stability of drug candidates. J Med Chem. 2003; 46(14): 3013–20.
13. Cano G, Garcia-Rodriguez J, Garcia-Garcia A, Perez-Sanchez H, Benediktsson JA, Thapa A, Barr A. Automatic selection of molecular descriptors using random forest: Application to drug discovery. Expert Syst Appl. 2017; 72 : 151–159.
14. Lagorce D, Bouslama L, Becot J, Miteva MA, Villoutreix BO. FAF-Drugs4: free ADME-tox filtering computations for chemical biology and early stages drug discovery. Bioinformatics. 2017; 15; 33(22):3658–3660.
15. Xiao Z, Morris-Natschke SL, Lee KH. Strategies for the Optimization of Natural Leads to Anticancer Drugs or Drug Candidates. Med Res Rev. 2016; 36(1): 32–91.
16. Sharifi-Rad J, Ozleyen A, Boyunegmez Tumer T, et al. Natural Products and Synthetic Analogs as a Source of Antitumor Drugs. Biomolecules. 2019; 9(11):679.
17. Sander T, Freyss J, von Korff M, Rufener C. DataWarrior: an open-source program for chemistry aware data visualization and analysis. J Chem Inf Model. 2015; 23; 55(2):460–73.
18. Testa, B, van der Waterbeemd, H, Folkers, G, Guy, RH & van de Waterbeemd, H 2001, Pharmacokinetic optimization in drug research. Wiley-VCH, Lausanne.
19. Estrada E, Molina E, Perdomo-López I. Can 3D structural parameters be predicted from 2D (topological) molecular descriptors? J Chem Inf Comput Sci. 2001; 41(4):1015–21.
20. Wegner JK, Sterling A, Guha R, et al. Cheminformatics. Communications of the ACM. 2012; 55(11), 65–75.
21. Miyao T, Kaneko H, Funatsu K. Ring system-based chemical graph generation for de novo molecular design. J Comput Aided Mol Des. 2016; 30(5):425–46.
22. Reutlinger M, Koch CP, Reker D, Todoroff N, Schneider P, Rodrigues T, Schneider G. Chemically Advanced Template Search (CATS) for Scaffold-Hopping and Prospective Target Prediction for 'Orphan' Molecules. Mol Inform. 2013; 32(2): 133–138.
23. Todeschini R, Ballabio D, Grisoni F. Beware of unreliable Q 2! A comparative study of regression metrics for predictivity assessment of QSAR models. J Chem Inf Model. 2016. 56(10), 1905–1913.
24. McKay BD, Practical Graph Isomorphism, CongressusNumerantium, 1981, 30, 45–87.
25. Reymond JL, Van Deursen R, Blum LC, Ruddigkeit L. Chemical space as a source for new drugs. MedChemComm. 2010; 1(1), 30–38.

26. Maggiora G, Gokhale V Non-specificity of drug-target interactions-consequences for drug discovery. In: Frontiers in Molecular Design and Chemical Information Science–Herman Skolnik Award Symposium 2015: Jürgen Bajorath. ACS Symposium Series. Vol.1222: Boston, MA: American Chemical Society; 2016. 91–142.

27. Maggiora G, Gokhale V. A simple mathematical approach to the analysis of polypharmacology and polyspecificity data. F1000Res.2017; 6:788.

28. Todeschini R, Valsecchi C. Activity cliffs and structural cliffs for categorical responses. MATCH Commun Math Comput Chem. 2018; 80:283–294.

29. Weininger D, Weininger A, Weininger JL. SMILES. 2. Algorithm for generation of unique SMILES notation. J Chem Inf Comput Sci. 1989; 29, 97–101.

30. Gómez-Bombarelli R, Wei JN, Duvenaud D, et al. Automatic Chemical Design Using a Data-Driven Continuous Representation of Molecules. ACS Cent Sci. 2018 Feb 28; 4(2):268–276.

31. Ma T, Chen J, Xiao C. Constrained generation of semantically valid graphs via regularizing variational autoencoders. 2018. arXiv:1809.02630.

32. O'Boyle N, Dalke A. DeepSMILES: An Adaptation of SMILES for Use in Machine-Learning of Chemical Structures. ChemRxiv. Cambridge: Cambridge Open Engage; 2018.

33. Schirris Y, Gavves E, Nederlof I, Horlings HM, Teuwen J. DeepSMILE: Contrastive self-supervised pre-training benefits MSI and HRD classification directly from H&E whole-slide images in colorectal and breast cancer. Med Image Anal. 2022.79:102464.

34. Landrum G. "RDKit: A software suite for cheminformatics, computational chemistry, and predictive modeling. Manual.2013.

35. Goodman JM, Pletnev I, Thiessen P, Bolton E, Heller SR. InChI version 1.06: now more than 99.99% reliable. J Cheminform. 2021;13(1):40.

36. Bento AP, Gaulton A, Hersey A, et al. The ChEMBL bioactivity database: an update. Nucleic Acids Res. 2014 Jan;42(Database issue):D1083-90.

37. Holliday JD, Salim N, Whittle M, Willett P. Analysis and display of the size dependence of chemical similarity coefficients. J Chem Inf Comput Sci. 2003;43(3):819–28.

38. Meanwell NA. Synopsis of some recent tactical application of bioisosteres in drug design. J Med Chem. 2011; 54:2529–2591.

39. Halgren TA. MMFF VI. MMFF94s option for energy minimization studies. J Comput Chem. 1999; 20:720–729.

40. Williams A, Tkachenko V. The Royal Society of Chemistry and the delivery of chemistry data repositories for the community. J Comput Aided Mol Des. 2014; 28(10):1023–30.

41. Ghani SS. A comprehensive review of database resources in chemistry. Eclética Química. 2020; 45, (3):57–68.

42. Bento AP, Gaulton A, Hersey A, et al. The ChEMBL bioactivity database: an update. Nucleic Acids Res. 2014 Jan;42(Database issue): D1083–90.

43. Kawabata T. Build-up algorithm for atomic correspondence between chemical structures. J Chem Info Model. 2011; 51:1775–1787.

44. Wang H, Klinginsmith J, Dong X, et al. Chemical data mining of the NCI human tumor cell line database. J Chem Inf Model. 2007; 47(6):2063–76.

45. Dutta D, Guha R, Wild D, Chen T. Ensemble feature selection: consistent descriptor subsets for multiple QSAR models. J Chem Inf Model. 2007;47(3):989–97.

46. Guha R, Van Drie JH. Structure—activity landscape index: identifying and quantifying activity cliffs. J Chem Inf Model. 2008; 48(3):646–58.

47. Murray-Rust P, Rzepa HS, Williamson MJ, Willighagen EL. Chemical markup, XML, and the World Wide Web. 5. Applications of chemical metadata in RSS aggregators. J Chem Inf Comput Sci. 2004; 44(2):462–9.

48. Frey JG, Bird CL. Cheminformatics and the Semantic Web: adding value with linked data and enhanced provenance. Wiley Interdiscip Rev Comput Mol Sci. 2013;3(5):465–481.

49. Feigenbaum L, Herman I, Hongsermeier T, Neumann E, Stephens S. The Semantic Web in action. Sci Am. 2007; 297:90–97.

50. Frey JG, Bird CL. Web-based services for drug design and discovery. Expert Opin Drug Discov. 2011; 6:885–895.

51. Curcin V, Ghanem M, Guo Y. Web services in the life sciences. Drug Discov Today. 2005; 10:865–871.

52. Tetko IV. Computing chemistry on the web. Drug Discov Today. 2005; 10:1497–500.

53. Frey JG. The value of the Semantic Web in the laboratory. Drug Discov Today.2009; 14:552–561.

54. Reese A. Databases and documenting data. Significance. 2007; 4:184–186.

55. Slater T, Bouton C, Huang E. Beyond data integration. Drug Discov Today. 2008; 13:584–589.

56. Chen B, Dong X, Jiao D, Wang H, Zhu Q, Ding Y, Wild DJ. Chem2Bio2RDF: a semantic framework for linking and data mining chemogenomic and systems chemical biology data. BMC Bioinformatics. 2010;11:255.

57. Hastings J, Chepelev L, Willighagen E, Adams N, Steinbeck C, Dumontier M. The chemical information ontology: provenance and disambiguation for chemical data on the biological semantic web. PLoS ONE. 2011; 6:e25513.

58. Chepelev L, Dumontier M. Chemical entity semantic specification: knowledge representation for efficient semantic cheminformatics and facile data integration. J Cheminformatics. 2011;3:20.

59. Curcin V, Ghanem M, Guo Y. Web services in the life sciences. Drug Discov Today. 2005; 10:865–871.

60. Frey J. Curation of laboratory experimental data as part of the overall data lifecycle. Int J Digital Curation. 2008. 3:44–62.

61. Bhat T. Chemical Taxonomies and Ontologiers for Semantic Web. Available at: semanticweb.com/chemical-taxonomies-and-ontologies-for-semanticweb b10926. (Accessed Oct 18, 2022).

62. Chemical Markup Language. Available at: www.xml-cml.org/documentation/biblio.html. (Accessed Oct 18, 2022).

63. Casher O, Rzepa HS. Planned Research Serendipity: Exploiting Web 3.0. Symplectic User Community Conference, May 5, 2009. Available at: www.symplectic.co.uk/assets/files/omercasher.ppt.(Accessed Oct 18, 2022).

64. Todor A, Paschke A, Heineke S. ChemCloud: chemical e-Science information cloud. Nat Preceding 2011. (Accessed Oct 18, 2022).

65. de Matos P, Alc ´antara R, Dekker A, Ennis M, Hastings J, Haug K, Spiteri I, Turner S, Steinbeck C. Chemical entities of biological interest: an update. Nucl Acids Res 2010 38(suppl 1):D249–D254.

66. Adams N, Cannon EO, Murray-Rust P. ChemAxiom—an ontological framework for chemistry in science. Nat Precedings 2009.

67. Borkum M, Lagoze C, Frey J, Coles S. A semantic eScience platform for chemistry. In: IEEE Sixth International Conference on e-Science. 2010; 316–323.

68. Chen M, Stott AC, Li S, Dixon DA. Construction of a robust, large-scale, collaborative database for raw data in computational chemistry: the Collaborative Chemistry Database Tool (CCDBT). J Mol Graphics Modelling. 2012; 34:67–75.

69. Taylor KR, Gledhill RJ, Essex JW, Frey JG, Harris SW, De Roure DC. Bringing chemical data onto the Semantic Web. J ChemInf Model. 2006; 46: 939–952.

70. Downs GM, Barnard JM. Chemical patent information systems. WIREs Comput Mol Sci 2011, 1:727–741.

71. Park JS. Towards secure collaboration on the Semantic Web. ACM SIGCAS Comput Soc. 2003; 33.

72. Berners-Lee T. Linked Data. Available at: www.w3.org/DesignIssues/LinkedData. (Accessed Oct 18, 2022).

73. eCrystals. Available at: ecrystals.chem.soton .ac.uk/. (Accessed Oct 18, 2022).

74. Coles SJ, Frey JG, Hursthouse MB, Light ME, Milsted AJ, Carr LA, De Roure D, Gutteridge CJ, Mills HR, Meacham KE, et al. An E-science environment for service crystallography—from submission to dissemination. J Chem Inf Model 2006, 46:1006–1016.

75. W3C. SWEO Community Project: Linking Open Data on the Semantic Web. Available at: www w3.org/wiki/TaskForces/CommunityProjects/Linking OpenData/DataSets. (Accessed Oct 18, 2022).

76. Todor A, Paschke A, Heineke S. ChemCloud: chemical e-Science information cloud. Nat Preceding 2011. (Accessed Oct 18, 2022).

77. Stephens S, LaVigna D, DiLascio M, Luciano J. Aggregation of bioinformatics data using Semantic Web technology. J Web Semantics 2006, 4:216–221. Journal of Cheminformatics, Thematic Series.

78. RDF technologies in chemistry. Available at: www.jcheminf.com/series/acsrdf2010. (Accessed Oct 18, 2022).

79. Willighagen EL, Br¨andle MP. Resource description framework technologies in chemistry. J Cheminform. 2011; 3:15.

80. Bachrach SM. Chemistry publication–making the revolution. J Cheminform. 2009; 1:2.

81. Bachrach SM, Krassavine A, Burleigh DC. End-user customized chemistry journal articles. J Chem Inf Comput Sci. 1999; 39:81–85.

82. Smith B, Ashburner M, Rosse C, et al. The OBO Foundry: coordinated evolution of ontologies to support biomedical data integration. Nat Biotechnol. 2007; 25:1252–1255.

83. Choi JY, Davis MJ, Newman AF, Ragan MA. A Semantic Web ontology for small molecules and their biological targets. J Chem Inf Model. 2010; 50: 732–741.

84. Chen H, Xie G. The use of web ontology languages and other semantic web tools in drug discovery. Expert Opin Drug Discov. 2010; 5: 413–423.

85. Quinnell R, Hibbert DB, Milsted A. eScience: evaluating electronic laboratory notebooks in chemistry research. In: Proceedings Ascilite Auckland; 2009.

3 An Introduction to Basic and Advanced Bioinformatics Computational Tools

Jay R. Anand, Manishkumar S. Patel, Manish Shah,
Bhavinkumar Gayakvad, Dignesh Khunt

1 INTRODUCTION

Bioinformatics can be defined as an interdisciplinary field for creating methods and software tools with the primary aim of biological investigation. It connects diverse field of basic sciences (biology, chemistry, physics) and application sciences (pharmacy, medicine, healthcare, etc.) with computer science and quantitative science (mathematic and statistics). Bioinformatics is an integral part of biological experimentation and interpretation of large quantify of data generated. Computer programming is the basis for the bioinformatic approach, which often creates a set of instruction called a 'pipeline' that are repeatedly used as a standalone program or as part of a software tool. A plethora of bioinformatics tools are available for researchers to use, and the tools of the trade can differ slightly or dramatically based on the field of research. Bioinformatics tools can broadly be divided into three categories: a) databases, b) data presentation, and c) data analysis. This chapter is focused on bioinformatic tools for data analysis and is divided into sections that enlist and discuss the key areas of bioinformatics analysis and common tools used (Figure 3.1). There are other more specialized areas of bioinformatics that deal in other quantitative factors (image analysis and signaling processing), which are covered elsewhere in this book.

2 SEQUENCE ANALYSIS TOOLS

The digital nature of DNA (i.e., organization as A, T, G, C) has several biological advantages such as DNA replication with high accuracy and high certainty in protein coding. It also allows us to easily store DNA as a string of information and helps us make meaningful interpretations about the genetic makeup of an organism. Since the inception of bioinformatics, sequence analysis has been a cornerstone of bioinformatic research, and several sequence analysis tools are developed to utilize the digital nature of DNA, RNA, and protein. The following section introduces different areas of sequence analysis and common bioinformatic tools.

2.1 SEQUENCE ALIGNMENT OR REFERENCE-BASED MAPPING/VISUALIZATION TOOLS

Sequence alignment tools are the primary example of this class and include database search tools (e.g., BLAST (1)); pairwise alignment tools (e.g., Bioconductor →Bistrings: pairwiseAlignment); multiple sequence alignment tools (e.g., Bali-Phy (2)); genomic analysis tools (e.g., DECIPHER (3); EAGLE (4)); motif finding tools (e.g., Phyloscan (5), PMS (6)); benchmarking tools (SMART (7)); alignment viewers/editors (e.g., Integrated Genome Browse IGB (8)); and short-read sequence

DOI: 10.1201/9781003353768-3

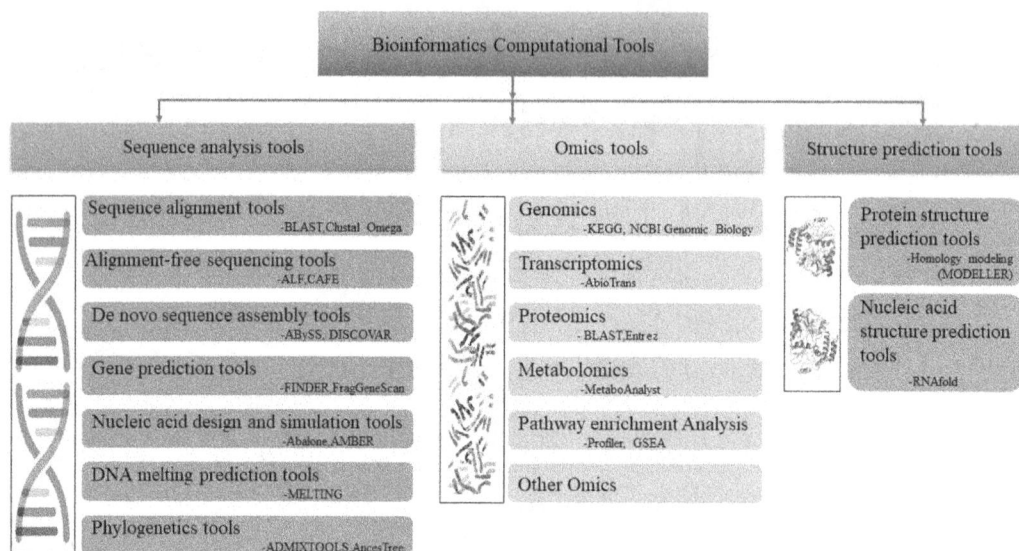

FIGURE 3.1 Graphical representation of types of bioinformatics computational tools with examples.

TABLE 3.1
Different types of BLAST programs for sequence analysis

Sr. No	Types of BLAST	Specification (Query vs. Database)
1	BLASTn	Nucleotides vs. nucleotides
2	BLASTp	Protein vs. protein
3	BLASTx	Translated nucleotide vs. Protein
4	tBLASTn	Protein vs. translated nucleotide
5	tBLASTx	Translated nucleotide vs. translated nucleotide

alignment (Bowtie (9)). Different types of BLAST analysis can be done based on query and target database, as shown in Table 3.1.

2.2 ALIGNMENT-FREE SEQUENCE ANALYSIS TOOLS

Sequences generated in the course of biological research can often be divergent, and a sequence alignment-based approach will be less practical and unreliable. Further, application of sequence-alignment tools for genomic data can be complex and slow. Alignment-free sequence analysis provides a reliable alternative in such cases. Alignment-free sequence analysis approaches are generally based on k-mer/word frequency, the length of common substrings, number of word matches, micro-alignments, information theory, and graphical representation. AFproject is a global community-based effort to standardize, enhance, and streamline alignment-free sequencing approaches.

2.3 DE NOVO SEQUENCE ASSEMBLY TOOLS

DNS assemblers allow construction of large nucleotide sequences from shorter sequences independent of a reference sequence. Major DNS assembly approaches include greedy assembly (find all overlaps between reads) (e.g., SEQAID (10) and CAP (11)) and de Bruijin graphs (break reads into k-mer and connect overlapping k-mer nodes; important in next-generation sequencing) (e.g., Abyss (12) and Trinity). Other noteworthy approaches are overlap-layout-consensus (build an overlap graph, create contigs, and pick most likely sequences) and string graphs (traverse a graph to produce a consensus) assemblers.

2.4　Gene Prediction Tools

A class of sequencing tools that allows searching for and analysis of sequences that encode for genes, RNAs, or other genomic elements such as regulatory regions are collectively referred to as gene prediction or finder tools. These tools utilize either sequence similarity or ab initio gene prediction approach (gene structure as a reference for gene identification). Similarity sequence search-based tools can be further classified based on whether they employ local alignment (e.g., BLAST and related group of gene tools) or global alignment approaches (e.g., PROCRUSTES and GeneWise). Ab initio gene predictions utilize signal sensors and content sensors for prediction. Signal sensors are short sequence motifs such as start codon, stop codon, splice sites, branch points, and polypyrimidine tracts. Content sensors denote species-specific codon usage pattern and help in exon detection. Commonly used tools based on ab initio gene prediction include GeneID (13), FINDER (14), GeneParser (15), GrailEX (16), and EuGene (17).

2.5　Nucleic Acid Design and Simulation Tools

Nucleic acid design is a growing field of creating a collection of nucleic acid sequences that will assemble into a chosen configuration. It is key to the area of DNA and RNA nanotechnology with several diverse applications, especially in the field of precision drug delivery. Three main approaches commonly used for nucleic acid design are heuristic methods, thermodynamics models, and geometric models. Heuristic methods involve the use of straightforward criteria based on previous knowledge to rapidly predict appropriateness of diverse nucleic acid sequences for a particular secondary structure. Thermodynamic models for nucleic acid design are based on the rationale that thermodynamic properties can be predicted from insight about nucleic acid sequences and their secondary complex. Tools for thermodynamic modeling of nucleic acid are Nupack (18), mfold/UNAFOLD (19), and Vienna (20). Geometric models in nucleic acid design are used to predict tertiary structure. Geometric modeling involves understanding constrains arising from misalignments in the nucleic acid structure and addressing them by modifying nucleic acid design. Commonly used geometric modeling tools applied to nucleic acid design include GIDEON (21), Tiamat, Nanoengineer-1, and UNIQUIMER 3D (21). DNA origami is a special case of nucleic acid design, and dedicated bioinformatic tools have been developed for this (e.g., caDNAno (22) and SARSE (23)). Finally, tools also exist for nucleic acid simulation, which involves nucleic acid structure modeling and justifying investigational outcomes with resolution at nucleotide-level of nucleic acid structure/complex behaviors. Popular nucleic acid simulation tools are oxDNA (24) and oxRNA (25).

2.6　DNA Melting (Denaturation) Prediction Tools

DNA melting reference to separation of DNA double helix into single-stranded DNAs or a DNA bubble (for DNA nucleotide molecules). DNA melting is usually studied as a function of the temperature and is a key area of research in the field of nucleic acid thermodynamics. DNA melting prediction is important for several molecular biology applications such as PCR and DNA hybridization as well as for studying biological processes such as DNA replication, transcription, recombination, and repair. Simple DNA melting prediction tools are based on either basic method (Marmur–Doty equation) or nearest neighbor method. The basic method for melting temperature depends on G:C content of the DNA molecules, do not take heterogenous base-pair stacking into calculation, and is good for predicting melting temperature of short nucleotides. The nearest neighbor method takes into account base pair interactions as well as effect of neighboring bases on the interaction between two DNA strands. It is a more realistic way of calculating melting temperature and is useful for large nucleotide molecules, but hardly provides additional benefit for short oligonucleotides. Oligo Calc (26) is an example of a simple DNA melting tool commonly used by scientists. Advanced DNA melting prediction tools employ statistical mechanical methods that are based on statistical calculation and probability instead of predetermined assumptions of the DNA molecules. Commonly used advanced DNA melting prediction tools include DNAmelt (26), MeltSim (26), and uMelt (27).

2.7 PHYLOGENETICS TOOLS

Phylogenetics tools are used either to generate phylogenetics or view phylogenetic trees. They are primarily applied in the fields of cladistics (biological classification of organisms) and comparative genomics. Early tools were based on multiple sequence alignment for generating phylogenies. Modern tools utilize a single method or a combination of methods for generating phylogenetic trees including likelihood-based methods (maximum likelihood or Bayesian phylogenetic interface), neighbor-joining, maximum parsimony, distance matrix methods, and unweighted pair group method with arithmetic mean (UPGMA). Commonly used phylogenetic tools include PHYLIP (28), ClustalW (29), PAUP (30), Dendroscope (31), MrBayes (32) and Geneious, iTOL (interactive Tree Of Life) (33), T-REX (34), and MEGA (35). A major challenge in phylogenetics is generating large phylogenetic tress from large datasets of sequences. Some of the advanced tools developed to overcome these challenges include ASTRA-PRO (36), DupTree (37), DISCO (38), APPLES (39), EPA-ng (40), pplacer-SCAMPP (41), and INSTRAL (42).

3 STRUCTURE PREDICTION TOOLS

Structure prediction is the inverse of nucleic acid or protein sequence design. In sequence design, a sequence is created that will produce a desired structure, while in structure prediction, the structure is predicted from a known sequence.

3.1 PROTEIN STRUCTURE PREDICTION TOOLS

Protein structure prediction involves suggesting a secondary and tertiary structure of a protein from its primary structure (i.e., amino acid sequences). The application of these tools is broad ranging from basic biology to applications in drug discovery and biotechnology (e.g., design novel and better enzymes). Experimental approaches for protein structure prediction such as X-ray crystallography, cyrogenic electron microscopy (cyro-EM), and NMR spectroscopy can be long, laborious, and costly. Bioinformatic tools provide an attractive alternative to these techniques, and have high throughput and are fast and inexpensive. Some of the common bioinformatic approaches used in protein prediction are homology modeling (predict target protein structure using target protein amino acid sequences and known structure of a homologous protein; examples include MODELLER (43), MaxMod (44), PyMOD (45), PRIMO (46) SWISS-MODEL (47)); protein threading (also referred to as fold recognition) predicts target protein structure for target proteins that do not have known homologous protein structure but have folds that are similar to other known protein structures; examples include HHPred (48), MUSTER (49), RaptorX (50), and PHYRE2 (51); Ab initio structure prediction (a de novo protein structure method that utilizes physicochemical parameters and algorithms to predict secondary structure and tertiary structures, respectively; examples include trRosetta (52), ROBETTA (53), Abalone, and C-Quark (54). The field of protein structure prediction is rapidly evolving and has fruitfully utilized advanced tools based on artificial intelligence and machine learning to improve protein structure prediction accuracy rapidly. One such example is Alphafold 2 (55, 56). It is an artificial intelligence (AI)-based protein structure prediction tool developed by DeepMind group at Alphabet. It made headlines when it ranked at the top of the 14th Critical Assessment of protein Structure Prediction (CASP) organized in 2020 with more than 90 percent accuracy for predicting two-thirds of the proteins in the CASP's global distance test (GDT), which are very hard to predict proteins. Of key recent importance, AlphaFold was helpful in inferring the structure of SARS-CoV-2.

3.2   Nucleic Acid Structure Prediction Tools

In structure prediction, we aim to predict nucleic acid structure (secondary and tertiary) from nucleotide sequences. Thus, the steps involved in predicting nuclei acid structure is opposite of that involved in nucleic acid design. Methods for DNA and RNA structure prediction are similar with minor differences in approaches. This is due to the fact that DNA generally base pair with complementary strands and exist in vivo as a duplex. In contrast, RNAs usually form complex secondary and higher-order structures that are difficult to accurately predict by same approaches as that used for DNA duplexes. A key factor governing differences in the approaches for DNA and RNA structure prediction is the differences in DNA and RNA thermodynamics. Examples of complex structures formed by RNAs include molecular machines such as ribosomes and spliceosomes. We are focusing our discussion on RNA structure prediction since it is a major field in bioinformatics with wide applications in the fields of biology, pathology, and medicine. A key promising application of RNA structure prediction is in studying biological functions of non-coding RNAs. Like protein structure prediction, experimental methods such as crystallography and NMR are also used for predicting RNA structure but are not always feasible due to technical difficulties and cost. Important structure prediction tools include RNAfold (57), RNAsoft (58), MXfold2 (59), RNA-SSPT (60), RNAstructure (61), and ValFold (62).

4   OMICS TOOLS

4.1   Genomics

Genomics involves analysis of all the genes of an organism including interaction between genes and the environment. The major subgroups of genomics are functional genomics, structural genomics, epigenomics, and metagenomics. Genomic analysis typical involves three steps. First, sequencing of the genome (i.e., genomic DNA); second, sequence assembly where all the sequenced DNA are aligned and merged to reconstruct the original sequence; and third, gene annotation, which is the process of assigning biological knowledge to the sequence. Common DNA sequencing methods for genome-level analysis include whole-genome sequencing (WGS), whole-exome sequencing (WES), methylation sequencing, targeting resequencing, and chromatin immunoprecipitation (ChIP) sequencing. There is a plethora of genomic tools, but some commonly used genomic databases and tools across different disciplines include NCBI genome, KEGG (63), database of essential genes (DEG) (64), and OrthoDB (65). The key areas that utilize genomic tools include genomic medicine, synthetic biology, and population genomics.

4.2   Transcriptomics

Transcriptomics is the study of all the RNA molecules in a cell. Because DNA is transcribed into RNA, which in turn is translated to protein, RNA serves as an intermediate information molecule. Transcriptomics analyze changes in RNA molecules in a cell and give us information regarding changes occurring across different biological processes in a cell under different conditions. It also provides information regarding non-coding RNAs that have diverse biological functions. There are two widely used methods for measuring RNA molecules: microarray and RNA-Seq, which is based on high-throughput sequencing and is popular these days. The RNA-seq method can be further classified into mRNA-seq, total RNA-seq, targeted RNA-Seq, single-cell RNA-Seq, RNA exome capture, small RNA-Seq, and ribosome profiling. RNA-seq generates raw reads of RNA in the sample, which is similar to DNA sequencing and undergoes different steps to generate meaningful biological information. The steps for RNA-seq include a) quality control (i.e., trimming off adapters, removing errors, or low-quality reads and filtering data; b) alignment (i.e., align the sequenced RNA reads against reference); c) quantification (i.e., calculating how often a read is present in a sample); and d) differential expression (i.e., comparing the RNA expression between groups). Common bioinformatics tools used for differential gene expression analysis are edgeR (66), LIMMA (66), and DESeq (67). Widely used databases for gene expression analysis include the Cancer Genome Atlas (TCGA) and the International Cancer Genome Consortium (ICGC) data portals.

4.3 PROTEOMICS

Proteomics is the comprehensive study of proteomes (i.e., all the proteins present in a organisms or sample), their structure, their modifications, their prevalence, and their functions under physiological conditions and in disease. Proteomic experiments involve three key steps. First is the purification of proteins, second is quantifying the protein, and third is the bioinformatic analysis of the generated protein data (often referred to as proteome informatics). Proteome informatic tools are generally employed for identifying and quantifying the proteins by aligning the protein fragments similar to DNA fragments against reference protein sequences using databases such as UniProt. PyMOL is graphic tool widely used for visualtion and protein structures. Other advanced tools such as AlphaFold can help decipher their structure from sequences. In the case of X-ray crystallography a range of advanced bioinformatic tools are used for different purposes. The X-ray Detector Software (XDS) program package (68) is a command line-based software used for processing/ refining single-crystal monochromatic diffraction data. CCP4 (69) is graphical interface-based software for processing/refining protein structure data obtained by X-ray crystallography. Cyroelectromicrosopy (Cyro-EM) has emerged as a major technique for protein structure predictions. Bioinformatics tools such as RELION (structure refinement) (70), EMAN2 (data processing) (71), Scipion (image processing), and (72) help decipher cyro-EM data to protein structure. Finally, tools such as PEIMAN (post-translational modifications term extraction from UniProtKB database) (73) and MusiteDeep (post-translational modifications site prediction) (74) are used to understand posttranslational modifications (PTMs).

4.4 METABOLOMICS

Metabolome includes metabolites in a sample, which may be a cell, tissue, biological fluid, or an organism, and their interactions with biological processes. Metabolomics in simple terms is the comprehensive analysis of metabolomes. The metabolomics workflow has three basic steps. First, is the separation of metabolites from the sample, followed by quantification of metabolites, and finally characterization of the metabolomes including changes in metabolites and their relationship to disease and outcomes. Metaboanalyst (general purpose) (75) and SIRIUS (*de novo* metabolite identification) (76) are some of the most commonly used bioinformatic tools in metabolomics.

4.5 OTHER OMICS

Several newer branches of omics are emerging. Phenomics is the large-scale analysis of phenotypes. Epigenomics, which overlaps with genomics, is the comprehensive analysis of epigenetic modifications such as DNA methylation and histone modifications. Interactomics, which overlaps with proteomics, is the all-encompassing analysis of proteins and their interactions with other proteins and biological molecules. Reactomics is a special case of metabolomics that utilize mass spectrometry data of small molecules to predict their chemical reactions.

4.6 PATHWAY ENRICHMENT ANALYSIS

Omics experiments of DNA, RNA, and protein generate big data, which is expanding dramatically. A general challenge for most researchers is analyzing and interpreting this data. Analysis of omics data by bioinformatics as discussed in previous sections usually generates lists or ranked lists of genes, RNAs, or proteins. The major challenge is to interpret these lists in biologically meaningful ways and extract useful information, which requires large literature reviews. A standard methodology employed to deal with this challenge is pathway enrichment analysis (PEA) using advanced bioinformatic tools, where gene lists are truncated into smaller biological or physiologically relevant and understandable pathways based on previous knowledge. PEA involves statistical assessment of probability of genes on those lists as compared to chance occurrences. Some of the common factors that are considered for such analysis are number of genes on the list of interest, their ranked order, and how representative they are of a pathway of interest. For example, if 35% of the genes on the list annotate for cell cycle pathway, which under normal conditions code for only 8% of the genes, the cell cycle pathway gens will be considered enriched. Two key steps involved in PEA are: a) generate list or ranked list of interest from the omics data analysis, and b) pathway enrichment analysis. PEA can be carried out in two ways. First, using select enriched gene list or ranked list, which meets the threshold for significance and fold change (using g:profiler bioinformatic tool (77)), and second, using all the ranked list of genes in the genome or the dataset, often referred to as gene set enrichment analysis (GSEA) (using GSEA software (78)). The final step in PEA is the graphical representation and interpretation of the enrichment analysis, which can be carried out by tools such as ENrichmentMap, STRING, KEGG, Reactome, and ClueGO among others.

5 CHALLENGES AND PERSPECTIVE

Given the explosion of biological data in recent times, data analysis has become the key bottleneck for advancement of bioinformatics. Most of the big advancements in genomics were accompanied by innovative advances in statistics and computational approaches. Thus, it is not surprising that future progress in life sciences will depend on technical and methodological advances in bioinformatics. Large-scale data increases the complexity of data analysis and creates new challenges for data visualization. Newer graphical and programming tools will help address these challenges while improving data simplification and insightful presentation. Another major challenge bioinformaticians face is unrelated to technical advancement but with education and culture. Bioinformaticians are increasingly becoming part of multidisciplinary teams from bench to bedside; however, interdisciplinary collaborative work is sometimes hampered due to the technical language barriers and differences in the way life scientists/clinicals differ from bioinformaticians in adopting/defining methodology and significance. This challenge has partly been addressed by inclusion of bioinformatics and statistics as a core course by universities training life scientists. Finally, specific areas of research require reliable databases to mine as well as deep knowledge regarding the subjects. For example, phytochemicals such as curcumin, esculetin, and many others have shown biological activities including anticancer activity but reliable databases that provide this information are lacking (79, 80). Similarly, bioinformatic models that consider unique properties of proteins are lacking. For example, topoisomerase-targeting anticancer agents generate anticancer activities by converting topoisomerase proteins into DNA lesion; however, bioinformatic tools that can screen for such chemical or DNA repair proteins that can resolve DNA-protein covalent complexes are lacking (81, 82). Despite all the challenges, bioinformatics will be a major contributor to our understanding of biological systems and will improve healthcare in the coming years.

REFERENCES

1. Johnson M, Zaretskaya I, Raytselis Y, Merezhuk Y, McGinnis S, Madden TL. NCBI BLAST: a better web interface. Nucleic Acids Res. 2008;36(Web Server issue):W5-9.
2. Redelings BD. Bali-Phy version 3: Model-based co-estimation of alignment and phylogeny. Bioinformatics. 2021.
3. Wright ES. DECIPHER: harnessing local sequence context to improve protein multiple sequence alignment. BMC Bioinformatics. 2015;16:322.
4. Pratas D, Silva JM. Persistent minimal sequences of SARS-CoV-2. Bioinformatics. 2021; 36(21):5129–32.
5. Carmack CS, McCue LA, Newberg LA, Lawrence CE. PhyloScan: identification of transcription factor binding sites using cross-species evidence. Algorithms Mol Biol. 2007;2:1.
6. Dinh H, Rajasekaran S. PMS: a panoptic motif search tool. PLoS One. 2013;8(12):e80660.
7. Letunic I, Khedkar S, Bork P. SMART: recent updates, new developments and status in 2020. Nucleic Acids Res. 2021;49(D1):D458-D60.
8. Freese NH, Norris DC, Loraine AE. Integrated genome browser: visual analytics platform for genomics. Bioinformatics. 2016;32(14):2089–95.
9. Langmead B, Salzberg SL. Fast gapped-read alignment with Bowtie 2. Nat Methods. 2012;9(4):357–9.
10. Peltola H, Soderlund H, Ukkonen E. SEQAID: a DNA sequence assembling program based on a mathematical model. Nucleic Acids Res. 1984;12(1 Pt 1):307–21.
11. Huang X. A contig assembly program based on sensitive detection of fragment overlaps. Genomics. 1992;14(1):18–25.
12. Jackman SD, Vandervalk BP, Mohamadi H, Chu J, Yeo S, Hammond SA, et al. ABySS 2.0: resource-efficient assembly of large genomes using a Bloom filter. Genome Res. 2017;27(5):768–77.
13. Blanco E, Parra G, Guigo R. Using geneid to identify genes. Curr Protoc Bioinformatics. 2007;Chapter 4:Unit 4 3.
14. Banerjee S, Bhandary P, Woodhouse M, Sen TZ, Wise RP, Andorf CM. FINDER: an automated software package to annotate eukaryotic genes from RNA-Seq data and associated protein sequences. BMC Bioinformatics. 2021;22(1):205.
15. Snyder EE, Stormo GD. Identification of protein coding regions in genomic DNA. J Mol Biol. 1995;248(1):1–18.
16. Uberbacher EC, Hyatt D, Shah M. GrailEXP and Genome Analysis Pipeline for genome annotation. Curr Protoc Bioinformatics. 2004;Chapter 4:Unit4 9.
17. Sallet E, Gouzy J, Schiex T. EuGene: An Automated Integrative Gene Finder for Eukaryotes and Prokaryotes. Methods Mol Biol. 2019;1962:97–120.
18. Zadeh JN, Steenberg CD, Bois JS, Wolfe BR, Pierce MB, Khan AR, et al. NUPACK: Analysis and design of nucleic acid systems. J Comput Chem. 2011;32(1):170–3.
19. Markham NR, Zuker M. UNAFold: software for nucleic acid folding and hybridization. Methods Mol Biol. 2008;453:3–31.
20. Lorenz R, Bernhart SH, Honer Zu Siederdissen C, Tafer H, Flamm C, Stadler PF, et al. ViennaRNA Package 2.0. Algorithms Mol Biol. 2011;6:26.
21. Birac JJ, Sherman WB, Kopatsch J, Constantinou PE, Seeman NC. Architecture with GIDEON, a program for design in structural DNA nanotechnology. J Mol Graph Model. 2006;25(4):470–80.
22. Douglas SM, Marblestone AH, Teerapittayanon S, Vazquez A, Church GM, Shih WM. Rapid prototyping of 3D DNA-origami shapes with caDNAno. Nucleic Acids Res. 2009;37(15):5001–6.
23. Andersen ES, Dong M, Nielsen MM, Jahn K, Lind-Thomsen A, Mamdouh W, et al. DNA origami design of dolphin-shaped structures with flexible tails. ACS Nano. 2008;2(6):1213–8.
24. Ouldridge TE, Louis AA, Doye JP. Structural, mechanical, and thermodynamic properties of a coarse-grained DNA model. J Chem Phys. 2011;134(8):085101.
25. Sulc P, Romano F, Ouldridge TE, Doye JP, Louis AA. A nucleotide-level coarse-grained model of RNA. J Chem Phys. 2014;140(23):235102.
26. Kibbe WA. OligoCalc: an online oligonucleotide properties calculator. Nucleic Acids Res. 2007;35(Web Server issue):W43–6.
27. Dwight Z, Palais R, Wittwer CT. uMELT: prediction of high-resolution melting curves and dynamic melting profiles of PCR products in a rich web application. Bioinformatics. 2011;27(7):1019–20.

28. Retief JD. Phylogenetic analysis using PHYLIP. Methods Mol Biol. 2000;132:243–58.

29. Larkin MA, Blackshields G, Brown NP, Chenna R, McGettigan PA, McWilliam H, et al. Clustal W and Clustal X version 2.0. Bioinformatics. 2007;23(21):2947–8.

30. Wilgenbusch JC, Swofford D. Inferring evolutionary trees with PAUP*. Curr Protoc Bioinformatics. 2003;Chapter 6:Unit 6 4.

31. Huson DH, Richter DC, Rausch C, Dezulian T, Franz M, Rupp R. Dendroscope: An interactive viewer for large phylogenetic trees. BMC Bioinformatics. 2007;8:460.

32. Huelsenbeck JP, Ronquist F. MRBAYES: Bayesian inference of phylogenetic trees. Bioinformatics. 2001;17(8):754–5.

33. Letunic I, Bork P. Interactive Tree Of Life (iTOL) v5: an online tool for phylogenetic tree display and annotation. Nucleic Acids Res. 2021;49(W1):W293-W6.

34. Boc A, Diallo AB, Makarenkov V. T-REX: a web server for inferring, validating and visualizing phylogenetic trees and networks. Nucleic Acids Res. 2012;40(Web Server issue):W573-9.

35. Tamura K, Stecher G, Kumar S. MEGA11: Molecular Evolutionary Genetics Analysis Version 11. Mol Biol Evol. 2021;38(7):3022–7.

36. Zhang C, Scornavacca C, Molloy EK, Mirarab S. ASTRAL-Pro: Quartet-Based Species-Tree Inference despite Paralogy. Mol Biol Evol. 2020;37(11):3292–307.

37. Wehe A, Bansal MS, Burleigh JG, Eulenstein O. DupTree: a program for large-scale phylogenetic analyses using gene tree parsimony. Bioinformatics. 2008;24(13):1540–1.

38. Willson J, Roddur MS, Liu B, Zaharias P, Warnow T. DISCO: Species Tree Inference using Multicopy Gene Family Tree Decomposition. Syst Biol. 2022;71(3):610–29.

39. Balaban M, Sarmashghi S, Mirarab S. APPLES: Scalable Distance-Based Phylogenetic Placement with or without Alignments. Syst Biol. 2020;69(3):566–78.

40. Barbera P, Kozlov AM, Czech L, Morel B, Darriba D, Flouri T, et al. EPA-ng: Massively Parallel Evolutionary Placement of Genetic Sequences. Syst Biol. 2019;68(2):365–9.

41. Wedell E, Cai Y, Warnow T. SCAMPP: Scaling Alignment-based Phylogenetic Placement to Large Trees. IEEE/ACM Trans Comput Biol Bioinform. 2022;PP.

42. Rabiee M, Mirarab S. INSTRAL: Discordance-Aware Phylogenetic Placement Using Quartet Scores. Syst Biol. 2020;69(2):384–91.

43. Webb B, Sali A. Comparative Protein Structure Modeling Using MODELLER. Curr Protoc Bioinformatics. 2016;54:5 6 1–5 6 37.

44. Parida BK, Panda PK, Misra N, Mishra BK. MaxMod: a hidden Markov model based novel interface to MODELLER for improved prediction of protein 3D models. J Mol Model. 2015;21(2):30.

45. Janson G, Paiardini A. PyMod 3: a complete suite for structural bioinformatics in PyMOL. Bioinformatics. 2021;37(10):1471–2.

46. Hatherley R, Brown DK, Glenister M, Tastan Bishop O. PRIMO: An Interactive Homology Modeling Pipeline. PLoS One. 2016;11(11):e0166698.

47. Waterhouse A, Bertoni M, Bienert S, Studer G, Tauriello G, Gumienny R, et al. SWISS-MODEL: homology modelling of protein structures and complexes. Nucleic Acids Res. 2018;46(W1):W296-W303.

48. Soding J, Biegert A, Lupas AN. The HHpred interactive server for protein homology detection and structure prediction. Nucleic Acids Res. 2005;33(Web Server issue):W244-8.

49. Wu S, Zhang Y. MUSTER: Improving protein sequence profile-profile alignments by using multiple sources of structure information. Proteins. 2008;72(2):547–56.

50. Kallberg M, Wang H, Wang S, Peng J, Wang Z, Lu H, et al. Template-based protein structure modeling using the RaptorX web server. Nat Protoc. 2012;7(8):1511–22.

51. Kelley LA, Mezulis S, Yates CM, Wass MN, Sternberg MJ. The Phyre2 web portal for protein modeling, prediction and analysis. Nat Protoc. 2015;10(6):845–58.

52. Du Z, Su H, Wang W, Ye L, Wei H, Peng Z, et al. The trRosetta server for fast and accurate protein structure prediction. Nat Protoc. 2021;16(12):5634–51.

53. Kim DE, Chivian D, Baker D. Protein structure prediction and analysis using the Robetta server. Nucleic Acids Res. 2004;32(Web Server issue):W526-31.

54. Mortuza SM, Zheng W, Zhang C, Li Y, Pearce R, Zhang Y. Improving fragment-based ab initio protein structure assembly using low-accuracy contact-map predictions. Nat Commun. 2021;12(1):5011.

55. Jumper J, Evans R, Pritzel A, Green T, Figurnov M, Ronneberger O, et al. Highly accurate protein structure prediction with AlphaFold. Nature. 2021;596(7873):583–9.

56. Skolnick J, Gao M, Zhou H, Singh S. AlphaFold 2: Why It Works and Its Implications for Understanding the Relationships of Protein Sequence, Structure, and Function. J Chem Inf Model. 2021;61(10):4827–31.

57. Denman RB. Using RNAFOLD to predict the activity of small catalytic RNAs. Biotechniques. 1993;15(6):1090–5.

58. Andronescu M, Aguirre-Hernandez R, Condon A, Hoos HH. RNAsoft: A suite of RNA secondary structure prediction and design software tools. Nucleic Acids Res. 2003;31(13):3416–22.

59. Sato K, Akiyama M, Sakakibara Y. RNA secondary structure prediction using deep learning with thermodynamic integration. Nat Commun. 2021;12(1):941.

60. Ahmad F, Mahboob S, Gulzar T, Din SU, Hanif T, Ahmad H, et al. RNA-SSPT: RNA Secondary Structure Prediction Tools. Bioinformation. 2013;9(17):873–8.

61. Bellaousov S, Reuter JS, Seetin MG, Mathews DH. RNAstructure: Web servers for RNA secondary structure prediction and analysis. Nucleic Acids Res. 2013;41(Web Server issue):W471-4.

62. Akitomi J, Kato S, Yoshida Y, Horii K, Furuichi M, Waga I. ValFold: Program for the aptamer truncation process. Bioinformation. 2011;7(1):38–40.

63. Kanehisa M, Goto S. KEGG: kyoto encyclopedia of genes and genomes. Nucleic Acids Res. 2000;28(1):27–30.

64. Zhang R, Ou HY, Zhang CT. DEG: a database of essential genes. Nucleic Acids Res. 2004;32(Database issue):D271-2.

65. Kriventseva EV, Kuznetsov D, Tegenfeldt F, Manni M, Dias R, Simao FA, et al. OrthoDB v10: sampling the diversity of animal, plant, fungal, protist, bacterial and viral genomes for evolutionary and functional annotations of orthologs. Nucleic Acids Res. 2019;47(D1):D807–D11.

66. Robinson MD, McCarthy DJ, Smyth GK. edgeR: a Bioconductor package for differential expression analysis of digital gene expression data. Bioinformatics. 2010;26(1):139–40.

67. Anders S, Huber W. Differential expression analysis for sequence count data. Genome Biol. 2010;11(10):R106.

68. Kabsch W. Xds. Acta Crystallogr D Biol Crystallogr. 2010;66(Pt 2):125–32.

69. Krissinel E, Lebedev AA, Uski V, Ballard CB, Keegan RM, Kovalevskiy O, et al. CCP4 Cloud for structure determination and project management in macromolecular crystallography. Acta Crystallogr D Struct Biol. 2022;78(Pt 9):1079–89.

70. Scheres SH. RELION: implementation of a Bayesian approach to cryo-EM structure determination. J Struct Biol. 2012;180(3):519–30.

71. Tang G, Peng L, Baldwin PR, Mann DS, Jiang W, Rees I, et al. EMAN2: an extensible image processing suite for electron microscopy. J Struct Biol. 2007;157(1):38–46.

72. de la Rosa-Trevin JM, Quintana A, Del Cano L, Zaldivar A, Foche I, Gutierrez J, et al. Scipion: A software framework toward integration, reproducibility and validation in 3D electron microscopy. J Struct Biol. 2016;195(1):93–9.

73. Nickchi P, Jafari M, Kalantari S. PEIMAN 1.0: Post-translational modification Enrichment, Integration and Matching ANalysis. Database (Oxford). 2015;2015:bav037.

74. Wang D, Liu D, Yuchi J, He F, Jiang Y, Cai S, et al. MusiteDeep: a deep-learning based webserver for protein post-translational modification site prediction and visualization. Nucleic Acids Res. 2020;48(W1):W140-W6.

75. Xia J, Psychogios N, Young N, Wishart DS. MetaboAnalyst: a web server for metabolomic data analysis and interpretation. Nucleic Acids Res. 2009;37(Web Server issue):W652–60.

76. Bocker S, Rasche F. Towards de novo identification of metabolites by analyzing tandem mass spectra. Bioinformatics. 2008;24(16):i49–i55.

77. Raudvere U, Kolberg L, Kuzmin I, Arak T, Adler P, Peterson H, et al. g:Profiler: a web server for functional enrichment analysis and conversions of gene lists (2019 update). Nucleic Acids Res. 2019;47(W1):W191–W8.

78. Subramanian A, Tamayo P, Mootha VK, Mukherjee S, Ebert BL, Gillette MA, et al. Gene set enrichment analysis: a knowledge-based approach for interpreting genome-wide expression profiles. Proc Natl Acad Sci U S A. 2005;102(43):15545–50.

79. Anand JR, Rijhwani H, Malapati K, Kumar P, Saikia K, Lakhar M. Anticancer activity of esculetin via-modulation of Bcl-2 and NF-κB expression in benzo [a] pyrene induced lung carcinogenesis in mice. Biomedicine & preventive nutrition. 2013;3(2):107–12.

80. Anand JR, Dandotiya R, Rijhwani H, Ranotkar S, Malapati K, Lahkar M. Protective Effect of Esculetin against Cyclophosphamide Induced Chromosomal Aberration, Micronuclei Formation and Oxidative Stress in Swiss Albino Mice.
81. Anand J, Sun Y, Zhao Y, Nitiss KC, Nitiss JL. Detection of Topoisomerase Covalent Complexes in Eukaryotic Cells. Methods Mol Biol. 2018;1703:283–99.
82. Menendez D, Anand JR, Murphy CC, Bell WJ, Fu J, Slepushkina N, et al. Etoposide-induced DNA damage is increased in p53 mutants: identification of ATR and other genes that influence effects of p53 mutations on Top2-induced cytotoxicity. Oncotarget. 2022;13:332–46.

4 Computational Methods in Cancer, Genome Mapping, and Stem Cell Research

*Sudha Vengurlekar, Piyush Trivedi, Harshada Shewale,
Rushabh Desarda, Bhupendra Gopalbhai Prajapati*

1 INTRODUCTION

Cancer is one of the leading causes of death worldwide and a huge burden to medicine discovery progress. Cancer is mainly a genetic disease, and vast studies on mutations in chromosomes and genes causing tumerogenesis have been done to date, which can be exhausted using novel techniques of drug design (1).

Computer-aided drug discovery (CADD) contributes as a powerful tool to achieve drug design in a cheaper and faster way. Recently, the faster growth of tools of computational methods for the discovery of new anticancer drugs has provided fruitful results for cancer therapies. Some techniques such as computational medicine and systems biology are powerful techniques in analyses of single-cell in certain areas like data science, machine learning, network theory, statistical, and multivariate information (1).

One successful development of a drug that was first designed based on target structure as an anti-HIV drug in the United States in 1997 (2) led to a very strong foundation in drug discovery through computational methods. Recently, advancements in graphical processing (GPUs) and in artificial intelligence (AI) (3,4), have created options in fundamental research (5, 6, 7) in the drug discovery field.

This chapter aims to provide an overview of various computational methods used in the development of anticancer agents, with special emphasis on genome mapping and stem cell research. We reviewed a few of the most representative studies performed for anticancer drug development utilizing genome mapping and stem cell techniques. A workflow of computational drug discovery is shown in Figure 4.1.

Computational methods are exploring new areas in research related to cancer. Certain novel applications of computational methods such as micro RNA (8,9) genome sequencing (10), text mining (11–13), and pathway analysis are based on fine information from patients to achieve personalized medical treatments.

For improved medical practices in cancer, understanding the alterations in pathways for a given cancer and reasons for alteration and identification of various targets in the pathway is important to reverse the damage. Exploring cancer bioinformatics opens various ways of finding novel mechanisms of action that may help researchers develop new molecules (14).

This is the era of big data and several fields such as biomedical science have been explored in search of genomic variants in humans. Some other data sets on genomics and proteomics offer big possibilities in research on identifying various mechanisms. The availability of abundant data related to cancer such as methylation levels and somatic mutations and their expression in cancer cells has opened doors widely for cancer research (15).

DOI: 10.1201/9781003353768-4

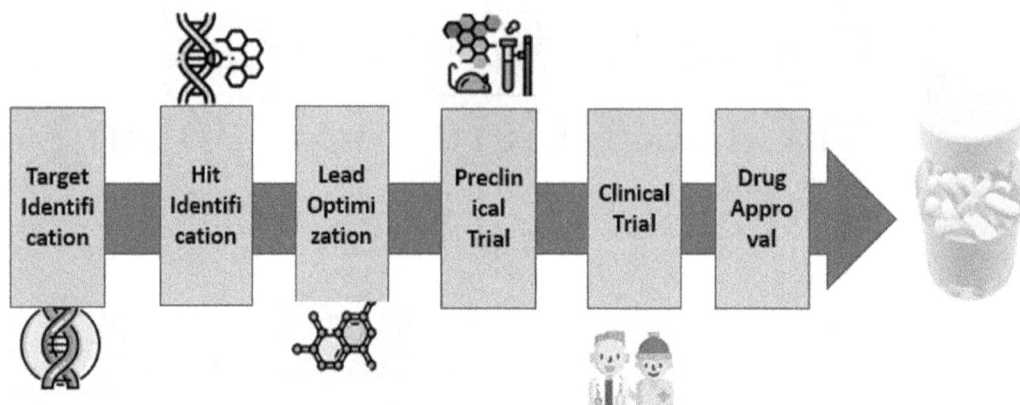

FIGURE 4.1 Computational drug discovery process.

2 GENOME MAPPING AS A TOOL FOR CANCER RESEARCH

Mapping the human genome sequence is an important aspect to explore cancer biology and treatment. Information on the complete sequence of DNA nucleotides and the classification of genes according to their functions reveals a clear role of genes in cancer. Genome mapping has emerged as an urgent need to overcome cancer's growing graph.

The idea of genome mapping started in the year 1981 with the discovery of a cancer-promoting human oncogene and with this discovery scientists were able to understand the primary cause as a mutation in the genes. Such mutations can occur through exposure to various radiations/toxins, or by errors in DNA repair. These kinds of mutations are generally inherited from one generation to the other. Independent of the origin of mutation alteration in the pathways of cells occurs that results in uncontrolled cell growth and replication, which eventually results in cancer (16).

Some of the mutations cause the inactivity of genes that are responsible for the protection of abnormal cell behavior whereas others cause an increase in the activity of disturbed genes. In recent years many scientific groups have utilized technologies based on molecular biology for finding the mutations in genes that are the main cause of altering the normal growth and behaviour of cells (17). Genome mapping revealed 350 genes related to cancer and provided significant information on cancer diseases. This technique is also helpful in some other types of cancers such as colon cancer, breast cancer, lymphoma, and leukemia.

Genome sequencing of circulating tumor DNA helps in tracking cancers with very resolution. A report published by George et al. (2020) indicated that complete genome sequencing of circulating tumor DNA provides scope to learn about how cancer is generated in a particular patient by providing information on evolutionary alterations in circulating tumor DNA with higher resolution (18).

To find out cancer genetics analysis of genomic structural variation (SV) is important as it utilizes optical genome mapping, which provides ample information on chromosomal microarrays. This technique involves extraction of data on the bases of DNA, then labels specific points and linearization for imaging. These moieties are then aligned with the human standard DNA assembly to find out large SVs. This type of genome mapping gives a fundamental understanding of cancer genetics (19).

Nucleotide-mapped genome excision repair methods are effective for finding the damaged DNA in cancer cells. This technique involves adaptor ligation sequencing that without using immunoprecipitation captures the repair substance by 3′-dA-tailing and 5′-adapter ligation and avoids the formation of dimers during PCR (20).

Some other techniques such as shallow whole genome sequencing (sWGS) are used to determine alterations in cell-free DNA (cfDNA). This is a PCR-free way of cancer research and uses peripheral bone marrow and blood or plasma samples of patients with lung cancer, lymphoma, and myeloid leukemia. In sWGS the mapping quality is remarkably higher and the percentage of unique reads and genome coverage is significant in comparison to the PCR method (21).

In continuation of efforts to sequence and map cancer genomes some protein–protein interactions of cancer drivers and additional cancer drivers are used to explore the possibilities of searching for novel disease-related pathways. Such mapping of the protein–protein and genetic interactions creates cell maps in cancer, which interpret cancer patients' mutations into pathways and thereby identify the targeted matching of genomes (22).

3   STEM CELLS AND CANCER RESEARCH

Tumors are different in terms of their morphology, kinetics of cell proliferation, and therapy response irrespective of whether the patient has the leukemic or malignant type of cancer. Most of the cells of any tumor exhibit a common genetic pattern that reflects their origin; analysis of single-cell indicates have shown the abnormal genetic or epigenetic pattern between the cells in a tumor. This heterogeneity and its molecular basis represent a challenge that has attracted many researchers. One explanation of such problems could be that most of the cells are biologically the same and heterogeneity arises from some internal or external influences that result in variable responses. Alternatively, cancer is a simulation of normal cell development and keeps a hierarchical system with stem cells at the apex (23).

Studies based on the purification of cells indicated that a set of cells called cancer stem cells (CSCs) or tumor-initiating cells are governing the growth maintenance of tumor cells long term in various cancers (24). The existence of CSCs is strongly evident in severe leukemias, although some current reports have shown its existence also in tumors of breast, brain, and colon.[3] This concept helped many researchers explore mechanisms of initiation and progression of tumor towards development of effective therapy against cancer. There are many reports with strong statements that targeting CSCs is more effectively involved in patient outcomes. Although the application of stem cells to cancer is a little older a number of obstacles still remain in its application for effective cancer therapies (24).

3.1   CANCER STEM CELL RESEARCH AND HETEROGENEITY OF TUMOR

Molecular basis of cancer cleared ways to search novel cancer therapies by focusing on inhibition of various molecular drivers of cancer such as erlotonib, imatinib-bcr/abl, trastuzumab-Her2/Neu, epidermal growth factor receptor, etc. Other factors favor cancer growth like vasculature, which is an extrinsic microenvironmental factor, have also been explored to find approaches that interfere with such factors (25).

Some reports published indicated some population of leukemic stem cell (LSC), which exhibits its cytokinetic properties similar to the hematopoietic stem cells (HSCs). (26) In these reports it was explained that the dormant cells have a small size with poor granularity as compared to a fraction that is rapidly proliferating leading to speculation that such cells represent a peculiar type of cellular fraction. Based on these observations, reviews of this era predicted that the limitations in eradicating the LSCs were the major cause of relapse and failure of various chemotherapies of cancer (27, 28). It was also observed that LSCs affect leukaemia cell mass depletion that occurs in the case of the administration of anticancer drugs to patients. LSCs also move to the growth cycle and expand as normal cells do after chemotherapy-induced cytopenias. Therefore, it was suggested that dormant LSCs can be eliminated by finding the window when they are into such cells cycle to kill them in the phase, they are most vulnerable.

CSCs exhibited analogy with the normal cells and showed the capability to initiate *in vivo* tumors by means of first digesting all differentiated cellular groups belonging to the primary tumor, and then can be re-transplanted in a series without loss of tumorigenic properties, indicating self-renewal of CSCs. Some xenografts based on cell lines demonstrated various histological and molecular aspects of the patient tumors help in cancer therapy strategies (29).

One study on CSCs explained their part in drug resistance and recurrence of tumor highly relied on the utilization of specific markers found in CSCs such as ALDH, CD133, Nanog, and ABCG2. One more type of cell known as epithelial-to-mesenchymal transition (EMT) is also responsible for higher resistance to tumor treatments. During this process of increased resistance, these phenotypic EMT cells attain mesenchymal nature, which allows EMT cells to acquire drug resistance (30).

CSC hypothesis suggested that the arrangement of tumors is in a hierarchical pattern, and consists of a subset of stem-like cells responsible for the initiation and growth of tumors. CSCs exhibit the ability to divide asymmetrically and renewal by self in addition to higher resistance to cancer therapy (31).

CSCs possess indefinite growth ability and their transformation of these cells requires few genomic changes (32, 33). In the case of gastric cancer two types of stem cells have been researched: one is cells with a slow cycle with the power of expressing the transcription factor Mist1 present in gastric corpus and G-protein coupled receptor 5 rich in leucine (Lgr5) in gastric antrum (34, 35). In the case of colon cancers, some reports in mice indicated that intestinal epithelial cells are potential CSCs (36).

CSC population is maintained by the symmetric division of tumor stem cells at the time when tumor initiation occurs (Figure 4.2) (37). This kind of division produces transient proliferating cells, which then symmetrically divide and exhibit high proliferating ability (38,39).

Because CSCs are associated with the initiation and relapse of the tumor, biomarkers that characterize these CSCs are important in therapeutic predictions. Importantly, several biomarkers in

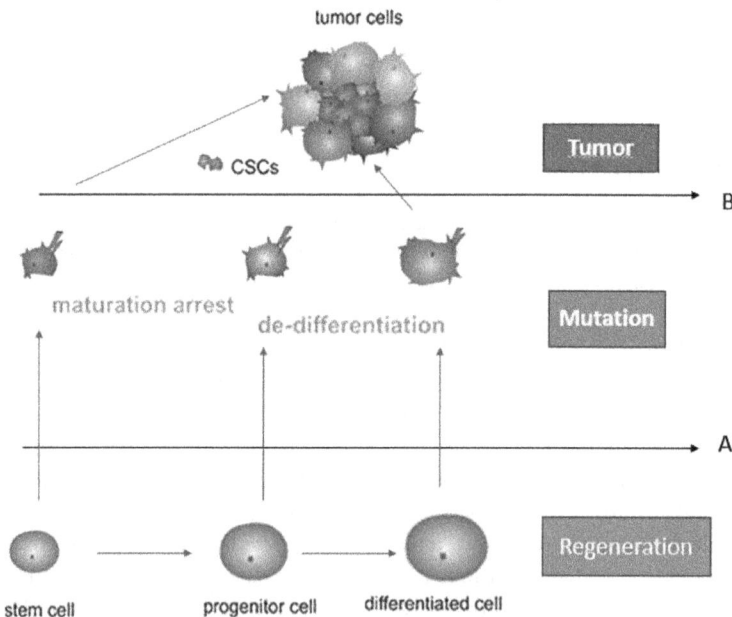

FIGURE 4.2 CSCs at Initiation of tumor: (A) Stem cells divide and produce amplifying cells with high proliferative ability. These cells differentiate terminally to help in organ homeostasis; (B) Tumors can be generated by step-wise accumulation of several mutations.

CSCs can be present in adult resident stem cell pools, human embryonic stem cells (hESC) (40). Stem cell research in cancer has been revolutionized by the search for pluripotent stem cells (iPSCs), which has opened new ways of exploring cancer diseases and designing techniques for the regeneration of tissues such as cell transplantation. Advancements in single-cell character determination allowed detailed knowledge of cellular phenotype in various conditions. Single-cell RNA sequencing (scRNA-seq) and generation of proteome and epigenome allowed systematic study and analysis of cell interaction, cell function, and type (41–43).

Research in the field of single-cell data opened many ways for generating computational models for the prediction of biological mechanisms. In particular, the gene regulatory network (GRN) gives more insight into cell differentiation and conversion of cells that helps to predict signaling molecules and transcription factors responsible for controlling these processes. Modeling the conversion of cells is possible with computational methods that are dependent on the analysis and reconstruction of intracellular GRNs, showing very vast utility in determining conversion factors (44, 45).

scRNA-seq is a technology that captures actual interactions between cells and genes by capturing the gene expression of many cells in a single experiment. Therefore, scRNA-seq allows inference of the type of cells or subtype of the cells. Stem cell research in cancer is revolutionized by the search for pluripotent stem cells (iPSCs) which has opened new ways of exploring cancer diseases, and designing techniques for the regeneration of tissues such as cell transplantation. Advancements in single-cell character determination allowed detailed knowledge of cellular phenotypes in various conditions. Single-cell RNA sequencing and generation of proteome and epigenome allowed systematic study and analysis of cell interaction, cell function, and type (46, 47).

The generation of the database on multi-omics single-cell helps in the development of computational models of high resolution, which can capture gene behavior at molecular level or tissue level. Some models that work on the cellular level such as gene regulatory network (GRN) explore cell differentiation and cell conversion through which various transcription factors can be predicted. Some other computational models based on cell-cell network interaction help in predicting homeostasis at the tissue level and the regeneration capacity of the tissues (48).

Breast cancer stem cells (BCSC) are responsible for the recurrence and resistance of tumors. Some factors present in human breast cancer cell line MCF7, RAC1B is helping BCSC to maintain its plasticity and in vitro chemoresistance towards the drug doxorubicin and in vivo tumor-initiating abilities. Two factors Rac1, Rac1b function is important for the growth and development of normal mammary glands and for the activity of mammary epithelial stem cell (MaSC). It was reported that loss of Rac1b activity in a model of mouse breast cancer diminishes the BCSC functions and promotes their doxorubicin chemosensitivity towards treatment (49).

Oncolytic viruses (OVs) is responsible for replication of cancer cells in selective manner lysis of tumor cells and produces some signals that activate the immune system to perform effective destruction of cancer cells (50). However, these OVs are identified easily by immune cells and removed from the body. Stem cells offer a promising carrier route to deliver OVs to the site of tumor with protection (51). Mesenchymal stem cells (MSCs) and neural stem cells (NSCs) were reported to load effectively the oncolytic HSV and attenuated oncolytic measles virus (OMV) to reduce the development of hepatocellular carcinoma in mice (52, 53). Some MSCs extracted from patients with ovarian cancer exhibited remarkable potential as carriers of OMV from healthy subjects (54). The role of stem cells as carriers is due to their therapeutic effects and combination of tumor and vehicle cells. For example, NSCs that are transduced show more potential towards tumour (55). Exosomes are natural carriers and have been used for encapsulation of therapeutic agents for cancer, such as small drugs, mi-RNAs, and proteins. These types of carriers exhibit stability, biocompatibility, and capacity to load cargo in tumor cells (56). Some genetic material such as siRNA is effectively packed in exosomes derived from stem cells for anticancer drug delivery through transfection technique. Some studies on exosome-based release from miR-146b stem cells of marrow (57). A few reports of brain tumor in a rat model indicated that these exosomes can directly be injected into tumors and can result in noticeable reduction in growth of glioma xenograft (57). One study showed that expression of MSCs occurs by exosomes released from miR-122 and causes significant increase in antitumor effect of sorafenib when tested on hepatocellular carcinoma tumor (58).

Some immune cells such as natural killer (NK) and chimeric antigen receptor (CAR) T cells are successfully used in cancer immunotherapy. These cells are obtained from patients and then further processed by activation and genetic transduction with CAR activated, genetically transduced with CAR, expanded, and finely charged to the patient (59). These cells are reported to be more beneficial in treatment of cancer if induction of CAR is done with HSCs and then transplanted in bone marrow. This results in continuous generation of a number of CAR expressing immune cells such as NK cells, T cells, neutrophils, and monocytes, which in combination generates improved immunity to kill cancer cells (60).

3.2 ANTICANCER VACCINES BASED ON STEM CELLS

Therapies based on CSCs to fight tumors are explored in earnest considering their roles in cancer formation and progress. Among various approaches of CSC targeting, anticancer vaccines shown tremendous potential to fight cancer due to their effect on the immune system (61, 62, 63). Some oncofetal peptides or cells based on ESC and CSC are utilized for the production of such anticancer vaccines, which involves charging of dendritic cells with antigens and then generates T cell responses for certain adoptive therapies (64).

The single use of oncofecal peptide-based vaccines cannot provide sufficient immune responses toward tumors due to tumor heterogeneity and rapid escape mechanisms. Anticancer vaccines derived from peptides can not render the required immune response for tumor due to its heterogenicity (64) but such vaccines provide better results on whole cell lysate (65). Anticancer vaccines may have some limitations such as possibility of formation of tetratoma and autoimmunity generation (66, 67). These vaccines are recommended to be used for prophylactic treatment not as therapeutic agent. Certain tumors that have high immunosuppressive nature may diminish the effect of treatment through vaccine. To achieve better therapeutic effect such anticancer vaccines should be combined with other therapies such as radiation, chemo, surgery, etc. (66).

4 RECENT ADVANCES IN STEM CELL RESEARCH AND GENOME MAPPING FOR CANCER

Stem cells and genome mapping are important areas of cancer research because they provide insights into the underlying molecular mechanisms of cancer as well as potential new treatments. Here are some recent developments in these areas.

Recent advances in single-cell analysis technologies have allowed the discovery of rare cancer stem cells within tumors. These cells are extremely difficult to detect and treat, but single-cell analysis enables researchers to better understand their molecular characteristics and develop targeted therapies. CRISPR-Cas9 genome editing techniques have been used to investigate the effects of genetic mutations on cancer stem cells. These techniques have also been used by researchers to create new cancer treatments that target specific mutations in cancer cells (68).

Epigenetic modifications can change the way genes are expressed and thus play a role in cancer development. To better understand cancer stem cells' biology and develop new treatments, researchers have been studying their epigenetic modifications. In liquid biopsy a non-invasive method of detecting cancer that analyses circulating tumour cells, DNA, and other biomarkers in the blood (69). This method can aid in the early detection of cancer and the monitoring of treatment response. Personalized medicine: Improvements in genome modeling and stem cell research have resulted in the development of personalized cancer medicine. Researchers can examine a patient's genome and cancer stem cells to define mutations and develop tailored treatments (70).

Drug resistance and cancer stem cells: Cancer stem cells are a small subpopulation of cells within a tumor that are thought to be responsible for tumor initiation, maintenance, and recurrence (70). These cells are particularly resistant to chemotherapy and radiation, making them a difficult treatment target. Stem cell research and genome mapping advances are assisting researchers in better understanding the mechanisms underlying drug resistance and developing new strategies to combat it (71).

Overall, stem cell research and genome mapping are providing important insights into cancer biology and helping to develop new treatments that are more effective and targeted. These advances are likely to continue to have a significant impact on the field of cancer research and treatment in the coming years.

5 LIMITATIONS OF STEM CELL RESEARCH AND GENOME MAPPING FOR CANCER RESEARCH

It is important to consider the limitations of stem cell research and genome mapping for cancer research, as these approaches are still in their early stages of development. Consider the following constraints.

Ethical concerns: The use of embryonic stem cells is a contentious issue, and the development of induced pluripotent stem cells (iPSCs) has raised its own ethical concerns. Furthermore, genome mapping raises ethical concerns about genetic privacy and discrimination (72).

Technical challenges: Identifying and isolating cancer stem cells can be difficult, and many current techniques have limitations in terms of sensitivity and specificity. Similarly, genome mapping technologies are still in their infancy and have limitations in terms of accuracy and cost. Despite showing promise in preclinical studies, stem cells and genome mapping have had limited clinical impact thus far. Many drugs developed based on stem cell research and genome mapping have failed in clinical trials.

Lack of standardization: Because the use of stem cells and genome mapping in cancer research is still new, there is a lack of standardization in experimental protocols, data analysis, and reporting. This makes comparing results across studies difficult and limits the ability to reach definitive conclusions.

Complexity of cancer biology: Cancer is a complex and heterogeneous disease, and it is unlikely that any single approach will be effective in all cases. Stem cells and genome mapping are just two of many tools that researchers are using to understand and treat cancer.

Sample scarcity: Stem cells and genome mapping require tissue samples, which can be scarce in some cases, particularly for rare or difficult-to-treat cancers (73).

High cost: The cost of stem cell research and genome mapping can be prohibitively expensive, particularly for developing countries or individual patients without health insurance.

Technical limitations of genome mapping: While genome mapping has advanced rapidly in recent years, there are still technical limitations, such as the inability to detect certain types of genetic alterations or the presence of somatic mutations in only a small subset of cells (74).

Potential for unintended consequences: As with any new technology, there is always the potential for unintended consequences or unforeseen risks associated with stem cell research and genome mapping. These risks will need to be carefully evaluated and monitored as these technologies continue to develop (75).

Overall, while stem cells and genome mapping hold great promise for cancer research and treatment, they are not without their limitations. It will be important for researchers to continue to address these limitations in order to maximize the potential of these approaches to benefit patients with cancer.

6 CONCLUSION

Genome mapping involves identifying and determining the sequence of genes in a DNA molecule. This technology has played an essential role in cancer research by helping to identify genetic mutations that can lead to cancer development, progression, and treatment response.

One area of cancer research where genome mapping has been beneficial is the study of cancer stem cells (CSCs). CSCs are a small population of cells within a tumor that have the ability to self-renew and differentiate into various cell types. They are thought to be responsible for tumor initiation, growth, metastasis, and treatment resistance.

By using genome mapping techniques, researchers have been able to identify specific mutations that are enriched in CSCs compared to non-CSCs. These mutations can provide insights into the underlying biology of CSCs and potential therapeutic targets.

Additionally, genome mapping has been used to develop personalized treatment strategies for cancer patients based on their genomic profiles. This approach, known as precision medicine, has shown promising results in treating various types of cancer.

In conclusion, genome mapping has proven invaluable in cancer research, particularly in studying CSCs. By identifying specific genetic mutations associated with these cells, researchers can better understand the biology of cancer and develop more targeted and effective treatments for patients.

REFERENCES

1. Zahir N, Sun R, Gallahan D, Gatenby RA, Curtis C. Characterizing the ecological and evolutionary dynamics of cancer. Nat Genet. 2020;52(8):759–767. doi:10.1038/s41588-020-0668-4
2. Kaldor, S. W., Kalish, V. J., Davies, J. F., Shetty, B. V., Fritz, J. E., Appelt, K., et al. (1997). Viracept (nelfinavir mesylate, AG1343): A potent, orally bioavailable inhibitor of HIV-1 protease. *J. Med. Chem.* 40, 3979–3985. doi: 10.1021/jm9704098
3. Chan, H. C. S., Shan, H., Dahoun, T., Vogel, H., Yuan, S. (2019). Advancing Drug Discovery via Artificial Intelligence. *Trends Pharmacol. Sci.* 40, 592–604. doi: 10.1016/j.tips.2019.06.004
4. Yang, X., Wang, Y., Byrne, R., Schneider, G., Yang, S. (2019). Concepts of Artificial Intelligence for Computer-Assisted Drug Discovery. *Chem. Rev.* 119, 10520–10594. doi: 10.1021/acs.chemrev.8b00728
5. Zhavoronkov, A., Ivanenkov, Y. A., Aliper, A., Veselov, M. S., Aladinskiy, V. A., Aladinskaya, A. V., et al. (2019). Deep learning enables rapid identification of potent DDR1 kinase inhibitors. *Nat. Biotechnol.* 37, 1038–103+. doi: 10.1038/s41587-019-0224-x
6. Cancer Biology Aspects of Computational Methods & Applications in Drug DiscoveryShang-Tao Chien [1], Ajay Kumar [23], Shifa Pandey [4], Chung-Kun Yen [3], Shao-Yu Wang [3], Zhi-Hong Wen [5], Aman C Kaushik [6], Yow-Ling Shiue [2], Cheng-Tang Pan [37]Curr Pharm Des 2018;24(32):3758-3766.doi: 10.2174/1381612824666181112104921.
7. NimritaKoulSunil Kumar S.Manvi, Recent Trends in Computational Intelligence Enabled Research, Theoretical Foundations and Applications, 2021, Pages 95–110, Chapter 6–Computational intelligence techniques for cancer diagnosis
8. Chen J, Zhang D, Zhang W, et al. Clear cell renal cell carcinoma associated microRNA expression signatures identified by an integrated bioinformatics analysis. J Transl Med. 2013;11:169.
9. Zhang W, Zang J, Jing X, et al. Identification of candidate miRNA biomarkers from miRNA regulatory network with application to prostate cancer. J Transl Med. 2014;12:66
10. Chen J, Zhang D, Zhang W, et al. Clear cell renal cell carcinoma associated microRNA expression signatures identified by an integrated bioinformatics analysis. J Transl Med. 2013;11:169.
11. Zhu F , Patumcharoenpol P, Zhang C, et al. Biomedical text mining and its applications in cancer research. J Biomed Inform. 2013;46:200–11.
12. Xu R, Wang Q. Automatic signal extraction, prioritizing and filtering approaches in detecting post-marketing cardiovascular events associated with targeted cancer drugs from the FDA Adverse Event Reporting System (FAERS). J Biomed Inform. 2014;47:171–7.
13. Xu R, Wang Q. Toward creation of a cancer drug toxicity knowledge base: automatically extracting cancer drug-side effect relationships from the literature. J Am Med Inform Assoc. 2013;21:90–6.
14. Tang Y, Yan W, Chen J, et al. Identification of novel microRNA regulatory pathways associated with heterogeneous prostate cancer. *BMC Syst Biol.* 2013;7(Suppl 3):S6. [PMC free article] [PubMed] [Google Scholar]
15. Wang Y, Chen J, Li Q, et al. Identifying novel prostate cancer associated pathways based on integrative microarray data analysis. *Comput Biol Chem.* 2011;35:151–8. [PubMed] [Google Scholar]

16. The New Era in Cancer Research. Harold Varmus in *Science*, Vol. 312, pages 1162–1165; May 26, 2006.

17. The Consensus Coding Sequences of Human Breast and Colorectal Cancers. Tobias Sj blom et al. in *Science*, Vol. 314, pages 268-274; October 13, 2006. (Published online September 7, 2006.)

18. George D Cresswell [#1], Daniel Nichol [#1], Inmaculada Spiteri [1], Haider Tari [1,2], Luis Zapata [1], Timon Heide [1], Carlo C Maley [3], Luca Magnani [4], Gaia Schiavon [5,6], Alan Ashworth [7], Peter Barry [8], Andrea Sottoriva [9] Mapping the breast cancer metastatic cascade onto ctDNA using genetic and epigenetic clonal tracking Nat Commun. 2020 Mar 27;11(1):1446. doi: 10.1038/s41467-020-15047-9.

19. Andy Wing ChunPangBenKellmanAlexHastieAlkaChaubey Use of Bionano Optical Genome Mapping in a multi-platform structural variation analysis of a cancer reference cell line, Cancer Genetics, Volumes 268–269, Supplement 1, November 2022, Pages 34–35.

20. SizhongWu[1]YanchaoHuang[1]Christopher P.Selby[2]MengGao[1]AzizSancar[2]JinchuanHu[1] A new technique for genome-wide mapping of nucleotide excision repair without immunopurification of damaged DNA Volume 298, Issue 5, May 2022, 101863.

21. Jamie J.Beagan[*]Esther E.E.Drees[*]PhyliciaStathi[*]Paul P.Eijk[*]LauraMeulenbroeks[*]FloortjeKessler[†]J aap M.Middeldorp[*]D. MichielPegtel[*]Josée M.Zijlstra[†]DaoudSie[‡]Daniëlle A.M.Heideman[*]ErikThun nissen[*]LindaSmit[†]Daphnede Jong[*]FlorentMouliere[*]BaukeYlstra[*]Margaretha G.M.Roemer[*]Erikvan Dijk[*] PCR-Free Shallow Whole Genome Sequencing for Chromosomal Copy Number Detection from Plasma of Cancer Patients Is an Efficient Alternative to the Conventional PCR-Based Approach The Journal of Molecular DiagnosticsVolume 23, Issue 11, November 2021, Pages 1553-1563

22. MehdiBouhaddou[123]ManonEckhardt[23]Zun ZarChi Naing[123]MinkyuKim[123]TreyIdeker[4]Nevan JKrogan Mapping the protein–protein and genetic interactions of cancer to guide precision medicineCurrent Opinion in Genetics & DevelopmentVolume 54, February 2019, Pages 110–117. doi.org/10.1016/j.gde.2019.04.005

23. JC Wang, JE Dick, Cancer stem cells: lessons from leukemia, Trends Cell Biol, 15 (2005), pp. 494–501, ArticleDownload PDFView Record in ScopusGoogle Scholar

24. 2 P Dalerba, RW Cho, MF Clarke, Cancer stem cells: models and concepts, Annu Rev Med, 58 (2007), pp. 267–284 View PDF CrossRefView Record in ScopusGoogle Scholar

25. John E.Dick[1,] Stem cell concepts renew cancer research Volume 112, Issue 13, 15 December 2008, Pages 4793-4807doi.org/10.1182/blood-2008-08-077941Get rights and content

26. B Clarkson, The survival value of the dormant state in neoplastic and normal populations.B Clarkson, R Baserga (Eds.), Control of Proliferation in Animal Cells, Cold Spring Harbor Laboratory, New York, NY (1974), pp. 945–972 View Record in ScopusGoogle Scholar

27. 14.BD Clarkson, J Fried, Changing concepts of treatment in acute leukemia, Med Clin North Am, 55 (1971), pp. 561–600, ArticleDownload PDFView Record in ScopusGoogle Scholar

28. 19. EP Cronkite, Acute leukemia: is there a relationship between cell growth kinetics and response to chemotherapy? Proc Natl Cancer Conf, 6 (1970), pp. 113–117, View Record in ScopusGoogle Scholar

29. MiriamLópez-Gómez[ab1]EnriqueCasado[ab1]MartaMuñoz[f]SoniaAlcalá[cde]JuanMoreno-Rubio[b]Gabriele D'Errico[c]Ana MaríaJiménez-Gordo[ab]SilviaSalinas[f]BrunoSainzJr[cde] Current evidence for cancer stem cells in gastrointestinal tumors and future research perspectives Critical Reviews in Oncology/Hematology, Volume 107, November 2016, Pages 54–7

30. doi: 10.1007/978-3-319-24932-2_4.The Role of Cancer Stem Cells in Recurrent and Drug-Resistant Lung CancerRaagini Suresh [1], Shadan Ali [2], Aamir Ahmad [3], Philip A Philip [2], Fazlul H Sarkar [4,5] Adv Exp Med Biol. 2016;890:57-74.

31. Cancer Lett. 2016 Mar 28;372(2):147–56. doi: 10.1016/j.canlet.2016.01.012. Epub 2016 Jan 18.Lung cancer stem cells: The root of resistanceLauren MacDonagh [1], Steven G Gray [1], Eamon Breen [2], Sinead Cuffe [1], Stephen P Finn [3], Kenneth J O'Byrne [4], Martin P Barr [5]

32. Li L, Neaves WB. Normal stem cells and cancer stem cells: the niche matters. *Cancer Res.* (2006) 66:4553–7. doi: 10.1158/0008-5472.CAN-05-3986

33. Sell S. Cellular origin of cancer: dedifferentiation or stem cell maturation arrest? *Environ Health Perspect.* (1993) 101(Suppl. 5):15–26. doi: 10.1289/ehp.93101s515

34. 25. Koulis A, Buckle A, Boussioutas A. Premalignant lesions and gastric cancer: current understanding. *World J Gastroentero Oncol.* (2019) 11:665–78. doi: 10.4251/wjgo.v11.i9.665

35. Hata M, Hayakawa Y, Koike K. Gastric stem cell and cellular origin of cancer. *Biomedicines.* (2018) 6:100. doi: 10.3390/biomedicines6040100

36. 28. Perekatt AO, Shah PP, Cheung S, Jariwala N, Wu A, Gandhi V, et al. SMAD4 suppresses WNT-driven dedifferentiation and oncogenesis in the differentiated gut epithelium. *Cancer Res.* (2018) 78: 4878–90. doi: 10.1158/0008-5472.CAN-18-0043

37. 2. Bu P, Chen K-Y, Lipkin SM, Shen X. Asymmetric division: a marker for cancer stem cells? *Oncotarget.* (2013) 4:950–1. doi: 10.18632/oncotarget.1029

38. Yamano S, Gi M, Tago Y, Doi K, Okada S, Hirayama Y, et al. Role of deltaNp63(pos)CD44v(pos) cells in the development of n-nitroso-tris-chloroethylurea-induced peripheral-type mouse lung squamous cell carcinomas. *Cancer Sci.* (2016) 107:123–32. doi: 10.1111/cas.12855

39. Hardavella G, George R, Sethi T. Lung cancer stem cells—characteristics, phenotype. *Transl Lung Cancer Res.* (2016) 5:272–9. doi: 10.21037/tlcr.2016.02.01

40. Kim W-T, Ryu CJ. Cancer stem cell surface markers on normal stem cells. *BMB Rep.* (2017) 50:285–98. doi: 10.5483/BMBRep.2017.50.6.039

41. T.E. Chan, *et al.* Gene regulatory network inference from single-cell data using multivariate information measures Cell Syst., 5 (2017), pp. 251–267.e3

42. Aibar, *et al.* SCENIC: single-cell regulatory network inference and clustering, Nat. Methods, 14 (2017), pp. 1083–1086

43. N. Papili Gao, *et al.* SINCERITIES: inferring gene regulatory networks from time-stamped single cell transcriptional expression profiles Bioinformatics, 34 (2018), pp. 258–266

44. P.S. Stumpf, B.D. MacArthur Machine learning of stem cell identities from single-cell expression data via regulatory network archetypesFront. Genet., 10 (2019), p. 2

45. H. Matsumoto, *et al.*, SCODE: an efficient regulatory network inference algorithm from single-cell RNA-seq during differentiationBioinformatics, 33 (2017), pp. 2314–2321

46. B.-R.; Kim, S.U.; Choi, K.-C. Co-treatment with therapeutic neural stem cells expressing carboxyl esterase and CPT-11 inhibit growth of primary and metastatic lung cancers in mice. *Oncotarget* 2014, *5*, 12835–12848. [Google Scholar] [CrossRef][Green Version]

47. Chang, J.C. Cancer stem cells: Role in tumor growth, recurrence, metastasis, and treatment resistance. *Medicine* 2016, *95* (Suppl. 1). [Google Scholar] [CrossRef]

48. Suryaprakash, S.; Lao, Y.-H.; Cho, H.-Y.; Li, M.; Ji, H.Y.; Shao, D.; Hu, H.; Quek, C.H.; Huang, D.; Mintz, R.L.; et al. Engineered Mesenchymal Stem Cell/Nanomedicine Spheroid as an Active Drug Delivery Platform for Combinational Glioblastoma Therapy. *Nano Lett.* 2019, *19*, 1701–1705. [Google Scholar] [CrossRef] [PubMed]

49. Miliotou, A.; Papadopoulou, L.C.; Androulla, M.N.; Lefkothea, P.C. CAR T-cell Therapy: A New Era in Cancer Immunotherapy. *Curr. Pharm. Biotechnol.* 2018, *19*, 5–18. [Google Scholar] [CrossRef]

50. Marelli, G.; Howells, A.; Lemoine, N.R.; Wang, Y. Oncolytic Viral Therapy and the Immune System: A Double-Edged Sword against Cancer. *Front. Immunol.* 2018, *9*, 866. [Google Scholar] [CrossRef] [PubMed][Green Version]

51. Tobias, A.L.; Thaci, B.; Auffinger, B.; Rincón, E.; Balyasnikova, I.V.; Kim, C.K.; Han, Y.; Zhang, L.; Aboody, K.S.; Ahmed, A.U.; et al. The timing of neural stem cell-based virotherapy is critical for optimal therapeutic efficacy when applied with radiation and chemotherapy for the treatment of glioblastoma. *Stem Cells Transl. Med.* 2013, *2*, 655–666. [Google Scholar] [CrossRef]

52. Ong, H.-T.; Federspiel, M.J.; Guo, C.M.; Ooi, L.L.; Russell, S.J.; Peng, K.-W.; Hui, K.M. Systemically delivered measles virus-infected mesenchymal stem cells can evade host immunity to inhibit liver cancer growth. *J. Hepatol.* 2013, *59*, 999–1006. [Google Scholar] [CrossRef][Green Version]

53. Duebgen, M.; Martinez-Quintanilla, J.; Tamura, K.; Hingtgen, S.; Redjal, N.; Shah, K.; Wakimoto, H. Stem Cells Loaded With Multimechanistic Oncolytic Herpes Simplex Virus Variants for Brain Tumor Therapy. *J. Natl. Cancer Inst.* 2014, *106*. [Google Scholar] [CrossRef]

54. Mader, E.K.; Butler, G.W.; Dowdy, S.C.; Mariani, A.; Knutson, K.L.; Federspiel, M.J.; Russell, S.J.; Galanis, E.; Dietz, A.B.; Peng, K.-W. Optimizing patient derived mesenchymal stem cells as virus carriers for a Phase I clinical trial in ovarian cancer. *J. Transl. Med.* 2013, *11*, 20. [Google Scholar] [CrossRef] [PubMed][Green Version]

55. Miska, J.; Lesniak, M.S. Neural Stem Cell Carriers for the Treatment of Glioblastoma Multiforme. *EBioMedicine* 2015, *2*, 774–775. [Google Scholar] [CrossRef] [PubMed][Green Version]

56. Fuhrmann, G.; Serio, A.; Mazo, M.M.; Nair, R.; Stevens, M.M. Active loading into extracellular vesicles significantly improves the cellular uptake and photodynamic effect of porphyrins. *J. Control. Release* 2015, *205*, 35–44. [Google Scholar] [CrossRef] [PubMed]

57. Katakowski, M.; Buller, B.; Zheng, X.; Lu, Y.; Rogers, T.; Osobamiro, O.; Shu, W.; Jiang, F.; Chopp, M. Exosomes from marrow stromal cells expressing miR-146b inhibit glioma growth. *Cancer Lett.* 2013, *335*, 201–204. [Google Scholar] [CrossRef] [PubMed][Green Version]

58. Lou, G.; Song, X.; Yang, F.; Wu, S.; Wang, J.; Chen, Z.; Liu, Y. Exosomes derived from miR-122-modified adipose tissue-derived MSCs increase chemosensitivity of hepatocellular carcinoma. *J. Hematol. Oncol.* 2015, *8*, 122. [Google Scholar] [CrossRef] [PubMed][Green Version]

59. Dolnikov, A.; Sylvie, S.; Xu, N.; O'Brien, T. Stem Cell Approach to Generate Chimeric Antigen Receptor Modified Immune Effector Cells to Treat Cancer. *Blood* 2014, *124*, 2437. [Google Scholar] [CrossRef]

60. Iriguchi, S.; Kaneko, S. Toward the development of true "off-the-shelf" synthetic T-cell immuno-therapy. *Cancer Sci.* 2019, *110*, 16–22. [Google Scholar] [CrossRef] [PubMed]

61. Chang, J.C. Cancer stem cells: Role in tumor growth, recurrence, metastasis, and treatment resistance. *Medicine* 2016, *95* (Suppl. 1). [Google Scholar] [CrossRef]

62. Batlle, E.; Clevers, H. Cancer stem cells revisited. *Nat. Med.* 2017, *23*, 1124–1134. [Google Scholar] [CrossRef] [PubMed]

63. Codd, A.S.; Kanaseki, T.; Torigo, T.; Tabi, Z. Cancer stem cells as targets for immunotherapy. *Immunology* 2017, *153*, 304–314. [Google Scholar] [CrossRef] [PubMed]

64. Ouyang, X.; Telli, M.L.; Wu, J.C. Induced Pluripotent Stem Cell-Based Cancer Vaccines. *Front. Immunol.* 2019, *10*, 1510. [Google Scholar] [CrossRef] [PubMed][Green Version]

65. Pattabiraman, D.; Weinberg, R.A. Tackling the cancer stem cells—What challenges do they pose? *Nat. Rev. Drug Discov.* 2014, *13*, 497–512. [Google Scholar] [CrossRef][Green Version]

66. Katsukawa, M.; Nakajima, Y.; Fukumoto, A.; Doi, D.; Takahashi, J.; Katsukawa, M.M. Fail-Safe Therapy by Gamma-Ray Irradiation Against Tumor Formation by Human-Induced Pluripotent Stem Cell-Derived Neural Progenitors. *Stem Cells Dev.* 2016, *25*, 815–825. [Google Scholar] [CrossRef] [PubMed]

67. Inui, S.; Minami, K.; Ito, E.; Imaizumi, H.; Mori, S.; Koizumi, M.; Fukushima, S.; Miyagawa, S.; Sawa, Y.; Matsuura, N. Irradiation strongly reduces tumorigenesis of human induced pluripotent stem cells. *J. Radiat. Res.* 2017, *58*, 430–438. [Google Scholar] [CrossRef] [PubMed][Green Version]

68. Rajabi A, Kayedi M, Rahimi S, et al. Non-coding RNAs and glioma: Focus on cancer stem cells. Mol Ther Oncolytics. 2022;27:100–123. Published 2022 Sep 17. doi:10.1016/j.omto.2022.09.005

69. Solier S, Müller S, Rodriguez R. Whole-genome mapping of small-molecule targets for cancer medicine. Curr Opin Chem Biol. 2020;56:42–50. doi:10.1016/j.cbpa.2019.12.005

70. Lee J, Bayarsaikhan D, Bayarsaikhan G, Kim JS, Schwarzbach E, Lee B. Recent advances in genome editing of stem cells for drug discovery and therapeutic application. Pharmacol Ther. 2020;209:107501. doi:10.1016/j.pharmthera.2020.107501

71. Wong DJ, Liu H, Ridky TW, Cassarino D, Segal E, Chang HY. Module map of stem cell genes guides creation of epithelial cancer stem cells. Cell Stem Cell. 2008;2(4):333–344. doi:10.1016/j.stem.2008.02.009

72. Santos F, Capela AM, Mateus F, Nóbrega-Pereira S, Bernardes de Jesus B. Non-coding antisense transcripts: fine regulation of gene expression in cancer. Comput Struct Biotechnol J. 2022;20:5652–5660. Published 2022 Oct 10. doi:10.1016/j.csbj.2022.10.009

73. Dashnau JL, Xue Q, Nelson M, Law E, Cao L, Hei D. A risk-based approach for cell line development, manufacturing and characterization of genetically engineered, induced pluripotent stem cell-derived allogeneic cell therapies. Cytotherapy. 2023;25(1):1–13. doi:10.1016/j.jcyt.2022.08.001

74. Bharti S, Anant PS, Kumar A. Nanotechnology in stem cell research and therapy. J Nanopart Res 2023;25:6. DOI: 10.1007/s11051-022-05654-6

75. Mortezaei Z. Computational methods for analyzing RNA-sequencing contaminated samples and its impact on cancer genome studies. Informatics in Medicine Unlocked. 2022 Aug 18:101054.

5 Using Chemistry to Understand Biology

*Mayank Bapna, Dhruv R. Parikh, Sumita Bardhan,
Bhupendra Gopalbhai Prajapati*

ABBREVIATIONS

AI	Artificial intelligence
DL	Deep learning
CAMD	Computational autonomous molecular design
ML	Machine learning
CRT	Conditional randomized transformer
QSAR	Quantitative structure activity relationship
SVM	Support vector machine
MD	Molecular docking
cAMP	Cyclic adenosine monophosphate
RBFNN	Radius basis function neural network
RNN	Recurrent neural networks
CNN	Convolutional neural networks
MPL	Multilayer perceptron networks

1 INTRODUCTION

1.1 ORIGIN OF SCIENCE

Science originated as early as the 6th century amongst Greek natural philosophers, B.C. and can be traced to the "scientific revolution" that occurred in the 17th century. Generally, whenever we attempt to improve and have mastery on or of the natural environment, we see science. With the professionalization of various scientific disciplines in the 19th century the application of science began. Today science is the knowledge of the world of nature. The term science comes from the Latin word *Scientia*, meaning "knowledge." Put simply, science is an attempt to discover specific facts and the ability to figure out patterns in which these facts are connected [1]. With the developments in science various branches originated amongst them incorporating life is Biology.

1.2 STUDYING BIOLOGY

The study of life processes, interactions between and among living matter, history and future of living things, interaction of human life with the external environmental, ecological interrelationship one organism has with another, etc., are possible only when we understand the basic concepts of biology. From the largest mammal to the microscopic DNA all have certain biological processes involved in evolution, maturation, functionality, and degradation, which are understood by studying the intricacies of the various biological systems.

1.3 ARTIFICIAL INTELLIGENCE

Artificial intelligence (AI) has increasingly gained attention in drug discovery, bioinformatics research, and computational molecular biology. Artificial intelligence, currently a cutting-edge concept, has the potential to improve the quality of life of human beings. Artificial intelligence is extremely useful in the process of synthesizing new molecules and chemical patterns. The fields of AI and biological research are becoming more intertwined, and methods for extracting and applying the information stored in live organisms are constantly being refined. As the field of AI matures with more trained algorithms, the potential of its application in epidemiology, the study of host–pathogen interactions and drug design, widens. Artificial intelligence is now being applied in several fields of drug discovery, customized medicine, gene editing, radiography, and image processing and medication management.

2 UNDERSTANDING BIOLOGY IN PURVIEW OF CHEMISTRY

While concentrating only on the understanding of modern biology and the various processes it encompasses (e.g., explaining the structure and function of all cellular processes at the molecular level), it is important to understand the facets of chemistry and hence biology.

Since the inception of life on the planet, there was biology. With full sovereignty over all other subject areas, biology was accepted and studied with grandeur and sophistication. It only defined and explained the living parts of the world and their environment, but also their interaction with others for proliferation and mortification. Therefore, to understand biology one requires a little knowledge of chemistry [2].

An incredibly fascinating field of study is chemistry as it is very much fundamental to our living world. It plays a role in everyone's lives and touches almost every aspect of our existence in some way [3]. For all our basic needs of food, clothing, shelter, health, energy, and clean air, water, soil, and other related aspirations of life we look forward to chemistry. Advancements in our solutions related to our health problems and enriching quality of life is achieved by using chemical technologies. Chemistry is the only bridge between biology and medicine and is the link between the earth and environmental sciences. Knowledge of chemical processes provides insight into a variety of physical and biological phenomena. Therefore, knowing something about chemistry is worthwhile because it provides an excellent basis for understanding the biological universe we live in [4]. For better or for worse, everything is chemical. In the sections to follow we will discuss the basics of chemistry and several important chemical concepts.

2.1 ATOMS AND MOLECULES

The smallest and the simplest form of a pure substance is known as an atom or as the smallest part that represents an element. An atom is composed of protons, neutrons, and electrons, which are positively charged, neutral, and negatively charged, respectively. Positively charged nucleus keeps the electrons from flying away from the orbit. A neutral atom after losing one or more electrons may gain a positive charge or a negative charge after gaining electron. The gaining and loosing of electrons results in the developent of ions that are responsible for formation of various types of bonds like covalent, ionic, coordinate, and pi bonds, which are all necessary for formation of a molecule. All biomolecules and dtheir functions can be explained in chemical terms.

Developing a small molecule in a laboratory is a very time-consuming task. It includes many steps like design of many experiments, execution of all experiments to synthesize the molecule,

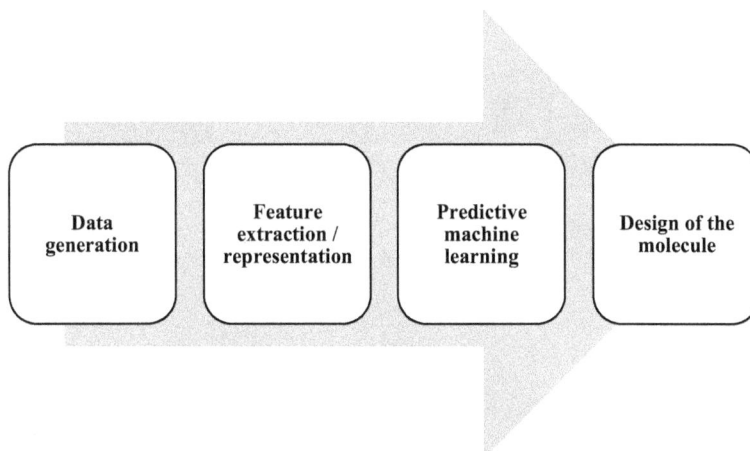

FIGURE 5.1   Steps for computational autonomous molecular design (CAMD).

characterization of the synthesized molecule, validating the complete process and decision making for the final developed molecule. Until recently most of these tasks were carried out in the laboratory manually, which slowed down the development of molecule. More recently, AI has come into sight as an approach of potential interest for this purpose. Deep learning-based models are widely used to develop a molecule. To proceed with the computational design of a small molecule is a very efficient alternative to the traditional manual laboratory work. [5]

Computational autonomous molecular design (CAMD) consists of the following steps as can be seen in Figure 5.1.

From the figure we can see that first step is data generation in autonomous molecular design (CAMD). For data generation high-throughput density functional theory (DFT) is used commonly because it is very accurate and efficient, followed by feature extraction/representation, predictive machine learning, and design of the molecule.

2.2   Biomolecules

Only about 30 of the more than 90 naturally occurring chemical elements are essential to organisms [6]. The four most abundant elements in living organisms are hydrogen, oxygen, nitrogen, and carbon. Carbon can form single bonds with hydrogen and both single and double bonds with oxygen and nitrogen [7] and of a greater significance is the carbon–carbon single bond. For a better understanding of the use of these bonds let us describe the structure of a biomolecule protein and understand it chemically.

2.2.1   Protein Architecture

The smallest unit of a protein is an amino acid (Figure 5.2A). When amino acid residue is covalently linked to another residue and forms an established chain, it is called as protein. Amino acids are also called α-amino acids due to linked carboxyl and amino functional groups linked to the same carbon, differing only in the side chain. Chemically a nitrogen atom shares its three valence electrons with carbon and hydrogen and forms a covalent bond; similarly when a carbon atom shares its valence electron with another carbon it forms a covalent bond.

Most of the time the structure of an amino acid is described as a Zwitterion (Figure 5.2B) or a dipolar ion, meaning that it possesses both positive and negative charge on amine and carboxyl functional groups. All amino acids have an amino group ($-NH_2$), which is basic, and a carboxylic

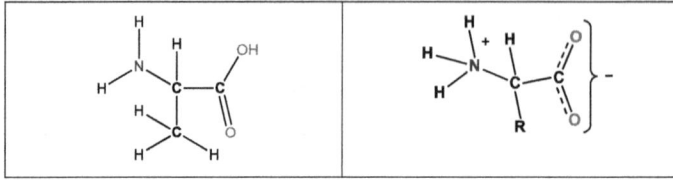

FIGURE 5.2A   Chemical structure of an amino acid.
FIGURE 5.2B   Amino acid as Zwitterion at isoelectric point.

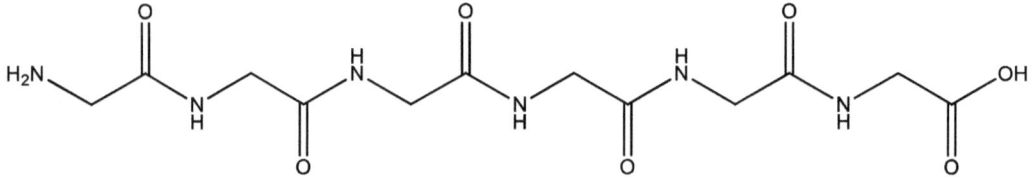

FIGURE 5.3   Polypeptide.

group (-COOH), which is acidic. In basic conditions, the –COOH group loses a H$^+$ to become –COO$^-$. By donating a H$^+$, the amino acid acts as an acid. When in acidic solutions, the –NH$_2$ group receives a H$^+$ ion and becomes –NH$_3^+$. By accepting a H$^+$ ion, the amino acid behaves as a base. Chemicals that behave as both acids and bases are called amphoteric. Depending on the pKa of amino acid when both the charges exist on amino acid at a particular pH then it is known as iso-electric point.

The amino acid molecule can be covalently joined through a substituted amide linkage, termed as a peptide bond to yield a dipeptide. Peptide bond formation is a type of condensation reaction, a common class of reaction common in living cells. Polypeptides have a characteristic amino acid composition leading to formation of a protein [8].

To understand how a long polypeptide chain of amino acid (Figure 5.3) (i.e., a protein) fits in a cell, we need to understand the chemistry behind bonding of amino acid sequences. The structure of a protein is often referred to as "folded." Once synthesized, proteins fold to a specific 3D conformation in a cellular compartment called the endoplasmic reticulum. This is a vital cellular process because proteins must be correctly folded into specific, 3D shapes in order to function correctly. Unfolded or misfolded proteins contribute to the pathology of many diseases [9].

A linear sequence of amino acids (polypeptide) is called the primary structure of amino acid. As seen in Figure 5.4 the secondary structure of amino acid is generated when intermolecular hydrogen bonds are formed between amino and carboxyl groups between two segments of the same chain and within the same segment giving β-sheets and α-helices. When both these interact in space through hydrophobic salt bridges (ionic bonds), hydrogen bonds, and disulphide bridges they give rise to the tertiary structure of protein. When several tertiary protein chains interact they build a multimeric protein. The interaction between these chains is similar to that seen in the tertiary structure.

Proteins play a part in many different functions from catalysis to specific recognition, transport, or regulation. Protein has direct application in areas such as green chemistry, health, and biotechnologies. Protein stability is of utmost importance, mainly due to its thermal stability. To get protein with new enhanced and desired property is the need of current century. Artificial intelligence can play a key role in developing protein molecule with new and enhanced properties. Computational protein design (CPD) can help in determining and producing protein sequence and protein structure of desired interest. Deep learning is very useful in structural biology. AlphaFold2 has shown success

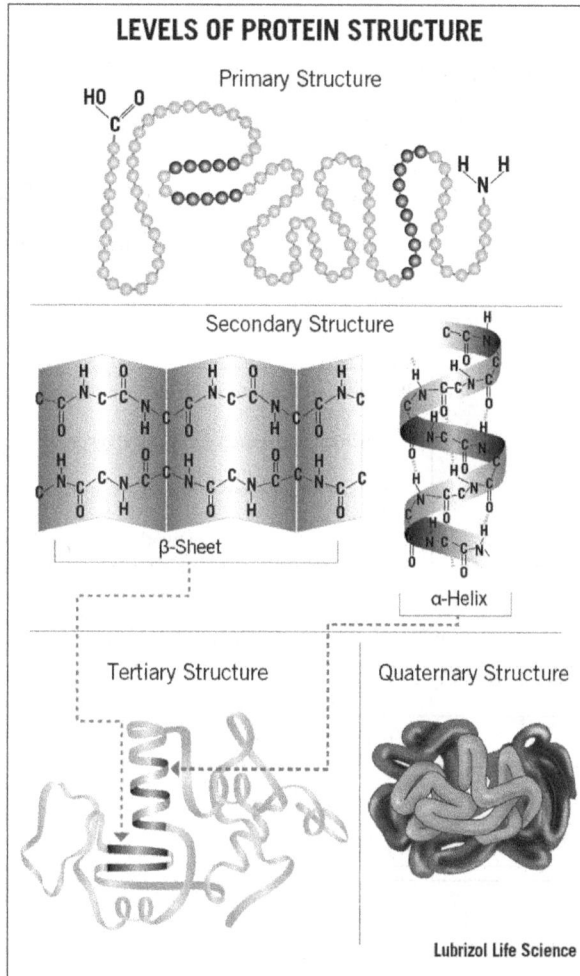

FIGURE 5.4 Primary, secondary, tertiary, and quaternary structures of proteins [10].

in structure prediction. Deep neural networks (architecture type in DL) are able to learn any function with enough data and computational power. Publically in the literature a large number of protein structures and sequences is available. These literature can help a lot in successful implementation of Deep Learning (DL). [11]

Deep learning consists of different steps like learning sequence, backbone generation, etc., as seen in Table 5.1.

2.3 WATER

Another example of where the biological properties can be understood by understanding the chemical behavior is water. It exists in all the three states of matter and is found in all the biological systems including earth. Water (Figure 5.5) is also called the "universal solvent" and the properties of water molecule that makes it unique can be understood by its chemical structure.

Water is a polar bent molecule and for the same reason it possesses the following biologically important characteristics critically important to creation and support of life on earth.

TABLE 5.1

Protein designing with the help of DL

Tasks dealing by DL		
Learning sequence embedding	Prediction of sequence with the help of other sequence, with the help of the sequence of the same family, from a random sequence or with the help of any other sequence	
	↓	
Backbone sequence	Prediction of sequence with the help of input structure, by reconstructing the native sequence, by predicting residues given their local environment, by inverting a folding network	
	↓	
Prediction of single point mutation		
	↓	
Prediction of side chain orientation		
	↓	
Foreword folding		

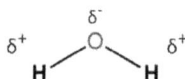

FIGURE 5.5   Water molecule.

2.3.1 Polarity

When a compound is formed with atoms differing in electronegativity then polarity arises, as seen in water molecule where due to electronegativity the electrons are shared more towards oxygen atom and less towards hydrogen atom. This polarity of water is responsible for effectively dissolving other polar molecules like sugars, ionic compounds, etc.

The bent shape of water molecule arises due to repulsion between the lone pair of electrons on oxygen atom. This bent shape is responsible for two dipoles as due to the shape the O-H dipoles do not cancel each other and the molecule becomes polar.

2.3.2 Hydrogen Bonding

The most favorable property of water is its self-bonding with other water molecule (Figure 5.6) through an unseen attraction between electronegative oxygen atom and hydrogen atom. Such types of bonds are commonly seen in bases of DNA.

2.3.3 Cohesion

Due to excessive hydrogen bonding between water molecules the water molecules tend to stick together. This is known as cohesion. The ascent of sap in the xylem tissue of plants is the upward

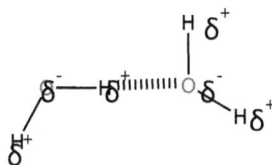

FIGURE 5.6 Intermolecular hydrogen bonding.

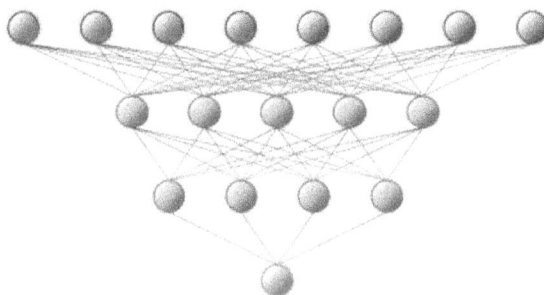

FIGURE 5.7 Bonds between atoms and molecules.

movement of water and minerals from the root to the aerial parts of the plant is an example of cohesion.

Bonding characters, such as length and strength, are of key importance for material structure and properties. Here, a machine learning (ML) model is used to predict the bonding properties from information pertaining to isolated systems before bonding.

Machine learning was used to predict the chemical bonding, binding energy, bond distance, covalent electron amount, and Fermi energy of bonding/adsorption systems. Machine learning is also helpful in identifying chemical bond type such as single, double, triple, and aromatic bonds based on spatial atomic coordinates. Quality of the predictions largely depends on the quality of the input stream and hence input data should be carefully selected in order to avoid misleading of the ML algorithm. [12]

2.3.4 Surface Tension

When the water molecules on the outside of the system align to form a strong intermolecular bond, a structure-like "atom net" is formed. Such layers allow water spiders to walk on the surface of water.

As water is a universal solvent it is very widely used as a solvent to dissolve many drugs to create formulations in the pharmaceutical industry. Artificial intelligence can help a lot in solubility prediction. Solubility prediction tools have been shown to be very efficient with the help of regression, ML, and DL methods. Solubility prediction can be improved more by going deeper with DL.

The published DL solubility prediction models are primarily shallow nets (3–7 layers). Deep learning performances are enhanced by going deeper (adding more layers to shallow nets). However, with going wider, performance can be enhanced but practically more convenient to develop deeper nets by tapping into the well-established architectures that do not require many parameters.

Some newly developed compounds were tested by different available tools and deeper-net models. Comparison of these different tools is given in Table 5.2.

As DL methods have been successfully implemented in other fields, the supreme learning capability of deeper-net models may be used successfully to improve solubility prediction of novel

TABLE 5.2
Different established tools

Model	R^2	RMSE	PCT-10-fold (%)
Established tools			
QikProp 2018-4 QP18	<0.2	0.926	69.4
Shallow-net deep learning model of a typically employed architecture for solubility prediction			
4-layer DNN mode	0.307	0.739	80.7
Shallow-net deep learning models developed			
8-layer ResNet-like model	<0.2	0.982	66.1
Deeper-net deep learning models developed			
20-layer ResNet-like model	0.412	0.681	82.3
26-layer ResNet-like model	0.075	0.854	77.4

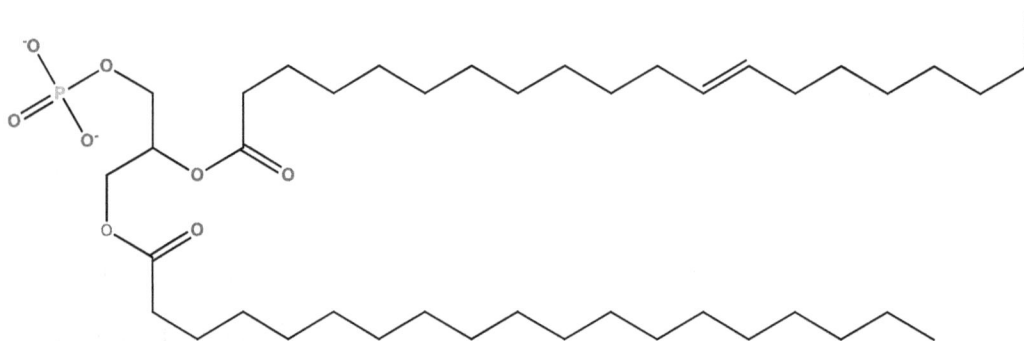

FIGURE 5.8 Phospholipid.

compounds. This can be very helpful for some compound solubility prediction, which can be very difficult for solubility prediction [13].

3 THE CHEMISTRY OF BIOLOGY

3.1 Knowing Living Molecules w.s.r. Chemistry

To understand biological systems it is necessary to understand the chemistry of carbon and its compounds, a special element in living systems. An atom with four valence electrons is capable of forming four bonds: single, double, and triple bonds. Many such compounds with carbon as the main element are known in biological systems. Long-chain molecules known as polymers are common in biological systems. When such polymers combine they form carbohydrates, amino acids, and lipids. Complex carbohydrates are formed from simple sugars, proteins are formed from amino acids, and lipids are long-chain fatty acids (Figure 5.8).

As mentioned above biomolecules are made of carbon atoms phosphorous, hydrogen, oxygen and some containing nitrogen as discussed previously in structure of proteins. Carbohydrates or hydrates of carbon (Figure 5.9) have a ratio of C:H as 2:1 as H_2O. Plants make carbohydrates and provide energy to all those who consume them. Another polymer lipid (Figure 5.8) is made of C, H, O. The C-H bond, unlike proteins, is non-polar with no lipid insolubility in water or a hydrophobic nature of the molecule. There are four types of lipids depending on their physical nature.

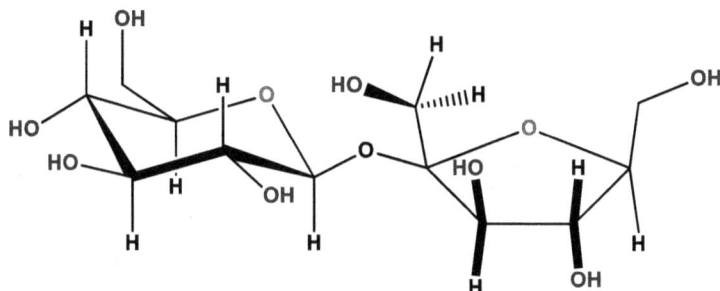

FIGURE 5.9   Sucrose.

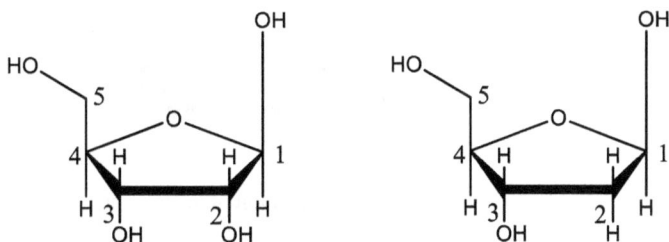

FIGURE 5.10A   β-D-Ribose.
FIGURE 5.10B   β-D-deoxyribose.

The first type, fats, is made of three fatty acid molecules linked to glycerol molecule. Fatty acids being long chain of covalently bonded carbon atoms with carboxyl group attached at one end. Typically each carbon in the glycerol molecule forms a bond with the first carbon atom from a fatty acid molecule resulting in a fat molecule. The second type of lipid (i.e., wax) is composed of one long-chain fatty acid bonded to a long-chain alcohol group. The long non-polar chain makes waxes highly hydrophobic. Such layers are used by plants as coating of stem, leaves, etc. Ear wax in humans protects the human ear. The third type is a phospholipid (Figure 5.8); it is similar to fat but has two fatty acid chains bonded to a glycerol and contains an element phosphorous. They are both hydrophilic and hydrophobic in nature and main constituents of cell membrane [14]. The last lipid is a steroid. It does not contain fatty acid. The most common example is cholesterol found in human beings.

What can be inferred from the above discussion is that simple molecules are made of carbon, hydrogen, oxygen, nitrogen, and other elements. The complex biomolecules that can be prepared using the above elements are best understood by explaining the structure of DNA and RNA. Both come under the category of nuclei acid and are composed of nucleotides. DNA and RNA are polymers of individual nucleotides. Each nucleotide is formed from three different components: (i) sugar, which can be ribose or deoxyribose; (ii) phosphate group; and (iii) nitrogen base.

The two main sugars are deoxyribose (Figure 5.10A, 5.10B) found in DNA and ribose in RNA. Since the inception of nature ribose was not randomly selected but was the only choice available from pentose sugars of D-arabinose, D-xylose, and D-lyxose as it fits best into the structure of physiological forms of nucleic acids. To understand this chemically it has been found that when replaced with these sugars in nucleotides, a lot of steric hindrance with the bulky nitrogenous base and C5-OH was observed due to the presence of –OH functional at C2-OH and/or the C3-OH above the furanose ring. Further, the selection of pentose (ribose) over hexose is due to the fact that fully hydroxylated six-carbon sugars probably would not have produced a stable base-pairing system capable of carrying genetic information as efficiently as DNA [15].

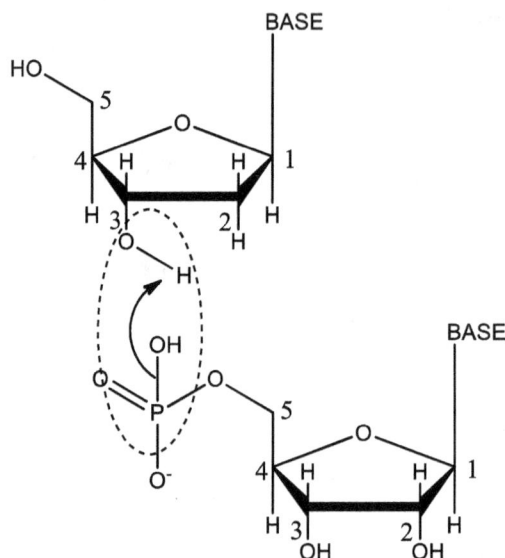

FIGURE 5.11   Formation of covalent bond and chain elongation in nucleic acid.

If it were only sugar selection by nature then it would have been easy to understand the structure of DNA or RNA. But the selection of phosphate group in nature over others can be explained chemically.

The role of phosphate during replication is that the phosphate group present on one nucleotide forms covalent bond with the sugar molecule of adjacent nucleotide, originally a triphosphate, to form a long chain of monomers also called as DNA polymerization. This sugar-phosphate group forms the backbone in a single strand of DNA. The heat energy released while removing two pyrophosphates is utilized for formation of bonds in chain in DNA polymerization [16].

In this reaction the cleavage of P-O bond occurs, which is one of the most significant reactions in biological systems, as it is involved in generation of energy, protein synthesis, maintaining integrity of genetic material, etc. Phosphoryl transfer reactions (Figure 5.11) are of S_N2 type and are very demanding as they need very high energy to overcome the repulsion between the negatively charged phosphates and the attacking nucleophile (in this case OH⁻). This makes phosphates one of the most inert compounds. This immediately gives phosphate esters an advantage as a biologically relevant compound, as the existence of life as we know it depends on the tight regulation of interrelated chemical reactions occurring in vivo, where a simple difference in rate of a factor of 2 can be the difference between life and death. Such tight regulation is impossible if the reactions being regulated could occur spontaneously, which is why we are also highly skeptical of the idea, for example, that arsenate can replace phosphate in biological systems [17].

Nitrogen bases: Both DNA and RNA have four nitrogen bases available to construct nucleotides (Figure 5.12). Three of the nitrogen bases are the same and the fourth one is Thymine in DNA and Uracil in RNA. Both DNA and RNA are large polymers synthesized by dehydration reaction during addition of nucleotides.

Nucleotides and nucleosides are the building blocks of DNA and RNA. They play a critical role in several biological processes including energy storage (ATP), signaling pathways (e.g., cyclic adenosine monophosphate, cAMP), innate immunity (e.g., cyclic guanosine monophosphate-adenosine monophosphate, cGAMP), as cofactors for specific enzymes (e.g., s-adenosylmethionine, SAM) and as neurotransmitters (e.g., adenosine). Chemical analogs of nucleotides and nucleosides therefore have a wide range of applications, such as as substrates for artificial synthesis of DNA/RNA sequences with increased resistance to nucleases or reduced effect on innate immune responses,

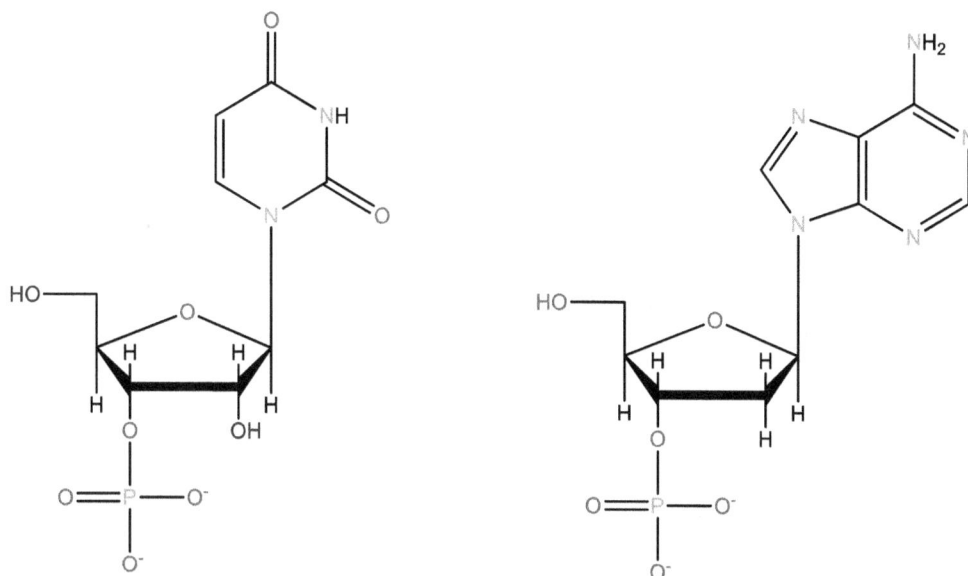

FIGURE 5.12A Uracil.
FIGURE 5.12B Adenine.

inhibition of viral DNA and RNA polymerases, inhibition of tumor cell replication, and treatment of autoimmune diseases.

Computational design of nucleoside analogs in which the resulting molecules are designed de novo using AI models, an approach that could potentially accelerate the design of new nucleoside analogs. To search chemical space(s) and identify promising drug candidates, deep generative models (by using neural networks) have been employed recently. The family of deep generative models consists of generative adversarial networks (GAN), adversarial auto encoders (AAE), variational autoencoders (VAE), Long Short-Term Memory (LSTM), and Gated Recurrent Unit (GRU) models.

Deep generative models are typically trained on a large entity of molecules. With the help of deep generative models, valid molecules that are divers can be generated. However, the molecules may not contain specific drug-like properties. Generating molecules with specific structural attributes is referred as conditional generation. The task of conditional molecule generation can be done by transfer learning, which is also called fine-tuning or, alternatively, direct steering of desired chemical properties. Transfer learning includes learning for model like to understand general rules of molecular construction from a large, widely available dataset and then transfer this general knowledge to a specific task. Direct steering simultaneously trains a model to learn the rules of valid chemical generation and to focus on a specific property, using a large dataset. To deal with the dual issues of generating focused, yet diverse molecules, development of a transformer-based model, which is called a Conditional Randomized Transformer (CRT), is required. [18]

Machine learning has helped immensely in difficult biological tasks like in automatically finding DNA patterns in different creatures. Artificial intelligence actually opened the door for biological sectors like vaccine preparation, drug production, study of virology, analysis process of antibody–antigen interactions and pathogens, etc.

Artificial intelligence does enable human-like reasoning. An AI-based engine can perform both complex tasks on the basis of reasoning, and repetitive tasks that can be executed over a large combinatorial space and can simulate billions of wet laboratory experimental research. In only one run, primary investigation is carried out of vaccine targets from greater than 20,000 flavivirus protein and in another only one single run screening vaccine targets from greater than 100,000 influenza protein.

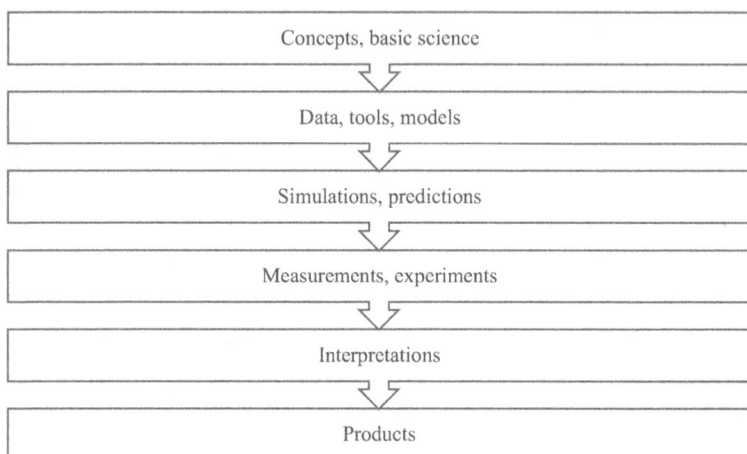

FIGURE 5.13 Well defined product developmental flow.

The results of runs are reports that have been used directly for patent applications. These automated applications need a well-defined workflow based on general science, and also well-defined sets for analytical tools combined to perform simulations and make predictions. These predictions may be validated into a number of well-chosen experiments. After accurate interpretation, then these results may be evolved into products – vaccine parts (components) and formulations of vaccine [19].

3.2   BIOCHEMICAL PROCESSES

In the previous section we discussed the chemistry of formation and functionality of biomolecules. In this section we discuss the biological processes w.s.r. to the chemical changes.

3.2.1   Formation of Lactic and Muscle Fatigue

Lactic acid formation and accumulation occurs in muscles when there is high demand of energy and insufficient supply of oxygen.

Myocytes are specialized cells that generate ATP as the immediate source of energy at times of skeletal muscle contraction. There are two types of muscle tissues: "Slow-twitch muscle" or Red muscle and "Fast-twitch muscle" or White muscle. Red muscles are highly resistant to fatigue, and are rich in mitochondria and served by dense network of blood vessels that supplies oxygen continuously. Due to this the ATP necessary for mechanical work is generated continuously. Red muscles are low tension muscles and due to non-stop supply of oxygen are resistant to fatigue.

$$C_6H_{12}O_6 + 6O_2 \rightarrow 6CO_2 + 6H_2O + ATP$$

3.2.2   Breakdown of Glucose in the Presence of Oxygen

Unlike Red muscle the White muscles have less mitochondria and are less well supplied with blood vessels. As the amount of oxygen supplied is less they cannot generate the required amount of energy in the form of ATP that is required. They are high tension muscles and can do so at faster rate, requiring high energy than it can replace and therefore they are liable to fatigue. From this description it is understood that fatigue develops in White, fast moving muscles. To understand the same we must go through the chemical changes that are taking place that lead to lactic acid formation and fatigue.

In maximally active fast-twitch muscle the need for ATP is much more than it can produce. This happens due to less oxygen supply and supply of fuels by aerobic respiration. From our food carbohydrates are broken down to glucose, a fuel that generates energy for our cells. When not needed, it is stored in the liver and muscles. This stored form of glucose is made up of many connected glucose molecules and is called glycogen.

Under the above conditions, when the Red muscles are unable to generate enough ATP during excessive mechanical work due to lack of oxygen, the stored muscle glycogen is broken down to lactate by fermentation. "Lactic acid fermentation" (anaerobic) thus responds to increased energy demands more so than oxidative phosphorylaton (aerobic).

Fermentation is a general term for anaerobic respiration, which is the most primitive way to generate energy from glucose in living cells in the absence of oxygen. In the process of glycolysis one glucose molecule generates two pyruvate molecules and ATP. This process occurs both in Red and White muscles, in the presence or absence of oxygen.

3.2.3   The Fate of Pyruvate

Under aerobic conditions pyruvate is further oxidized to carbon dioxide and water in citric acid cycle and oxidative phosphorylation.

Under anaerobic conditions (hypoxia) or in vigorously contracting muscles glucose is converted to two pyruvate molecules but the NADH generated in the glycolysis cannot be oxidized as there is a lack of oxygen. It can be recalled that NADH and NAD^+ are reduced and oxidized forms of coenzyme. The coenzyme undergoes reversible reduction when a substrate is oxidized giving up two hydrogens in the form of hydride. Reduction of NAD^+ converts the benzenoid ring of nicotinamide moiety to the quinonoid form without any charge on nitrogen.

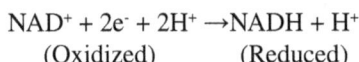

$$NAD^+ + 2e^- + 2H^+ \rightarrow NADH + H^+$$
$$\text{(Oxidized)} \qquad \text{(Reduced)}$$

The above reaction occurs in the presence of O_2, where it serves as an electron carries. In Fast-twitch muscles when there is no oxygen the electron cannot be given to a carrier for NADH to oxidize. In such cases NAD^+ cannot be generated. As NAD^+ is an electron acceptor, failure to generate it stops the conversion of glyceraldyhyde-3-phosphate to 1,3-diphosphosphate in the "payoff phase" of glycolysis (Figure 5.14). The energy yielding of glycolysis will then stop. It is therefore necessary to regenerate NAD^+ for the process of energy generation.

Many advanced living organisms including human beings have developed a way to generate NAD^+ by lactic acid fermentation. In such organisms the pyruvate is reduced to lactate as per the reaction shown in Figure 5.15, finally generation NAD^+.

When lactate is formed in large quantities during vigorous exercise, the acidification that results from ionization of lactic acid in muscle and blood limits the period of vigorous exercise. In other words fatigue occurs [20], [21].

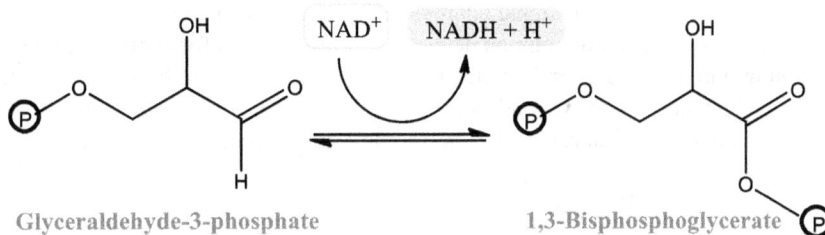

FIGURE 5.14   Reaction conversion of glyceraldyhyde-3-phosphate to 1,3-diphosphosphate.

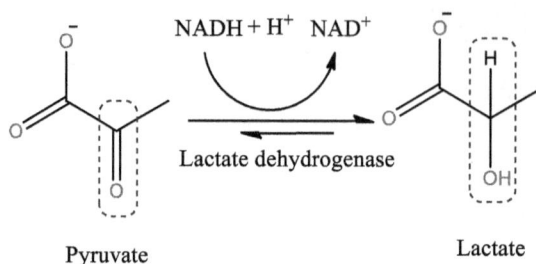

FIGURE 5.15 Conversion of pyruvate to lactate.

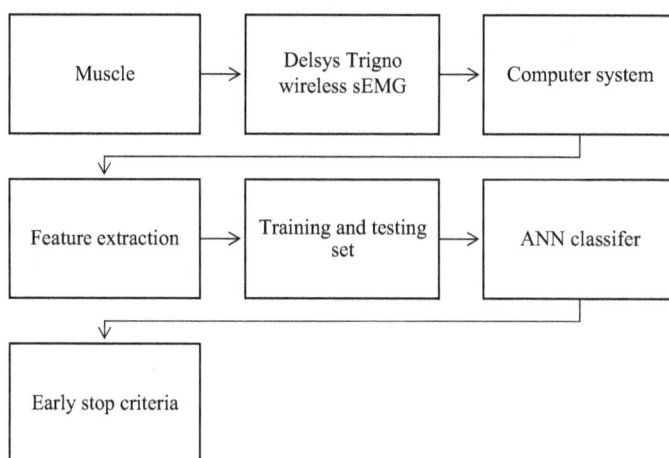

FIGURE 5.16 Use of AI in predicting muscle fatigue detection.

Muscle fatigue can be detected by AI. Anyone too enthusiastic about lifting weights on their first day in the gym is aware of muscle fatigue. It is a condition in which one's muscles are unable to generate enough force. It is observed not only in people who undergo vigorous exercise, like athletes, but is also a common syndrome in neurological disorders like multiple sclerosis, Parkinson's disease. and stroke.

Different methods and techniques can be tried that can determine whether a person has a normal or fatigued condition. The most accurate among the methods tried is the Support Vector Machine (SVM), which imagines the transformed signals from each bicep curl to be like a point in some space and draws an imaginary boundary that can best classify the points while keeping maximum distance from either type of point. Using SVM with four selected features gave a high (91%) accuracy in detecting muscle fatigue.

Fatigue prediction is a part of muscle endurance analysis, which is normally based on expert experience and guided by a muscle signal chart such as surface electromyography. Computational modeling is widely used in assisting human decision making, such as for sport coaching. In muscle endurance training, fatigue prediction is typically recommended either to boost the muscle strength or the muscle endurance for sport conditioning purposes. Many analysis tools use automated methods to assist human decision especially on muscle fatigue prediction. The onset of contractile fatigue was successfully predicted using Radius Basis Function Neural Network (RBFNN) model and Multilayer Perceptron (MLP) model [22].

3.3   PLANT KINGDOM

Plant science or biology is more of understanding of chemical biology. Plants are the very basic source of energy of all other living organisms on earth. Gaining knowledge about plant chemical biology is beginning to enhance our understanding of diverse cellular processes in plants, including endomembrane trafficking, hormone transport, and cell wall biosynthesis. All living things need energy to survive. Thus, let's first understand the chemistry behind the energy source of plants. Reactions that take place within living things are called biochemical reactions. Two of the most important are photosynthesis and cellular respiration. These two processes are responsible for providing energy to almost all of earth's organisms.

3.3.1   Photosynthesis

Plants get their energy by the process of photosynthesis. Sunlight provides energy to plants for photosynthesis. This is the process in which plants and certain other organisms synthesize glucose ($C_6H_{12}O_6$). The process of photosynthesis uses sunlight, carbon dioxide, and water and also produces oxygen along with glucose. The oxygen gets released in air and is taken by other organisms to carry out various oxygen-dependent chemical processes to get energy to survive. The overall chemical equation for photosynthesis is:

$$6CO_2 + 6H_2O + \text{Light Energy} \rightarrow C_6H_{12}O_6 + 6O_2$$

Photosynthesis changes light energy to chemical energy. The chemical energy is stored in the bonds of glucose molecules. Glucose is used for energy by the cells of almost all living things. Plants make their own glucose. Other organisms get glucose by consuming plants (or organisms that consume plants). This photosynthetic process is carried out in small organelles inside plant cell called chloroplast. Membranes of chloroplast contain pigment called chlorophyll, which is responsible for absorption of sunlight and reflecting green wave to make plants appear green in colour. This process of photosynthesis requires direct stream of sunlight and takes place within the thallykoid membrane of chloroplast. Apart from this sunlight-independent chemical reactions or other metabolic processes also occur within the plant cells.

3.3.2   Cellular Respiration

During dark times, plants cannot get the energy from sunlight but still must carry on basic metabolic processes using stored food. Plant cells, like animal cells, contain mitochondria in which stored food is converted to energy by cellular respiration.

Cellular respiration in terms of chemical reaction is the opposite of photosynthesis. Both processes are interlinked with each other and thus living organisms survive. The product of photosynthesis (i.e., oxygen) and glucose breaks down to produce energy in the form of ATP by the process of cellular respiration.

- Photosynthesis: $6CO_2 + 6H_2O \rightarrow C_6H_{12}O_6 + 6O_2$
- Cellular Respiration: $C_6H_{12}O_6 + 6O_2 \rightarrow 6CO_2 + 6H_2O$

Cellular respiration is a process of breakdown of glucose by living cells in the presence of oxygen to produce energy. Cellular respiration occurs in three main steps: glycolysis, transformation of pyruvate, the Krebs cycle (also called the citric acid cycle), and oxidative phosphorylation.

Animals who cannot use sunlight to produce food depend on organic matter produced by plants for their food and are said to be heterotrophic. They act as "middlemen" in the chemical reaction between oxygen and food material using the energy from the reaction to carry out their life processes. Plant cells, which use sunlight as a source of energy and CO_2 as a source of carbon, are

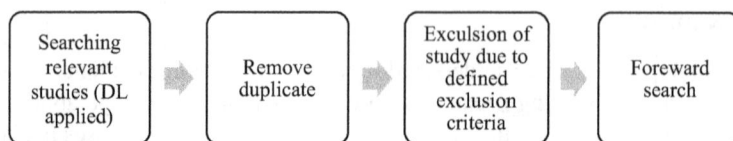

FIGURE 5.17   Deep learning.

TABLE 5.3
Different network architecture

Classification tasks	Network architecture
Flowers, fruits	ResNet 50
8 stages + harvest time	Cust. two-branch CNN processing images and temperature
10 stages, 9 stages	AlexNet

TABLE 5.4
Species summary

Total species	Possibly threatened	Not threatened
100%	44%	56%

---

classified as autotrophic. In contrast, animal cells must depend on organic material manufactured by plants for their food. These are called heterotrophic cells.

Biomass-produced biochemical conversions involving energy are very important in the practice of green chemistry and sustainability. The most obvious connection is the capture of solar energy as chemical energy by photosynthesis. Photosynthetically produced biomass can serve as a source of chemically fixed carbon for the synthesis of chemical fuels including synthetic natural gas, gasoline, diesel fuel, and ethanol.

Anoxic fermentation of biomass (abbreviated $\{CH_2O\}$) from sources such as sewage sludge or food wastes yields methane (natural gas), the cleanest burning of all hydrocarbon fuels:

$$2(CH_2O) \rightarrow CH_4 + CO_2$$

As plant is the very basic source of energy for all living organisms, study of the timing of seasonal events, such as budburst, flowering, fructification, and senescence, photosynthesis is of utmost important. Climatic conditions play an essential role in ecosystem processes, such as carbon and nutrient cycling. Study of the plant kingdom can be done in many ways such as by human observation, images captured by satellite, etc. But evaluating all data manually can be a very tedious task. Artificial intelligence can help to understand different matters related to the plant kingdom. Thus, there are many different studies in the literature including those using DL.

In the plant kingdom, classification tasks are done by following DL methods [23].

As plants are very necessary for human survival, their protection and conservation is also of utmost important. Currently, globally many tree species are threatened with extinction by human activities, and AI can be used to get near-to-accurate data about tree species that are prone to get extinction. Some approximate data are shown in Table 5.4.

Thus, DL helps us determine which species are possibly threatened so efforts can be made to conserve these threatened species and to further conserve the plant kingdom [24].

4   DRUG DISCOVERY AND DRUG–RECEPTOR INTERACTION

The above instances were some natural processes that are happening in a biological being. While these chemical concepts help us understand these processes they can also help us better understand various other concepts like drug discovery or drug–receptor interactions, etc.

Put simply, drug discovery is the process of finding a new chemical agent with the help of basic knowledge gained at the submolecular level. We all know whenever there is a signal for protein synthesis in the body, inside the cell or specifically nucleus (DNA & RNA) and ribosomes starts the process of translation. Similarly, whenever we administer a medicine, it binds to a receptor, which is linked to messenger systems, and finally there is an activity/action at the cellular level of a particular organ.

For the above drug–receptor action to happen the design of the drug is performed in such a manner that maximum interaction of drug and receptor takes place with the least amount of Gibb's free energy (more negative the better) and less spontaneity.

The process of drug discovery again starts with identification of biological receptor located on or spanning the semipermeable membrane. A receptor is a 3D chemical structure made of amino acid residues linked together. The term receptor or "receptive substance" was coined by Paul Ehrlich in the late 19th century. The interaction of drug with receptor is analogous to a "lock" (receptor) and "key" (the drug). Those ligands that fit properly and activate the receptor are considered for preparation of drug molecules. Figure 5.18 shows the 3D structure of receptor with amino acid residues and a drug molecule bound at a specific site on the receptor.

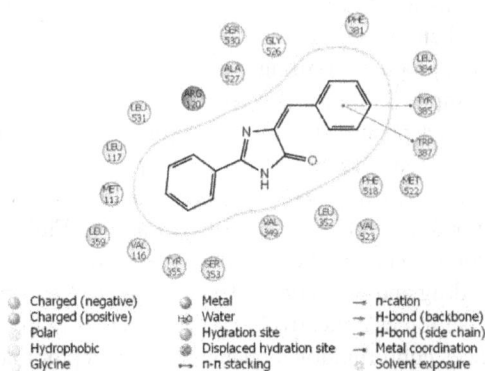

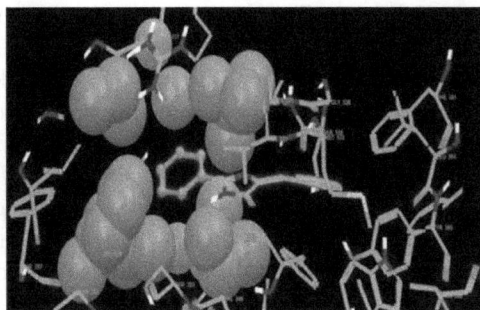

FIGURE 5.18   Ligand interaction diagram for receptor and ligand.

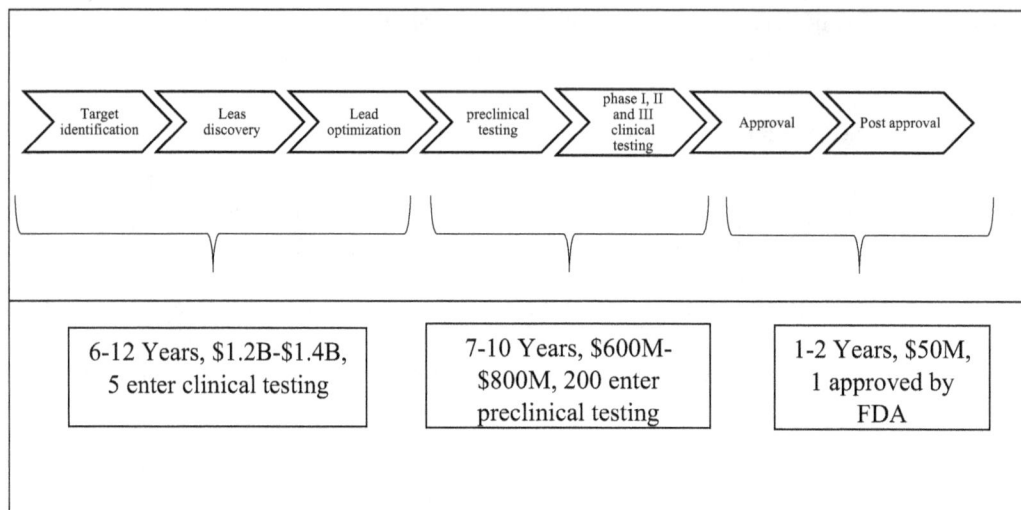

FIGURE 5.19 Typical drug developmental flow.

4.1 HOW DOES THIS INTERACTION OCCUR?

When a drug interacts with a receptor, multiple chemical interactive forces, both weak and strong, between drug and its receptor are believed to be responsible for initial interaction. Such interactions are covalent, ionic, and hydrogen bonds and hydrophobic interactions. The first three have already been discussed but the fourth type of interaction is seen in Figure 5.18 where the non-polar (aromatic) groups of ligands are enclosed in the non-polar cavity (grey circles) of the receptor. This will cause attraction between the ligand and the receptor. These interactions are called van der Waals forces and come into effect when two non-polar molecules come in close range to each other.

A covalent bond interaction produces a strong bonding, which is most of the time irreversible and destroys the receptor. For example, long-lasting blockade of α-adrenoceptors by phenoxybenzamine. Chemically, at physiological pH phenoxybenzamine forms a carbonium ion that links itself with amino, sulfhydryl, or carboxyl group of the α-adrenoceptors. A weaker bond than the covalent bond is the ionic bond, which is formed when two oppositely charged atoms approach each other. This tendency to form an ionic bond depends on the electronegativity. Fluorine, chlorine, hydroxy, and sulfhydryl groups form strong bonds as their tendency to attract electrons is greater. Hydrogen bond when formed between drug and receptor confers stability to the complex. The combination of variety of bonds and attractive forces collectively contributes to the strength of drug binding to the receptor. The binding produces conformational change in the receptor that initiates the activation of a biological response. Such forces also tend to align the atoms/molecules in order to increase their interaction and reduce the potential energy. The above discussion again shows that a biological phenomenon can be understood by understanding its chemistry [25].

Drug discovery, research, and development consist of many steps like design of experiments, execution of experiments, drug discovery and lead optimization, pre-clinical and clinical testing, and approval for commercial production. This process can take a very long time (approximately 10–15 years and high costs).

Artificial intelligence can help a lot by making this process faster and with lower amount of investment required. Computer-aided design can predict the required structure with the required properties and can screen new drugs from large databases. These discoveries are done using various methodologies like molecular docking, pharmacophore modeling, VS, and quantitative structure activity relationship (QSAR). Among this group VS is highly successful. It accelerates the drug discovery and lowers the number of experiments to be done in the laboratory.

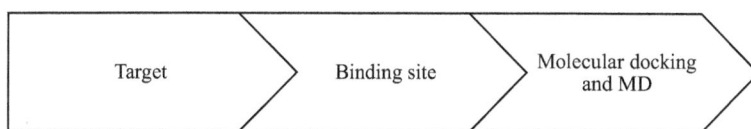

FIGURE 5.20 Molecular docking.

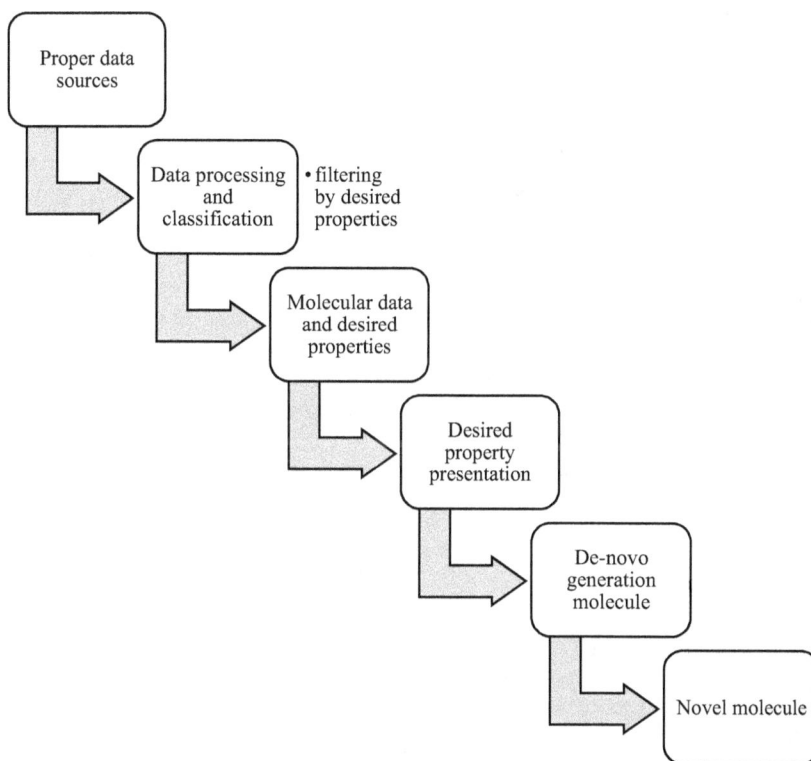

FIGURE 5.21 Machine learning.

4.2 TYPICALLY, COMPUTER-AIDED DRUG DESIGN CONSISTS OF THE FOLLOWING PROCESS

In computer-aided drug design, target identification is the first step followed by binding site and compound library preparation. Then, put in each compound from the library into the identified binding site for the evaluation of the score. However, in molecular docking (MD) simulations help obtain more flexible targets using ligand-based drug design in which as per the activity of ligand virtual screening is performed [26].

4.3 APART FROM THIS, MACHINE LEARNING-BASED PROCEDURE IS ALSO POPULAR. IT CONSISTS OF THE FOLLOWING STEPS.

Different deep learning techniques are as per below for molecule generation

- Recurrent neural networks (RNN)
- Gated Recurrent Neural Network (Gated RNN)
- Convolutional neural networks (CNN)
- Multilayer perceptron networks (MPL)

5   CONCLUSION

All biological processes have a chemical basis. From the above discussion it can be inferred that changes at the atomic level (i.e., electrons in the presence of favourable energy) lead to flow of a process and formation of a product, for which chemical bonding is required. Any process that occurs in a biological system has a chemical basis and therefore to understand such a process chemistry all the states of matter involved must be thoroughly studied. Artificial intelligence can immensely help humankind since many different fields related to chemistry and biological things. Humans are incapable of breaking free of their biologically determined limits. But with the evolution of AI, humans are capable of transcending those limits [27].

REFERENCES

1. Nelson, D. L., & Cox, M. M., Lehninger Principles of Biochemistry. 7th ed. WH Freeman. Chicago (notes-bibliography), 2014. 523–526 and 898.
2. Hruby, V. J. Organic Chemistry and Biology: Chemical Biology through the Eyes of Collaboration. J. Org. Chem. 2009. 74, 9245–9264.
3. Mattaini, K. (2020). Chapter 1. The study of life. rwu.pressbooks.pub/bio103/chapter/1-the-study-of-life/
4. Molnar, C., & Gair, J. (2015). 5. 1: Overview of photosynthesis. opentextbc.ca/biology/chapter/5-1-overview-of-photosynthesis/
5. Joshi, R.P.; Kumar, N. Artificial Intelligence for Autonomous Molecular Design: A Perspective. Molecules 2021, 26, 6761.
6. 2. 1: The elements. (2017, June 1). Chemistry LibreTexts. chem.libretexts.org/Courses/University_of_South_Carolina__Upstate/USC_Upstate%3A_CHEM_U109_-_Chemistry_of_Living_Things_(Mueller)/02%3A_Elements%2C_Atoms%2C_and_the_Periodic_Table/2.1%3A_The_Elements
7. Chemical elements for life. (n.d.). Retrieved December 14, 2022, from hyperphysics.phy-astr.gsu.edu/hbase/Chemical/chonps.html
8. Protein structure: Primary, secondary, tertiary, quaternary structures I lls health cdmo. (2019, October 28). lubrizolcdmo.com/technical-briefs/protein-structure/
9. Protein folding—An overview I sciencedirect topics. (n.d.). Retrieved December 14, 2022, from www.sciencedirect.com/topics/immunology-and-microbiology/protein-folding#:~:text=Unfolded%20or%20misfolded%20proteins%20contribute,secrete%20proteins%20is%20working%20properly
10. Banfalvi G. Why ribose was selected as the sugar component of nucleic acids. DNA and cell biology. 2006 Mar 1;25(3):189–96.
11. Defresne M, Barbe S, Schiex T. Protein Design with Deep Learning. International Journal of Molecular Sciences. 2021 Oct 29;22(21):11741.
12. Suzuki E, Shibata K, Mizoguchi T. Accurate prediction of bonding properties by a machine learning–based model using isolated states before bonding. Applied Physics Express. 2021 Jul 19;14(8):085503.
13. Cui Q, Lu S, Ni B, Zeng X, Tan Y, Chen YD, Zhao H. Improved prediction of aqueous solubility of novel compounds by going deeper with deep learning. Frontiers in oncology. 2020 Feb 11;10:121.
14. Kamerlin SC, Sharma PK, Prasad RB, Warshel A. Why nature really chose phosphate. Quarterly reviews of biophysics. 2013 Feb;46(1):1–32.
15. Egli M, Pallan PS, Pattanayek R, Wilds CJ, Lubini P, Minasov G, Dobler M, Leumann CJ, Eschenmoser A. Crystal structure of homo-DNA and nature's choice of pentose over hexose in the genetic system. Journal of the American Chemical Society. 2006 Aug 23;128(33):10847–56.
16. Lactic acid fermentation. (n.d.). Retrieved December 14, 2022, from www.tempeh.info/fermentation/lactic-acid-fermentation.php
17. Kamerlin SC, Sharma PK, Prasad RB, Warshel A. Why nature really chose phosphate. Quarterly reviews of biophysics. 2013 Feb;46(1):1–32.
18. Damien A. Dablain, Geoffrey H. Siwo, Nitesh V. Chawla. Generative AI Design and Exploration of Nucleoside Analogs
19. Bera M. Artificial Intelligence in Bioinformatics. International Journal of Innovative Science and Research Technology, 2021; 6 (2), 433–36.

20. Sahlin K. Muscle fatigue and lactic acid accumulation. Acta physiologica Scandinavica. Supplementum. 1986 Jan 1;556:83–91.

21. Hultman E, Spriet LL, Söderlund K. Biochemistry of muscle fatigue. Biomed Biochim Acta. 1986; 45(1–2), S97–106. PMID: 3964254.

22. Nur Shidah Ahmad Sharawardi, Yun-Huoy Choo, Shin-Horng Chong, and Nur Ikhwan Mohamad. Isotonic Muscle Fatigue Prediction for Sport Training using Artificial Neural Network Modelling. Springer-Verlag Berlin Heidelberg 2011.

23. Katal N, Rzanny M, Mäder P and Wäldchen J (2022) Deep Learning in Plant Phenological Research: A Systematic Literature Review. Front. Plant Sci. 13:805738

24. Silva SV, Andermann T, Zizka A, Kozlowski G and Silvestro D (2022) Global Estimation and Mapping of the Conservation Status of Tree Species Using Artificial Intelligence. Front. Plant Sci. 13:839792

25. Bapna M, Chauhan LS. Microwave Assisted Synthesis and In-silico Molecular Modeling, Toxicity Studies of Some New Derivatives of (E)–(Substituted)Benzylidene-1-Isonicotinoylpyrazolidine3,5-Dione as Lead Compounds. Inventi Impact: Molecular Modeling, 2014(4):179–184, 2014.

26. Zhang, Y.; Luo, M.; Wu, P.; Wu, S.; Lee, T.-Y.; Bai, C. Application of Computational Biology and Artificial Intelligence in Drug Design. Int. J. Mol. Sci. 2022, 23, 13568.

27. Harari YN. Sapiens: a brief history of humankind by Yuval Noah Harari. The Guardian. 2014.

6 Chemical Biology-based Toolset for Artificial Intelligence Usage in Drug Design

Ashutosh Agarwal, Parixit Prajapati,
Bhupendra Gopalbhai Prajapati

1 INTRODUCTION

1.1 CHEMICAL BIOLOGY

The study of molecular events in biological systems with the help of chemistry is known as chemical biology. It includes the subdivision of biological chemistry like biochemistry, biophysics, bioanalytic, bioinorganic, and bioorganic (1). The role of chemical biology in drug design is to identify the new target molecule or the lead compound that can modulate the targeted protein in the human body (2). The basic aim is to study the living system with active groups and not to treat and cure of the disease by these chemical probes (3). To achieve the above application, there are some toolsets in chemical biology toolset such as chemical probes, antisense and RNAi, protein degradation studies, phenotypic screening, chemical genomics, and many more (4,5). Speaking of chemical probes, these are the compounds or small molecules that modulate the targeted protein function; more specifically they can reversibly or irreversibly inhibit the target protein or protein domain in human cells. There are some requirements for being a highly qualified probe such as defined mechanism of action, selectivity toward targeted protein, high bioavailability, aqueous solubility, potency, and being freely available. Basically, chemical probes are employed in the interpretation of biological experiments required for target validation. Also, since current methods for target validation are biased, developing a chemical probe with a framework for its use can play a central role in identification of targets fulfilling the basic criteria of chemical biology (6). In the past few decades, RNAi interferences have been widely used for reducing gene expression. By taking use of oligonucleotide's capacity to bind to targeted RNAs via Watson-Crick hybridization, antisense oligonucleotide agents cause the suppression of target gene expression in a sequence-specific manner (7). These technologies are robust and used to identify and validate potential drug targets. Similar to the above two chemical tools, phenotypic screening and chemical genomics are also used for the same purpose (i.e., the identification of lead molecule with therapeutic efficacy). Many compounds have been discovered and optimized through phenotypic screening like Daptomycin, linezolid, lacosamide, trametinib, bedaquiline, etc., having good efficacy towards their target (8). Affinity-based methods, computational approaches, resistance selection, genetic modifier screening, etc., are some of the techniques used to understand the small molecule mechanism of action in phenotypic screening. Among all the above, the affinity-based method is well

DOI: 10.1201/9781003353768-6

known for the identification of the lead molecule, although the other methods are equally effective (9). Similarly, in chemical genomics, a target-based chemical ligand is added to the cell of interest where it binds to the target and causes either the loss of function or gain of function of its target. The study of these chemical ligands can be further aided by the use of chemical genomic tools like gene expression profiling and protein profiling, which form the basis of validation studies (10). Today, many of the targeted proteins lack the active site, which limits the discovery of new compounds or lead molecules. To overcome the above problem, protein degradation or proteolytic targeting chimera method has been proven to be a technology that can specifically target these kinds of proteins (11).

1.2 DRUG DESIGN

As we all know, human lives are always in threat due to new diseases like coronavirus, monkeypox virus, and many more. Drugs are the compounds or molecule used for the diagnosis, prevention, and treatment of disease in humans and animals (12). For the design of these drugs there are several stages to be worked on (13). The traditional method of drug design is very time consuming as well costly and sometimes leads to failure. A simple process of traditional drug design has been explained and later compared with the new available technologies like computer-aided drug design and role of artificial intelligence in drug design further in this chapter. The basic drug design process involves many steps, among which the first step is selection of drug target and to obtain a pure solution of it. Further, the structure of the target is determined by the use of crystallographic techniques as well as NMR studies. The analysis of the structure thus obtained results in information about the possible inhibitory binding sites of target in the human body. Then from the stored databases the list of available compounds is obtained and from the selected list, top compounds are picked out that have good binding affinity and selectivity against the targeted protein (14). After this, the lead selected is synthesised and checked for its micromolar inhibition in solution. Next, the structure of this lead compound is determined by the use of crystallographic techniques and many analytical techniques like NMR, Mass, etc. Information regarding whether the lead is a Nm inhibitor or not is further obtained. If yes then the potency is determined and the drug molecule is sent for further clinical trials (15). The process stated above takes nearly 12 yrs. For its completion as well as its estimated cost is around $2.6 billion (16). Today, to increase the efficiency of drug design, computational methods are used (17). Basically, this method consists of two forms of computer-aided drug design (CADD), namely structure-based drug design and ligand-based drug design. (18). In this design process, first the target is selected, and then availability of 3D structure of target is checked. If the structure is available, a structure-based drug design (SBDD) approach is used and if the structure remains unknown, ligand-based drug design (LBDD) approach is applied (19). When a completely novel molecule or lead is discovered throughout the SBDD process, database screening is required. Utilizing this screening technique, novel lead compounds with active sites and potential pharmacological activity are found. There are also docking programs, such as DOCK and DockSearch, which allow researchers to screen a lot of leads in a database (20). Following this, the best compound having active site is selected for further drug design procedures like experimental testing and pharmacology testing (21). In contrast, the LBDD technique analyzes ligands with known pharmacological activities even while the structure of the targeted protein is not known. This can be done by using database screening and models like Pharmacore models, QSAR models, and QSPR models (22,23). Currently, 3D QSAR models are also utilized for screening, which aids in figuring out how ligand features are distributed in 3D space (24). Following the determination of the structure, docking tests are conducted to choose the lead molecule that fits its group with the target's active site. The remaining steps follow the same pattern as those described earlier. While traditional approaches make it very difficult to understand the properties of a drug and a ligand, computational methods are quick, inexpensive, and simple for the estimation of biomolecular systems, particularly the determination of structure of targeted ligands using CADD (25).

1.3 ARTIFICIAL INTELLIGENCE

Artificial intelligence is used to make intelligent devices, particularly intelligent computer programs, which is a scientific and engineering endeavour. Although AI is related to the relevant work of utilizing computers to comprehend human intelligence, AI should not be limited to techniques that can be observed biologically (26). In 1956, John McCarthy coined the term Artificial Intelligence. In 1961 the first industrial robot was utilized for normal work at General Motors. In medicine AI was first used by the Saul Amarel at Rutgers University for the development of Research Resources in Computers in 1971. Whereas in 1972 the first molecule to be synthesized by the AI was MYCIN (27). As time passed, AI advances. Currently there are four types of AI: Machine Learning (ML), Natural Language Processing (NLP), Fuzzy Logic (FL), and Computer Vision (CV) (28). Machine learning is generally based on the analysis of information provided using some algorithms by machines leading to the development of learning techniques (29). The use of computers for understanding the languages of humans, retrieval of information, analysis of lexical, question answering, etc., falls under NLP (30). Fuzzy logic is a type of AI used for studying human reasoning (31). Due to the fuzziness of the majority of natural classes and thoughts, fuzzy set theory was developed. Fuzzy logic has been majorly used in dental medicine, but can be extended to other fields of medicines like otology, dermatology, gynecology, rhinology, ophthalmology, and many more (32). In order to create software algorithms, CV is frequently used. Currently in medicine CV software is used in diagnostic laboratories. For example, Cellavision is a software used for the determination of morphology of white blood cells (33). A subclass of AI called "machine learning" aims to have computers automatically learn new skills and get better over time without explicit programming. Deep learning is a subset of ML in which the computer uses the input data to automatically find the features needed for the task. Convolution neural network (CNN) and recurrent neural network (RNN) are two widely used deep learning models (34). Convolution neural networks are used for text detection, visual saliency detection, action identification, scene labeling, object tracking, position estimation, picture classification, audio recognition, and NLP (35). Whereas RNNs are used for sequential data, such text, audio, and video, and have also become widespread (36). In the field of medicine, AI is used in radiology, pathology, endoscopy, ultrasonography, diagnosis, surgery, 3D printing, anaesthesiology assistance, rehabilitation assistance, drug production, drug design, etc. (37).

2   ARTIFICIAL INTELLIGENCE IN DRUG DESIGN

Artificial intelligence plays a crucial part in drug development. As such, there are many applications of AI in drug design processes, some of which include understanding pathways and finding molecular targets, finding leads, prediction of mode of action of active molecule, selection of population for clinical trials, in polypharmacology, etc. (38). Recent developments in physico-chemical and ADMET (Absorption, Distribution, Metabolism, Elimination, Toxicity) properties, as well as other applications in property or activity estimations, support the strength of this technology in Quantitative Structure-Activity Relationships (QSAR) (39). In order to predict the properties of molecules, the molecules are first represented mathematically, and thereafter various ML techniques are used to ascertain how those representations connect to the activity or value values that have been observed. The molecular structure is the starting point for the majority of molecular representations, which can be further translated into SMILES (Simplified Molecular-Input Line-Entry System), Graph Model, Fingerprint, and Molecular Descriptors. After these molecules have been properly represented, molecular property prediction algorithms can be created using ML techniques. These algorithms include DNN, Random Forest, LightGBM, and Ensemble for the

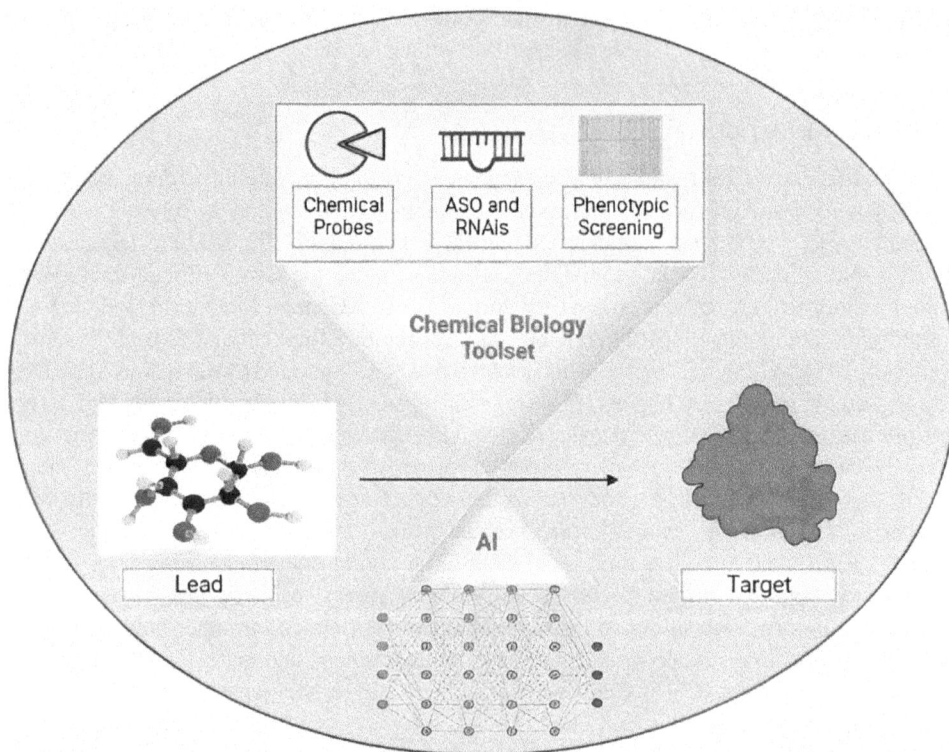

FIGURE 6.1   Role of chemical biology toolset in drug design.

estimation of ADME, ADME and potency, toxicity, and physicochemical property determination, respectively (40–43). Currently, a revolutionary idea for de novo molecular design that makes use of generative AI has been put forward. It a technique for learning from already-known bioactive chemicals that creates unique molecules on its own that inherit bioactivity and synthesizability (44). In 2021, using a similar method 33 active compounds were synthesized against SARS-CoV-2. For this a dataset of 7665 compounds were collected that were active 3CL protease. Further, the screening was performed using pChEMBL and 2515 molecules passed the test. For pretraining the generative mode, a dataset of 1.6 million drug-like small compounds in SMILES format was used. On a Tesla V100 graphics processing unit (GPU), the model was trained for 500 epochs, and the weights from the trained model were used for the pipeline's subsequent tasks. A collection of target-specific small molecules was then used to include the target information into the model and the pretrained generative model was retrained. Utilizing transfer learning and reinforcement learning, the generative model was enhanced to focus on the chemical space relating to the protease inhibitors. The last screening was conducted using a number of physicochemical property filters and a virtual screening score, which resulted in new 33 compound (45). Figure 6.1 illustrates the role of the chemical biology toolset in drug design.

3   CHEMICAL BIOLOGY TOOLSET FOR AI IN DRUG DESIGN

For drug design some of the chemical biology toolsets have been reported including the use of chemical probes, phenotype screening, antisense oligonucleotides and RNAis, chemical genomics, and protein degradation. Figure 6.2 illustrates the types of chemical biology toolsets for AI in drug design. The first three toolsets are discussed in the following sections.

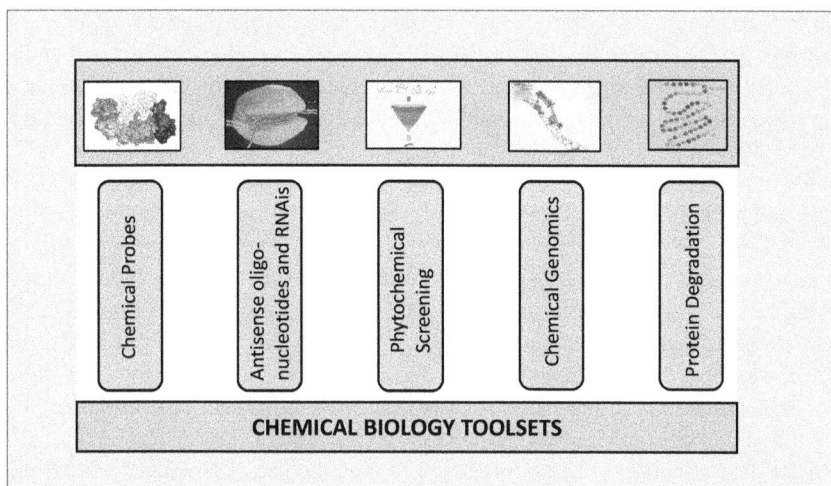

FIGURE 6.2 Types of chemical biology toolsets for AI in drug design.

3.1 CHEMICAL PROBES

A chemical probe is a small molecule with a well-defined and targeted biological response. It, in a qualitative and quantitative manner, contains strong affinity, high efficacy, and a clearly defined pharmacological and biochemical profile (46). Chemical probes are used to investigate how a target's phenotype and the capacity to modify that phenotype by a small molecule are related. Use of a high-quality chemical probes may provide information about the tractability and translatability of a specific target (47). Chemical probes are used for cell-based target validation in drug design and have four main pillars: exposure of chemical probe at the site of action, target engagement, functional pharmacology, and relevant phenotype. The exposure of chemical probe at the site of action means the availability of small molecule probe where it has to act in a suitable concentration. For this to happen, cell permeability of the probes is a prerequisite. Various analytical techniques like LC-MS can help in estimating the concentration of probe in cells. Exciting developments in methodologies like activity-based proteomics have made it possible to quantify target engagement. A more physiologically appropriate environment can be created by using advanced functional probes to assess availability within the cell and enable impartial selectivity assessment. Assays that test the pharmacology of the probe can be developed for the expression of functional pharmacology, frequently evaluating a proximal biomarker for activity. For the proof of phenotype perturbation, cell biologists must develop tests with a high degree of confidence in their "translatability" and that capture the phenotypic alterations that are most pertinent to human disease. It is important to rule out potential false positives as soon as possible, including nonspecific cell death (48). The design of these chemical probes can be done with the help of virtual screening web servers like ChEMBL, DrugBank, hmdb, PubChem, SuperDRUG 2, SureCheMBL, ZINC, etc. Chemical or text queries serve as the "Input" for these models, while a similar molecule serves as the "Output" (49).

3.2 PHENOTYPIC SCREENING

In drug discovery, phenotypic screening is a method for finding molecules that can change a cell's phenotype. Without knowing the pharmacological target beforehand, Phenotypic Drug Discovery (PDD) refers to the screening and selection of hit or lead molecules based on measurable phenotypic endpoints using cell-based assays (69). The PDD approach is also known as "forward pharmacology," "forward chemical biology," or "classical pharmacology." The benefit of phenotypic screening in drug development is that it is not mandatory to identify a disease's molecular target, and it is also possible to target signalling pathways with a variety of biological domains and complexes (70). As said earlier the mechanism of action for PDD can be done by various methods. Here only two of them are explained: affinity-based method and computational method. In the affinity-based method the SM-derived affinity probe is generated first with an appropriate linker and a synthetic moiety that acts as a contact to enable its adherence to a solid support ("bead") at an ideal regulated concentration and packing density. Then a cell or tissue lysate is prepared using physical means such as mechanical disruption, liquid homogenization, or reagent-based techniques (often detergents), etc. Afterwards, the non-covalent SM-protein interactions are brought by incubating the affinity probe-carrying beads with a cell or tissue lysate. Then the washing of beads is gently carried out to get rid of any proteins that were randomly bound. Strong washing is done to release proteins from the beads, followed by either "on-bead" trypsin digestion or trypsin digestion to produce peptides. Furthermore, utilizing software programs such as Mascot, Andromeda, the enriched peptides that arise are separated by HPLC, identified, and quantified by MS (71–73). A computational method or AI approach for determination of MoA for PDD involves the use of annotated cluster multidimensional enrichment (ACME) method for the identification of possible relationships between specific CCL (Cancer Cell Lines) dependencies and small-molecule MoA. Here a large number of CCL are treated with individual many small molecules, some of them with known mechanism of action. After that, "hot spots" are mined from the resulting matrix. It is reasonable to speculate that unidentified compounds in the same cluster will have a comparable MoA if one or more molecules in the hot spot have a known MoA. Another method under AI is called "Cell Painting," whereby five microscopy channels and six fluorescent dyes are used to label seven biological components. More than 800 cellular characteristics are generated by CellProfiler (Software) analysis, which may be grouped and examined similarly to gene expression or CCL. The method for assessing MoA compounds was validated by the large number of compounds with comparable MoA clustering (74–76).

3.3 ANTISENSE OLIGONUCLEOTIDE (ASO) AND RNAIS

In 1978, the term "antisense" and the concept of developing oligonucleotides to bind to certain regions in target RNAs via Watson–Crick hydrogen bonding were developed (90). Oligonucleotides are the true form of molecule utilized for inhibiting protein synthesis of the target mRNA, resulting in the knockdown of the target gene. They are identical to a section of a target mRNA molecule (91). The two basic mechanisms of ASO-based methods are occupancy-only mediated and occupancy-mediated RNA cleavage. It was discovered in 2004 that the use of ASO to regulate RNase H1 results in the reduction of RNA in human cells. Human RNase H1 interacts with the duplex produced by ASO-containing DNA and a target RNA via its N-terminal. The RNase H1 catalytic domain requires at least five consecutive DNA/RNA base pairs to cleave the RNA in the duplex, and cleavage normally happens between 7–10 nucleotides. Following cleavage, the affected phosphate group on the 5′-end and the hydroxyl group on the 3′-end of the RNA are identified, and the RNA is destroyed by cellular nuclease enzymes. After being liberated, ASO is likely ready to participate in another transcript. This was all about the underlying mechanism of occupancy-mediated RNA cleavage whereas in occupancy-only mediated method the use splice-switching oligonucleotide (a type of ASO), leads to modulate splicing in the nucleus and on chromatin by halting splice sites and regions that link enhancers and silencers that are present in both exons and introns. Additionally, by skipping particular exons, these ASOs can potentially be employed to inactivate proteins. Additionally, by skipping particular exons, these ASOs can potentially be employed to inactivate proteins (92–95). TargetFinder is a tool for interactively choosing effective antisense oligonucleotides using target mRNA secondary structures and accessible site tagging. The target mRNA's secondary structure and single-strand likelihood profile are also shown for reference purposes. The alignment method used to display all of these sequences and profiles makes it easier to spot the target mRNA's accessible

TABLE 6.1

List of chemical probes with their probable targeted proteins

Sr No.	Chemical Probes	Targeted Protein	References
1	Fluorosulfates	tyrosine, lysine, serine, and histidine	(50)
2	Rapamycin	mTOR	(51)
3	Cyclopamine	Smoothened	(51)
4	GSK4112	REV-ERBα	(52)
5	ω-azido-fatty acids	fatty-acylated proteins	(53)
6	Rohinitib	Heat shock factor 1 activation inhibitor	(54)
7	Robotnikinin	sonic hedgehog binder	(55)
8	Congo red, thioflavin T	amyloid-β (Aβ) peptide	(56)
9	5-arylidene-2-thioxo-4-thiazolidinone	galactofuranose	(57)
10	triacetyloleandomycin (TAO), α-nasymphthoflavone (ANF), and diethyldithiocarbamate (DDC)	cytochrome P450 (CYP)	(58)
11	trichostatin A and trapoxin	histone deacetylases	(59)
12	2,4-dinitrophenylhydrazine	Protein carbonyls	(60)
13	Tritiated sodium borohydride	Protein carbonyls	(61)
14	(2-substituted)-piperidyl-1,2,3–triazole ureas	diacylglycerol lipase-β	(62)
15	[36S]dithiobis(suooinimidyl propionate), a	haemoglobin	(63)
16	Dimedone	Sulfenic Acid	(64)
17	Conus peptides	Ion Channels	(65)
18	PROteolysis TArgeting Chimeras (PROTAC)	Protein Of Interest (POI)	(66)
19	Hexanediol	-	(67)
20	Cysteine sulfenic acids	sulfenic acid	(68)

TABLE 6.2

List of molecules screened with their targeted disease or parameters

Sr No.	Disease/Parameters	Molecules	References
1	Cancer	Elesclomol	(77)
2	Cancer	Quinacrine	(78)
3	gene expression analysis	ethyl methane sulfonate	(79)
4	Cancer	Zybrestat	(80)
5	Tuberculosis	Bedaquiline, Delamanid, Pretomanid	(81)
6	Parasitic disease	Sirtinol	(82)
8	Cancer	CRISPR-Cas9	(83)
9	Tuberculosis	NITD-529	(84)
10	schistosomiasis	Anisomycin, Gambogic acid	(85)
11	Cancer	Tasquinimod	(86)
12	Antimicrobial thearpy	carbapenemase	(87)
13	Diarrhoea	Cedrelone and baicalein	(88)
14	Cancer	Zalypsis	(89)

TABLE 6.3
List of ASO/RNAis molecule with their probable targeted

Sr No.	Molecules	Target	References
1	Fomivirsen	CMV mRNA	(104)
2	Milasen	intron 6 spice acceptor cryptic site	(105)
3	Patisiran	TTR mRNA	(106)
4	Golodirsen	DMD pre-mRNA	(107)
5	Nusinersen	SMN2 pre-mRNA	(108)
6	RG-012 03373786 *	miR-21	(109)
7	LY900003*	protein kinase C-alpha (PKC-α)	(110)
8	LErafAON*	c-raf kinase	(111)
9	OGX-011*	Clusterin	(112)
10	HsPCSK9-1811*	*PCSK9* mRNA	(113)
11	QPI-1002*	p53 mediated apoptosis	(114)
12	PF-04523655*	RTP801	(115)
13	Apo-B SNALP*	liver ApoB mRNA	(116)

(*: under clinical trial)

locations (96). eSkip-Finder (97) and PFRED (98) are some of the other computation technologies used for the design of the above ASOs.

RNA interference (RNAi) is a strong biological mechanism that selectively silences gene expression in cells. It has enormous potential for functional genomics, drug discovery via in vivo target validation, and drug development (99). Small interfering RNA (siRNA) is employed as a therapeutic component for the silencing of genes, according to Ryan et al. Pre-miRNA is created in the nucleus from pri-miRNA as part of the overall pathway by a microprocessor complex (Drosha–DGCR8). With the assistance of Exportin 5, this pre-miRNA is delivered to the cytoplasm. After that it binds to the TAR RNA-binding proteins (TRBP) and Dicer. The Dicer causes the development of an RNA-induced silencing complex (RISC)-loading complex (RLC) with an Argonaute (Ago1-Ago4) protein by cleaving the terminal loop of pre-miRNA. The sense strand is deleted while the antisense strand is put into Ago1-Ago4 protein. By preventing mRNA translation, encouraging mRNA breakdown, and guiding transcriptional gene silence of the target gene, the mature RISC may control gene expression. In order to create the RLC, the synthetic small interfering RNAs (siRNAs) now reach the cytosol via endocytosis and bind with a new Dicer and TRBP. Since this siRNA almost exactly resembles the targeted mRNA, it is quite simple for the siRNA to cause powerful and precisely targeted gene silence (100–102). Similar to ASO determination, ATARiS (103) and PFRED are the two AI-based software tools used for RNAi screening.

REFERENCES

1. Begley TP. Chemical biology: an educational challenge for chemistry departments. Nature Chemical Biology 2005 1:5 [Internet]. 2005 [cited 2022 Dec 25];1(5):236–8. Available from: www.nature.com/articles/nchembio1005-236

2. Schenone M, Dančík V, biology BW… chemical, 2013 undefined. Target identification and mechanism of action in chemical biology and drug discovery. nature.com [Internet]. [cited 2022 Dec 25]; Available from: www.nature.com/articles/nchembio.1199?message-global=remove&page=2

3. Schmidt M. Chemical Biology: And Drug Discovery [Internet]. 2022 [cited 2022 Dec 25]. Available from: books.google.com/books?hl=en&lr=&id=xCZcEAAAQBAJ&oi=fnd&pg=PR5&dq=chemical+biology+toolset+in+drug+discovery&ots=aTk-ZQG-tb&sig=3j3hwrchYmBGFp-h3HQKMjqHAMA

4. Civjan N. Chemical biology: approaches to drug discovery and development to targeting disease [Internet]. 2012 [cited 2022 Dec 25]. Available from: books.google.com/books?hl=en&lr=&id=l8nZdmqtDnEC&oi=fnd&pg=PT11&dq=chemical+biology+tools+%22drug+candidates%22&ots=9DoNI0LD2E&sig=rie3EoL08jAIkkzMmh4dLQUxruY

5. Riaz A, Rasul A, Sarfraz I, Nawaz J, Sadiqa A, Zara R, et al. Chemical Biology Toolsets for Drug Discovery and Target Identification. Cheminformatics and its Applications. 2020 Jul 15;

6. Workman P, biology ICC&, 2010 undefined. Probing the probes: fitness factors for small molecule tools. Elsevier [Internet]. [cited 2023 Jan 2]; Available from: www.sciencedirect.com/science/article/pii/S1074552110001985

7. Dallas A, monitor AVM science, 2006 undefined. RNAi: a novel antisense technology and its therapeutic potential. didattica-2000.archived.uniroma2.it [Internet]. [cited 2022 Dec 25]; Available from: didattica-2000.archived.uniroma2.it/BiologiaApplicata/deposito/RNAinterf.pdf

8. Plowright A, Chemistry LDAR in M, 2017 undefined. Phenotypic screening. Elsevier [Internet]. [cited 2022 Dec 25]; Available from: www.sciencedirect.com/science/article/pii/S0065774317300015

9. Wagner B, biology SSC chemical, 2016 undefined. The power of sophisticated phenotypic screening and modern mechanism-of-action methods. Elsevier [Internet]. [cited 2022 Dec 25]; Available from: www.sciencedirect.com/science/article/pii/S2451945615004675

10. Steven Zheng XFS, Chan TF. Chemical Genomics: A Systematic Approach in Biological Research and Drug Discovery. Current Issues in Molecular Biology 2002, Vol 4, Pages 33–43 [Internet]. 2002 Apr 5 [cited 2022 Dec 25];4(2):33–43. Available from: www.mdpi.com/1467-3045/4/2/4

11. Li H, Dong J, Cai M, Xu Z, Cheng XD, Qin JJ. Protein degradation technology: a strategic paradigm shift in drug discovery. J Hematol Oncol. 2021 Dec 1;14(1).

12. Mandal S, pharmacology SME journal of, 2009 undefined. Rational drug design. Elsevier [Internet]. [cited 2022 Dec 26]; Available from: www.sciencedirect.com/science/article/pii/S0014299909008784

13. Martin Y. Quantitative drug design: a critical introduction [Internet]. 2010 [cited 2022 Dec 26]. Available from: www.taylorfrancis.com/books/mono/10.1201/9781420071009/quantitative-drug-design-yvonne-martin

14. Shoichet B, McGovern S, Wei B, chemical JIC opinion in, 2002 undefined. Lead discovery using molecular docking. Elsevier [Internet]. 2002 [cited 2022 Dec 26]; Available from: www.sciencedirect.com/science/article/pii/S1367593102003393

15. biology AAC&, 2003 undefined. The process of structure-based drug design. Elsevier [Internet]. [cited 2022 Dec 26]; Available from: www.sciencedirect.com/science/article/pii/S1074552103001947

16. Mohs R, & NGA& DTR, 2017 undefined. Drug discovery and development: Role of basic biological research. Elsevier [Internet]. [cited 2022 Dec 28]; Available from: www.sciencedirect.com/science/article/pii/S2352873717300653

17. Surabhi S, Therapeutics BSJ of D delivery and, 2018 undefined. Computer aided drug design: an overview. jddtonline.info [Internet]. [cited 2022 Dec 27]; Available from: www.jddtonline.info/index.php/jddt/article/view/1894

18. Yu W, Mackerell AD. Computer-aided drug design methods. Methods in Molecular Biology. 2017;1520:85–106.

19. Sharma V, Wakode S, Kumar H. Structure–and ligand-based drug design: concepts, approaches, and challenges. Chemoinformatics and Bioinformatics in the Pharmaceutical Sciences. 2021 Jan 1;27–53.

20. Ewing TJA, Makino S, Skillman AG, Kuntz ID. DOCK 4.0: Search strategies for automated molecular docking of flexible molecule databases. J Comput Aided Mol Des. 2001;15(5):411–28.

21. Veselovsky A, Disorders AIDTI, 2003 undefined. Strategy of computer-aided drug design. ingentaconnect.com [Internet]. [cited 2022 Dec 28]; Available from: www.ingentaconnect.com/content/ben/cdtid/2003/00000003/00000001/art00005

22. Acharya C, Coop A, design JP… aided drug, 2011 undefined. Recent advances in ligand-based drug design: relevance and utility of the conformationally sampled pharmacophore approach. ingentaconnect.com [Internet]. [cited 2022 Dec 28]; Available from: www.ingentaconnect.com/content/ben/cad/2011/00000007/00000001/art00002

23. Pérez-Castillo Y, Lazar C, Taminau J, Froeyen M, Cabrera-Pérez MÁ, Nowé A. GA(M)E-QSAR: A novel, fully automatic genetic-algorithm-(meta)-ensembles approach for binary classification in ligand-based drug design. J Chem Inf Model [Internet]. 2012 Sep 24 [cited 2022 Dec 28];52(9):2366–86. Available from: pubs.acs.org/doi/abs/10.1021/ci300146h

24. Kubinyi H, Folkers G, Martin Y. 3D QSAR in Drug Design: Volume 2: Ligand-Protein Interactions and Molecular Similarity [Internet]. 1998 [cited 2022 Dec 28]. Available from: books.google.com/books?hl= en&lr=&id=7JbSM718LX8C&oi=fnd&pg=PA1&dq=Ghose,+A.K.%3B+Wendoloski,+J.J.+In+ 3D+QSAR+in+Drug+Design%3B+Kubinyi,+H.+Ed.%3B+Kluwer+Academic+Publishers:+Great+ Britain,+1998,+Vol.+3,+pp.+253-271.&ots=qP-_SJW0mq&sig=6cpaoOlmvq_iCRv88sKeKkxfrlA

25. Faver J, Ucisik M, … WYA medicinal chemistry, 2013 undefined. Computer-aided drug design: using numbers to your advantage. ACS Publications [Internet]. 2013 Sep 12 [cited 2022 Dec 28];26(9):47. Available from: pubs.acs.org/doi/abs/10.1021/ml4002634

26. Mccarthy J. What is artificial intelligence? 2007 [cited 2022 Dec 28]; Available from: faculty.otterbein. edu/dstucki/inst4200/whatisai.pdf

27. Kaul V, Enslin S, endoscopy SGG, 2020 undefined. History of artificial intelligence in medicine. Elsevier [Internet]. [cited 2022 Dec 30]; Available from: www.sciencedirect.com/science/article/pii/ S0016510720344667

28. Bellini V, Cascella M, Cutugno F, … MRABM, 2022 undefined. Understanding basic principles of artificial intelligence: a practical guide for intensivists. ncbi.nlm.nih.gov [Internet]. [cited 2022 Dec 30]; Available from: www.ncbi.nlm.nih.gov/pmc/articles/PMC9686179/

29. García CG, … ENV… JOI, 2019 undefined. A review of artificial intelligence in the internet of things. digibuo.uniovi.es [Internet]. 2019 [cited 2022 Dec 30];5(4):9. Available from:digibuo.uniovi.es/ dspace/bitstream/handle/10651/52310/A%20Review.pdf?sequence=1

30. Deng L, Liu Y. Deep learning in natural language processing [Internet]. 2018 [cited 2022 Dec 30]. Available from: books.google.com/books?hl=en&lr=&id=y_lcDwAAQBAJ&oi=fnd&pg= PR5&dq=Natural+learning+processing+in+artificial+intelligence&ots=aiwrb2RUgd&sig= aVcTj5svms4SHQfSeM3pMSmNECw

31. Dernoncourt F, Paris ESPR, 2011 undefined. Fuzzy logic: between human reasoning and artificial intelligence. researchgate.net [Internet]. 2011 [cited 2022 Dec 30]; Available from:www.resea rchgate.net/profile/FranckDernoncourt/publication/235333084_Fuzzy_logic_between_human_ reasoning_and_artificial_intelligence/links/5444075d0cf2a6a049ab071e/Fuzzy-logic-between- human-reasoning-and-artificial-intelligence.pdf

32. Mahfouf M, Abbod M, medicine DLA intelligence in, 2001 undefined. A survey of fuzzy logic monitoring and control utilisation in medicine. Elsevier [Internet]. [cited 2022 Dec 30]; Available from: www.sciencedirect.com/science/article/pii/S0933365700000725

33. Rhoads DD. Computer vision and artificial intelligence are emerging diagnostic tools for the clinical microbiologist. J Clin Microbiol. 2020 Jun 1;58(6).

34. Das N, Topalovic M, in WJC opinion, 2018 undefined. Artificial intelligence in diagnosis of obstructive lung disease: current status and future potential. ingentaconnect.com [Internet]. [cited 2022 Dec 31]; Available from: www.ingentaconnect.com/content/wk/mcp/2018/00000024/00000002/art00004

35. Gu J, Wang Z, Kuen J, Ma L, Shahroudy A, recognition BSP, et al. Recent advances in convolutional neural networks. Elsevier [Internet]. [cited 2022 Dec 31]; Available from: www.sciencedirect.com/scie nce/article/pii/S0031320317304120

36. Yu Y, Si X, Hu C, computation JZN, 2019 undefined. A review of recurrent neural networks: LSTM cells and network architectures. direct.mit.edu [Internet]. 2019 [cited 2022 Dec 31]; Available from: direct. mit.edu/neco/article-abstract/31/7/1235/8500

37. Liu P ran, Lu L, Zhang J yao, Huo T tong, Liu S xiang, Ye Z wei. Application of Artificial Intelligence in Medicine: An Overview. Curr Med Sci. 2021 Dec 1;41(6):1105–15.

38. Mak K, today MPD discovery, 2019 undefined. Artificial intelligence in drug development: present status and future prospects. Elsevier [Internet]. 2019 [cited 2023 Jan 2];24. Available from: www. sciencedirect.com/science/article/pii/S1359644618300916

39. Hessler G, Molecules KB, 2018 undefined. Artificial intelligence in drug design. mdpi.com [Internet]. [cited 2023 Jan 1]; Available from: www.mdpi.com/346666

40. Shen J, Technologies CNDDT, 2019 undefined. Molecular property prediction: recent trends in the era of artificial intelligence. Elsevier [Internet]. [cited 2023 Jan 1]; Available from: www.sciencedirect. com/science/article/pii/S1740674920300032

41. Svetnik V, Liaw A, Tong C, Christopher Culberson J, Sheridan RP, Feuston BP. Random Forest: A Classification and Regression Tool for Compound Classification and QSAR Modeling. J Chem Inf Comput Sci. 2003 Nov;43(6):1947–58.

42. Gardiner L, Carrieri A, Wilshaw J, reports SCS, 2020 undefined. Using human in vitro transcriptome analysis to build trustworthy machine learning models for prediction of animal drug toxicity. nature. com [Internet]. [cited 2023 Jan 1]; Available from: www.nature.com/articles/s41598-020-66481-0

43. Limongelli V, Marinelli L, Cosconati S, Braun HA, Schmidt B, Novellino E. Ensemble-docking approach on BACE-1: pharmacophore perception and guidelines for drug design. Wiley Online Library [Internet]. 2007 May 14 [cited 2023 Jan 1];2(5):667–78. Available from: chemistry-europe. onlinelibrary.wiley.com/doi/abs/10.1002/cmdc.200600314

44. Merk D, Friedrich L, Grisoni F, Schneider G. De Novo Design of Bioactive Small Molecules by Artificial Intelligence. Wiley Online Library [Internet]. 2018 Jan 1 [cited 2023 Jan2];37(1). Available from: onlinelibrary.wiley.com/doi/abs/10.1002/minf.201700153

45. Bung N, Krishnan SR, Bulusu G, Roy A. De novo design of new chemical entities for SARS-CoV-2 using artificial intelligence. Future Med Chem. 2021 Mar 1;13(6):575–85.

46. Discovery HSEO on D, 2020 undefined. The chemical probe–scopes, limitations and challenges. Taylor & Francis [Internet]. 2020 Dec 1 [cited 2023 Jan 2];15(12):1365–7. Available from: www.tand fonline.com/doi/abs/10.1080/17460441.2020.1781086

47. Garbaccio R, biology EPC chemical, 2016 undefined. The impact of chemical probes in drug discovery: a pharmaceutical industry perspective. Elsevier [Internet]. [cited 2023 Jan 2]; Available from: www.sciencedirect.com/science/article/pii/S2451945615004729

48. Bunnage M, Chekler E, biology LJN chemical, 2013 undefined. Target validation using chemical probes. nature.com [Internet]. [cited 2022 Dec 25]; Available from: www.nature.com/articles/nchem bio.1197

49. Singh N, Chaput L, bioinformatics BVB in, 2021 undefined. Virtual screening web servers: designing chemical probes and drug candidates in the cyberspace. academic.oup.com [Internet]. [cited 2023 Jan 2]; Available from: academic.oup.com/bib/article-abstract/22/2/1790/5809605

50. Letters LJAMC, 2018 undefined. Emerging utility of fluorosulfate chemical probes. ACS Publications [Internet]. [cited 2023 Jan 3]; Available from: pubs.acs.org/doi/abs/10.1021/acsmedchemlett.8b00276

51. Arrowsmith C, Audia J, Austin C, … JBN chemical, 2015 undefined. The promise and peril of chemical probes. nature.com [Internet]. [cited 2023 Jan 3]; Available from: www.nature.com/articles/nchem bio.1867

52. Trump RP, Bresciani S, Cooper AWJ, Tellam JP, Wojno J, Blaikley J, et al. Optimized chemical probes for REV-ERBα. J Med Chem. 2013 Jun 13;56(11):4729–37.

53. Hang HC, Geutjes EJ, Grotenbreg G, Pollington AM, Bijlmakers MJ, Ploegh HL. Chemical probes for the rapid detection of fatty-acylated proteins in mammalian cells. J Am Chem Soc. 2007 Mar 14;129(10):2744–5.

54. Santagata S, Mendillo ML, Tang YC, Subramanian A, Perley CC, Roche SP, et al. Tight coordination of protein translation and HSF1 activation supports the anabolic malignant state. Science (1979). 2013;341(6143).

55. Stanton B, Peng L, Maloof N, … KNN chemical, 2009 undefined. A small molecule that binds Hedgehog and blocks its signaling in human cells. nature.com [Internet]. [cited 2023 Jan 3]; Available from: www.nature.com/articles/nchembio.142

56. Reinke A, design JGC biology & drug, 2011 undefined. Insight into amyloid structure using chemical probes. Wiley Online Library [Internet]. [cited 2023 Jan 3]; Available from: onlinelibrary.wiley.com/ doi/abs/10.1111/j.1747-0285.2011.01110.x

57. Carlson E, May J, biology LKC&, 2006 undefined. Chemical probes of UDP-galactopyranose mutase. Elsevier [Internet]. [cited 2023 Jan 3]; Available from: www.sciencedirect.com/science/article/pii/ S1074552106002171

58. Chang T, Gonzalez F, and DWA of B, 1994 undefined. Evaluation of triacetyloleandomycin, α-nasymphthoflavone and diethyldithiocarbamate as selective chemical probes for inhibition of human cytochromes P450. Elsevier [Internet]. [cited 2023 Jan3];Available from: www.sciencedirect.com/scie nce/article/abs/pii/S0003986184712598

59. Yoshida M, Horinouchi S, Beppu T. Trichostatin A and trapoxin: Novel chemical probes for the role of histone acetylation in chromatin structure and function. BioEssays. 1995;17(5):423–30.

60. Levine RL, Garland D, Oliver CN, Amici A, Climent I, Lenz AG, et al. Determination of carbonyl content in oxidatively modified proteins. Elsevier [Internet]. 1990 [cited 2023 Jan 3]; Available from: www. sciencedirect.com/science/article/pii/007668799086141H

61. Lenz AG, Costabel U, Shaltiel S, Levine RL. Determination of carbonyl groups in oxidatively modified proteins by reduction with tritiated sodium borohydride. Anal Biochem. 1989;177(2):419–25.

62. Hsu KL, Tsuboi K, Whitby LR, Speers AE, Pugh H, Inloes J, et al. Development and optimization of piperidyl-1,2,3-triazole ureas as selective chemical probes of endocannabinoid biosynthesis. J Med Chem. 2013 Nov 14;56(21):8257–69.

63. Lomant A, biology GFJ of molecular, 1976 undefined. Chemical probes of extended biological structures: synthesis and properties of the cleavable protein cross-linking reagent dithiobis (succinimidyl propionate). Elsevier [Internet]. [cited 2023 Jan 3]; Available from: www.sciencedirect.com/science/article/abs/pii/0022283676900115

64. Klomsiri C, Nelson K, Bechtold E, … LSM in, 2010 undefined. Use of dimedone-based chemical probes for sulfenic acid detection: evaluation of conditions affecting probe incorporation into redox-sensitive proteins. Elsevier [Internet]. [cited 2023 Jan 3]; Available from: www.sciencedirect.com/science/article/pii/S0076687910730032

65. Myers RA, Cruz LJ, Rivier JE, Olivera BM. Conus Peptides as Chemical Probes for Receptors and Ion Channels. Chem Rev. 1993;93(5):1923–36.

66. Němec V, Schwalm M, Reviews SMCS, 2022 undefined. PROTAC degraders as chemical probes for studying target biology and target validation. pubs.rsc.org [Internet]. [cited 2023 Jan 6]; Available from: pubs.rsc.org/en/content/articlehtml/2022/cs/d2cs00478j

67. Kroschwald S, Maharana S, Matters AS, 2017 undefined. Hexanediol: a chemical probe to investigate the material properties of membrane-less compartments. sosjournals.s3.amazonaws.com [Internet]. 2017 [cited 2023 Jan 6]; Available from: sosjournals.s3.amazonaws.com/Qm4Tb429k12BV4sA.pdf

68. Shi Y, biology KCR, 2021 undefined. Parallel evaluation of nucleophilic and electrophilic chemical probes for sulfenic acid: Reactivity, selectivity and biocompatibility. Elsevier [Internet]. [cited 2023 Jan 6]; Available from: www.sciencedirect.com/science/article/pii/S2213231721002317

69. Warchal SJ, Unciti-Broceta A, Carragher NO. Next-generation phenotypic screening. Future Med Chem. 2016 Jul 1;8(11):1331–47.

70. Zheng W, Thorne N, today JMD discovery, 2013 undefined. Phenotypic screens as a renewed approach for drug discovery. Elsevier [Internet]. [cited 2023 Jan 4]; Available from: www.sciencedirect.com/science/article/pii/S135964461300202X

71. Jenmalm Jensen A, Cornella Taracido I. Affinity-Based Chemoproteomics for Target Identification. 2019 Oct 14;25–49.

72. Perkins D, Pappin D, … DCE, 1999 undefined. Probability-based protein identification by searching sequence databases using mass spectrometry data. Wiley Online Library [Internet]. [cited 2023 Jan 4]; Available from: analyticalsciencejournals.onlinelibrary.wiley.com/doi/abs/10.1002/(SICI)1522-2683(19991201)20:18%3C3551::AID-ELPS3551%3E3.0.CO;2-2

73. Urgen Cox J, Neuhauser N, Michalski A, Scheltema RA, Olsen J v, Mann M. Andromeda: a peptide search engine integrated into the MaxQuant environment. ACS Publications [Internet]. 2011 Apr 1 [cited 2023 Jan 4];10(4):1794–805. Available from: pubs.acs.org/doi/abs/10.1021/pr101065j

74. Wagner B, biology SSC chemical, 2016 undefined. The power of sophisticated phenotypic screening and modern mechanism-of-action methods. Elsevier [Internet]. [cited 2023 Jan 5]; Available from: www.sciencedirect.com/science/article/pii/S2451945615004675

75. Seashore-Ludlow B, Rees M, Cheah J, discovery MCC, 2015 undefined. Harnessing Connectivity in a Large-Scale Small-Molecule Sensitivity DatasetHarnessing Connectivity in a Sensitivity Dataset. AACR [Internet]. [cited 2023 Jan 5]; Available from: aacrjournals.org/cancerdiscovery/article-abstract/5/11/1210/4735

76. Carpenter AE, Jones TR, Lamprecht MR, Clarke C, Kang IH, Friman O, et al. CellProfiler: Image analysis software for identifying and quantifying cell phenotypes. Genome Biol. 2006 Oct 31;7(10).

77. Zhang M, Serono E, Barsoum J, Therapeutics R, Kirshner JR, He S, et al. Elesclomol induces cancer cell apoptosis through oxidative stress. AACR [Internet]. 2008 [cited 2023 Jan 9]; Available from: aacrjournals.org/mct/article-abstract/7/8/2319/93240

78. Nature HL, 2012 undefined. Drug candidates derailed in case of mistaken identity. nature.com [Internet]. [cited 2023 Jan 9]; Available from: www.nature.com/articles/483519a

79. Espina MJ, Ahmed CMS, Bernardini A, Adeleke E, Yadegari Z, Arelli P, et al. Development and phenotypic screening of an ethyl methane sulfonate mutant population in Soybean. Front Plant Sci. 2018 Mar 29;9.

80. Tozer GM, Prise VE, Wilson J, Locke RJ, Vojnovic B, Stratford MRL, et al. Combretastatin A-4 phosphate as a tumor vascular-targeting agent: early effects in tumors and normal tissues. AACR [Internet]. 1999 [cited 2023 Jan 9];59:1626–34. Available from: aacrjournals.org/cancerres/article-abstract/59/7/1626/505976

81. Grzelak E, Choules M, Gao W, Cai G, … BWTJ of, 2019 undefined. Strategies in anti-Mycobacterium tuberculosis drug discovery based on phenotypic screening. nature.com [Internet]. [cited 2023 Jan 9]; Available from: www.nature.com/articles/s41429-019-0205-9

82. Grozinger C, Chao E, … HBJ of B, 2001 undefined. Identification of a class of small molecule inhibitors of the sirtuin family of NAD-dependent deacetylases by phenotypic screening. ASBMB [Internet]. [cited 2023 Jan 9]; Available from: www.jbc.org/article/S0021-9258(20)74143-8/abstract

83. Chiang T, Sage C le, Larrieu D, reports MDS, 2016 undefined. CRISPR-Cas9D10A nickase-based genotypic and phenotypic screening to enhance genome editing. nature.com [Internet]. [cited 2023 Jan 9]; Available from: www.nature.com/articles/srep24356

84. Manjunatha U, chemistry PSB& medicinal, 2015 undefined. Perspective: Challenges and opportunities in TB drug discovery from phenotypic screening. Elsevier [Internet]. [cited 2023 Jan 9]; Available from: www.sciencedirect.com/science/article/pii/S0968089614008797

85. Abdulla MH, Ruelas DS, Wolff B, Snedecor J, Lim KC, Xu F, et al. Drug discovery for schistosomiasis: Hit and lead compounds identified in a library of known drugs by medium-throughput phenotypic screening. PLoS Negl Trop Dis. 2009 Jul;3(7).

86. Isaacs JT. The long and winding road for the development of tasquinimod as an oral second-generation quinoline-3-carboxamide antiangiogenic drug for the treatment of prostate cancer. Expert Opin Investig Drugs. 2010 Oct;19(10):1235–43.

87. Stuart J, antimicrobial MLVHI journal of, 2010 undefined. Guideline for phenotypic screening and confirmation of carbapenemases in Enterobacteriaceae. Elsevier [Internet]. 2010 [cited 2023 Jan 9];36(3):205. Available from: www.sciencedirect.com/science/article/pii/S0924857910002323

88. Jin Z, Ma J, Zhu G, Zhang H. Discovery of novel anti-cryptosporidial activities from natural products by in vitro high-throughput phenotypic screening. Front Microbiol. 2019;10(AUG).

89. Guirouilh-Barbat J, Antony S, therapeutics YPM cancer, 2009 undefined. Zalypsis (PM00104) is a potent inducer of γ-H2AX foci and reveals the importance of the C ring of trabectedin for transcription-coupled repair inhibition. AACR [Internet]. [cited 2023 Jan 9]; Available from: aacrjournals.org/mct/article-abstract/8/7/2007/93569

90. Crooke S, Liang X, Baker B, Chemistry RC of B, 2021 undefined. Antisense technology: A review. ASBMB [Internet]. [cited 2023 Jan 5]; Available from: www.jbc.org/article/S0021-9258(21)00189-7/abstract

91. Wang S, Sim T, Kim Y, biology YCC opinion in chemical, 2004 undefined. Tools for target identification and validation. Elsevier [Internet]. [cited 2023 Jan 5]; Available from: www.sciencedirect.com/science/article/pii/S1367593104000778

92. Rigo F, Seth PP, Bennett CF. Antisense oligonucleotide-based therapies for diseases caused by pre-mRNA processing defects. Adv Exp Med Biol. 2014;825:303–52.

93. Zamecnik PC, Stephenson ML. Inhibition of Rous sarcoma virus replication and cell transformation by a specific oligodeoxynucleotide. Proc Natl Acad Sci U S A. 1978;75(1):280–4.

94. Polymenidou M, Lagier-Tourenne C, research KHB, 2012 undefined. Misregulated RNA processing in amyotrophic lateral sclerosis. Elsevier [Internet]. [cited 2023 Jan 5]; Available from: www.sciencedirect.com/science/article/pii/S0006899312003915

95. Lentz J, Pan F, Deininger P, Research/Fundamental BKM, 2007 undefined. Ush1c216A knock-in mouse survives Katrina. Elsevier [Internet]. [cited 2023 Jan 5]; Available from: www.sciencedirect.com/science/article/pii/S0027510706003204

96. Bo X, Bioinformatics SW, 2005 undefined. TargetFinder: a software for antisense oligonucleotide target site selection based on MAST and secondary structures of target mRNA. academic.oup.com [Internet]. [cited 2023 Jan 5]; Available from: academic.oup.com/bioinformatics/article-abstract/21/8/1401/250353

97. Chiba S, Lim K, Sheri N, … SAN acids, 2021 undefined. eSkip-Finder: A machine learning-based web application and database to identify the optimal sequences of antisense oligonucleotides for exon skipping. academic.oup.com [Internet]. [cited 2023 Jan 5]; Available from: academic.oup.com/nar/article-abstract/49/W1/W193/6294209

98. Sciabola S, Xi H, Cruz D, Cao Q, Lawrence C, Zhang T, et al. PFRED: A computational platform for siRNA and antisense oligonucleotides design. PLoS One. 2021 Jan 1;16(1 January).

99. biology MMC opinion in chemical, 2004 undefined. RNA interference and chemically modified small interfering RNAs. Elsevier [Internet]. [cited 2023 Jan 5]; Available from: www.sciencedirect.com/science/article/pii/S1367593104001383

100. Setten R, Rossi J, Discovery SHNRD, 2019 undefined. The current state and future directions of RNAi-based therapeutics. nature.com [Internet]. [cited 2023 Jan 6]; Available from: www.nature.com/articles/s41573-019-0017-4

101. Lee H, Zhou K, Smith A, … CNN acids, 2013 undefined. Differential roles of human Dicer-binding proteins TRBP and PACT in small RNA processing. academic.oup.com [Internet]. [cited 2023 Jan 6]; Available from: academic.oup.com/nar/article-abstract/41/13/6568/1125588

102. Morris K, Chan S, Jacobsen S, Science DL, 2004 undefined. Small interfering RNA-induced transcriptional gene silencing in human cells. science.org [Internet]. 2004 Aug 27 [cited 2023 Jan 6];305(5688):1289–92. Available from: www.science.org/doi/abs/10.1126/science.1101372

103. Shao DD, Tsherniak A, Gopal S, Weir BA, Tamayo P, Stransky N, et al. ATARiS: computational quantification of gene suppression phenotypes from multisample RNAi screens. genome.cshlp.org [Internet]. [cited 2023 Jan 6]; Available from: genome.cshlp.org/content/23/4/665.short

104. Azad RF, Driver VB, Tanaka K, Crooke RM, Anderson KP. Antiviral activity of a phosphorothioate oligonucleotide complementary to RNA of the human cytomegalovirus major immediate-early region. Antimicrob Agents Chemother. 1993;37(9):1945–54.

105. Kim J, Hu C, Moufawad El Achkar C, Black LE, Douville J, Larson A, et al. Patient-Customized Oligonucleotide Therapy for a Rare Genetic Disease. New England Journal of Medicine. 2019 Oct 24;381(17):1644–52.

106. Adams D, Gonzalez-Duarte A, O'Riordan WD, Yang CC, Ueda M, Kristen A v., et al. Patisiran, an RNAi therapeutic, for hereditary transthyretin amyloidosis. diposit.ub.edu [Internet]. 2018 Jul 5 [cited 2023 Jan 10];379(1):11–21. Available from: diposit.ub.edu/dspace/handle/2445/138257

107. Frank DE, Schnell FJ, Akana C, El-Husayni SH, Desjardins CA, Morgan J, et al. Increased dystrophin production with golodirsen in patients with Duchenne muscular dystrophy. AAN Enterprises [Internet]. 2020 [cited 2023 Jan 10]; Available from: n.neurology.org/content/94/21/e2270.abstract

108. Hua Y, Liu YH, Sahashi K, Rigo F, Bennett CF, Krainer AR. Motor neuron cell-nonautonomous rescue of spinal muscular atrophy phenotypes in mild and severe transgenic mouse models. genesdev.cshlp.org [Internet]. [cited 2023 Jan 10]; Available from: genesdev.cshlp.org/content/29/3/288.short

109. Hinkel R, Ramanujam D, Kaczmarek V, Howe A, Klett K, Beck C, et al. AntimiR-21 Prevents Myocardial Dysfunction in a Pig Model of Ischemia/Reperfusion Injury. J Am Coll Cardiol. 2020 Apr 21;75(15):1788–800.

110. Roychowdhury D, oncology MLS in, 2003 undefined. Antisense therapy directed to protein kinase C-alpha (Affinitak, LY900003/ISIS 3521): potential role in breast cancer. Elsevier [Internet]. [cited 2023 Jan 10]; Available from: www.sciencedirect.com/science/article/pii/S0093775403701432

111. Steinberg JL, Mendelson DS, Block H, Green SB, Shu VS, Parker K, et al. Phase I study of LErafAON-ETU, an easy-to-use formulation of liiposome entrapped c-raf antisense oligonucleotide, in advanced cancer patients. Journal of Clinical Oncology. 2005 Jun;23(16_suppl):3214–3214.

112. Chi K, Eisenhauer E, Fazli L, … EJJ of the, 2005 undefined. A phase I pharmacokinetic and pharmacodynamic study of OGX-011, a 2'-methoxyethyl antisense oligonucleotide to clusterin, in patients with localized prostate. academic.oup.com [Internet]. [cited 2023 Jan 10]; Available from: academic.oup.com/jnci/article-abstract/97/17/1287/2521380

113. Wada F, Yamamoto T, Acids TK… TN, 2021 undefined. Drug discovery and development scheme for liver-targeting bridged nucleic acid antisense oligonucleotides. Elsevier [Internet]. [cited 2023 Jan 10]; Available from: www.sciencedirect.com/science/article/pii/S2162253121002511

114. Peddi V, Ratner L, Cooper M, Gaber O, … SF, 2014 undefined. Treatment with QPI-1002, a Short Interfering (SI) RNA for the Prophylaxis of Delayed Graft Function.: Abstract# 2967. journals.lww.com [Internet]. [cited 2023 Jan 10]; Available from: journals.lww.com/transplantjournal/Fulltext/2014/07151/Treatment_with_QPI_1002,_a_Short_Interfering__SI_.467.aspx

115. Phase II Open Label Multicenter Study For Age Related Macular Degeneration Comparing PF-04523655 Versus Lucentis In The Treatment Of Subjects With CNV (MONET Study).–Full Text View–ClinicalTrials.gov [Internet]. [cited 2023 Jan 10]. Available from: clinicaltrials.gov/ct2/show/NCT00713518?term=PF-04523655

116. Tadin-Strapps M, Peterson LB, Cumiskey AM, Rosa RL, Mendoza VH, Castro-Perez J, et al. siRNA-induced liver ApoB knockdown lowers serum LDL-cholesterol in a mouse model with human-like serum lipids. J Lipid Res [Internet]. 2011 Jun 1 [cited 2023 Jan 10];52(6):1084–97. Available from: www.jlr.org/article/S0022227520313274/fulltext

7 Machine Learning and Data Mining

Uses and Challenges in Bioinformatics

Muhasina K.M., Puja Ghosh, Akey Krishna Swaroop, Jubie Selvaraj, Bhupendra Gopalbhai Prajapati, B. Duraiswamy, Dhanabal S. Palaniswamy

1 INTRODUCTION

Improvements and developments in machine learning and data mining have witnessed revolutionary changes over the last few decades in the biomedical research fields. The field of biomedicine is still in its infancy and is considered a subject that crosses disciplines that were developed from theories and approaches to life science, general medicine, and biology. Application of biology and engineering is the primary task in methods for examining and resolving issues in biological science, in particular in the field of medicine (Li et al., 2005). Academic research and the development of new biomedical information based on biomedicine, nanotechnology, gene chips, novel materials, and medical imaging technology. The current biomedical system evolved as a result of the development of systems biology and the social-psycho-biomedical paradigm. (Santosh et al., 2016). It is a crucial area of engineering that is linked to raising the bar for medical diagnosis and enhancing human health in the twenty-first century (Kharya, 2012). It is directly linked to formation of the biotechnology sector at the industry level. In addition, due to the multidisciplinary integration of biomedical information and computer science, bioinformatics is a brand-new, exciting field that is emerging and developing quickly as a result of the involvement of biotechnology and data analysis techniques.

On the other hand, because of the rapid growth of biotechnology over the course of history, the exponential growth rate of biomedical data over time can vary significantly depending on research and application fields such as micromolecular level (protein interactions, gene functions); level of biological tissue (brain connection map, MRIs, etc.); clinical patient level (intensive care unit, electronic medical record, etc.); and full population level (medical message board, social media, etc.) (Hsieh et al., 2012). It is an undeniable fact that handling biomedical data with such qualities is significantly more difficult than normal data due to growing speed and heterogeneous structure (Banaee et al., 2013). Consequently, it is desirable to develop more potent methodologies theoretically and powerful tools for analysing and collecting valuable points from the complicated bio-data indicated above. The fundamental ideas behind the wide range of scalable and effective techniques and tools, including but not limited to prediction, classification, outlier detection, grouping, creation of deep network architecture, sequential processing, frequent item query, analysis, and visualisation of spatial and temporal data.

Data mining methods with core methodology known as machine learning are used to extract usable knowledge from original raw data. They have been widely used and shown to be successful through creating models, formulating hypotheses, and classifying data, discovering linked rules, and by clustering. In contrast, deep learning is a relatively modern idea and framework that is significantly better at representing features at an abstract level than ordinary machine learning. In order to properly bridge the two domains and mine biological data, it is necessary to combine advanced data analysis techniques with bioinformatics. This is where the challenge arises. In this chapter the basic concept is offered with data analysis methods represented by machine learning and data mining techniques, which have been used in a variety of bioinformatics applications. The challenges while dealing with big data and the methods for bridging the gap between industry and academia are also covered.

2.1 MACHINE LEARNING

Since the beginning of time, people have used a wide range of tools to complete multiple roles more quickly. Human innovations have resulted in the development of machines and technologies that benefit human life in industry, computing, and travel. Machine learning (ML) is one such innnovation and is the systematic investigation of the mathematical models and tools that computer systems use to perform tasks without being specifically instructed as shown in Figure 7.1.

Machine learning is typically considered to be a branch of artificial intelligence. Today computers are able to make decisions on their own without any outside assistance support using ML algorithms. Finding valuable underlying patterns inside complex data allows for the making of such decisions (Sah, 2020). The importance of ML is the practise of using statistical methods to help computer systems "learn" and "improve" from data in order to make accurate predictions without pattern recognition (Aromolaran et al., 2021) as shown in Figure 7.2.

Machine learning techniques teach computers to manage data more productively. Sometimes even after obtaining the data, we are unable to elucidate or extrapolate the results. In that case, we apply ML. With the knowledge gained from these data, a range of application scenarios in the relevant disciplines can be built. For instance, the appropriate data in cybersecurity can be used to create an intelligent and automated cybersecurity system (Sarker et al., 2020). Using the pertinent mobile data, it is possible to create individualised, situationally smart mobile applications (Sarker et al., 2021). To lay the groundwork for practical applications, it is imperative to create tools for data management and procedures that can swiftly and shrewdly extract insights or valuable knowledge from data (Sarker, 2021). Industries including manufacturing, banking, healthcare, and countless

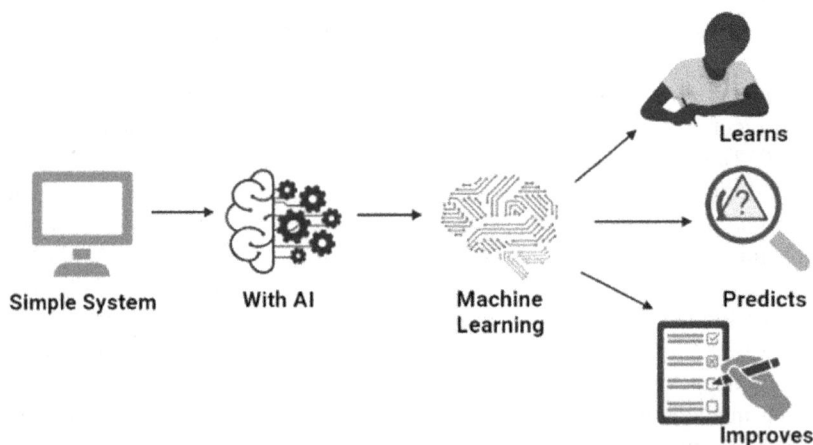

FIGURE 7.1 Introduction of machine learning.

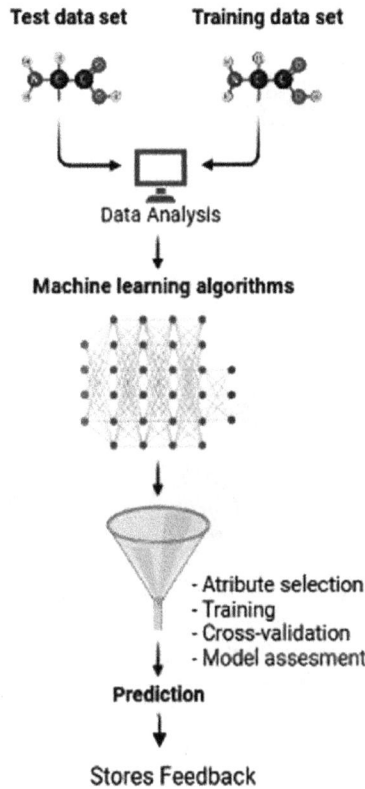

FIGURE 7.2   Importance of machine learning.

more have been transformed by ML. Machine learning has become a crucial component of contemporary industry. In several fields, such as speech recognition, medicine, computer vision, and email filtering, ML algorithms are used when it is difficult or impractical to develop conventional algorithms to perform the necessary tasks (Hu et al., 2020).

There are two major categories of ML: inductive and deductive. Deductive learning draws new knowledge from previously acquired facts and information. Through the extraction of rules and patterns from enormous data sets, inductive ML techniques are used to develop computer programmes (Singh et al., n.d.). However, simply making models is insufficient. You must also properly optimise and modify the model so that it produces correct results. Optimisation strategies entail fine-tuning the hyperparameters to achieve the best possible result. Machine learning is applied in all fields. It is utilised to give sophistication to static systems, and is employed to construct smart solutions using the information derived from data (Team, 2017). The basic ML model has four steps: environment, learning, repository, and execute (Figure 7.3). The main determinant of ML is the reliability of the data the system receives from the outside world. Learning is the conversion of external information into knowledge. Repository maintains numerous broad principles that guide a portion of the implementation activity. Because the environment offers all types of information for the learning system, the quality of information has a direct impact on whether learning is realised easily or in a disorganized manner (Wang et al., 2009).

Machine learning software makes it simple to use ML techniques. Rapid prototyping as well as tool capabilities introduced to various languages are made possible with the use of ML software (Team, 2017). A list of ML software is shown in Figure 7.4 and in Table 7.1.

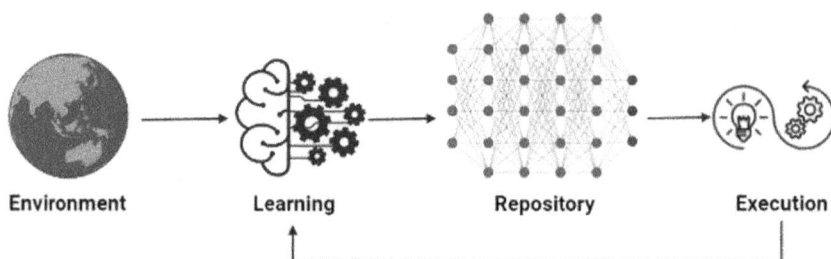

FIGURE 7.3   Basic model of machine learning.

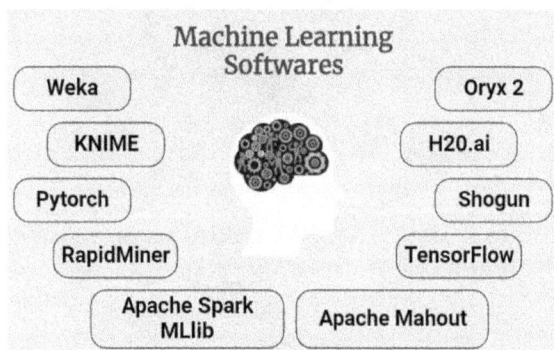

FIGURE 7.4   Different softwares for machine learning.

TABLE 7.1
Sources and ses of machine learning software

S. No	Software name	Source	Uses
1	TensorFlow	free and open-source software	Expertize in training and speculation of deep neural networks.
2	Shogun	free and open-source software	It assists various languages like R, Python, C++, Scala, Ruby, etc.
3	Apache Mahout	open-source software	Focused on collaborative filtering and its classification.
4	Apache Spark MLlib	Spark is a powerful data streaming platform	Provides a extensible machine learning platform with its several APIs that permit users to execute machine learning on real-time data.
5	Oryx 2	open-source software	It facilitates end to end model development for collaborative filtering, clustering operations, regression, and classification
6	H_2O.ai	fully open-source	H2O's deep learning platform provides an expandable multi-layer artificial neural network.
7	Pytorch	open-source software	PyTorch is a machine learning library assisted the Torch library, used for applications like computer vision and tongue processing.
8	RapidMiner	open-source software	RapidMiner facilitates an comprehensive and integrated environment for carrying out different tasks like data preparation, deep learning, machine learning, text mining, and predictive analytics.
9	Weka	open-source software	Few of these tools are including classification, clustering, regression, visualization as well as data preparation.
10	Knime	open-source data analytics	To perform data analysis, modeling, and visualization without the need for extensive programming.
11	Keras	open-source neural network	It is popular for its ease of use, modularity, and speed. Therefore, it can be used for rapid prototyping as well as fast experimentation.

2.2   REVIEW OF RESEARCH USING MACHINE LEARNING

Data are indeed the new DNA of the twenty-first century, holding significant knowledge, ideas, and possibilities, and they are now an essential part of all organisms that are powered by data. In many different industries, including sciences, medicine, industry, education, corporate finance, cybersecurity, data governance, law enforcement, and advertising, information can be gathered from data to develop a variety of smart applications. Various applications are discussed below and shown in Figure 7.5.

A)   CLINICAL RESEARCH

For managing massive and diverse data sources, spotting complex and hidden patterns, and forecasting difficult events, multiple ML techniques are available. Although when ML is not really engaged, bias issues are crucial because gender, racial, and socioeconomic discrimination in clinical studies have such a long history. It is essential to actively handle ML's capacity to amplify and maintain bias in clinical studies, even without such research group knowledge. If bias is detected, it can often frequently be rectified and reduced; the worst-case situation is when a system containing hidden bias is applied to a fresh cohort with important outcomes. A moral obligation to investigate, this prospect is created by the fact that ML-enabled clinical studies might enhance the effectiveness and calibre of biomedical findings, may save lives and lessen pain for humans (Weissler et al., 2021).

B)   ECOSYSTEM SERVICE RESEARCH

The investigation of complicated processes that entail connections involving biodiversity, human impacts, and the biophysical environment is a component of ecosystem service (ES) research. In ES studies, ML is utilised as a descriptive tool, where characteristics of mechanisation need to process

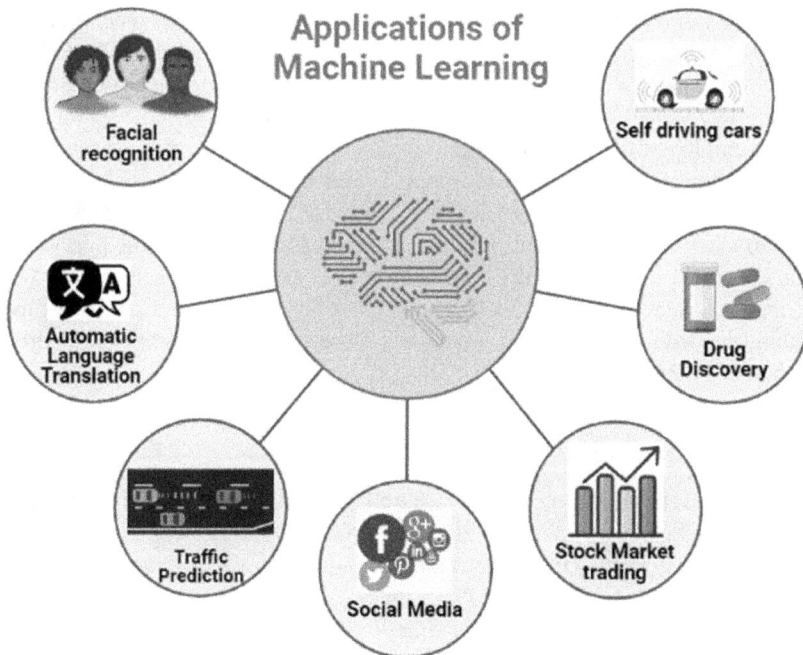

FIGURE 7.5   Applications of machine learning.

massive amounts of complicated data, and as a parameter estimation tool, where precise judgments regarding ES may be formed (Scowen et al., 2021).

c) HEALTHCARE

Machine learning is a statistical tool for evaluating the model to information and 'learning' from data through training sets. Among the most popular ones of AI is machine learning. The neural network is an even more advanced type of ML, which has existed since the 1960s and has been widely employed in medical research and for addressing problems in the context of inputs, outputs, and variable weights or "features" that link inputs with outputs. Using population health, ML models are employed to identify the number of people at danger of illnesses or accidents, as well as hospital readmission. Machine learning algorithms in medical care might also be subjected to statistical bias, such as forecasting high chance of illness based on ethnicity or gender when such elements are not really responsible (Davenport & Kalakota, 2019).

d) TRAFFIC PREDICTION

To predict traffic flow, three ML ideas are used: DAN (neural network technique), DBN (Deep Belief Network, which is perhaps the most recognisable and effective method across all deep learning techniques), and RF (RANDOM FOREST, an ensemble technique used to handle continuous reaction and label values). Climate, temperature, area name, and day are all key factors in predicting traffic flow in a specific zone. A system's effectiveness can be evaluated via accurate, precise, RMSE, and MSE values (Ramchandra & Rajabhushanam, 2021).

e) AGRICULTURE

Agricultural activities are divided into three key categories: pre-harvesting, harvesting, and post-harvesting. Machine learning is cutting-edge technique utilised to tackle difficult agricultural problems and assist farmers in reducing losses. Today's requirement is for precise and personalised ML models that can operate rapidly, dynamically evaluate larger, more complicated data sets, and aid in the optimisation of agricultural operations such as categorisation, suggestions, and predictions. To increase the effectiveness of the classification approach a multilayer perceptron neural network design has been employed for separating high-quality vs. less-quality pepper seeds, which uses a variety of ML, computer vision, and convolution neural network (CNN) techniques. It classifies Cavendish bananas according to their grades by using a deep learning system and computer vision. The algorithm to categorise the bananas into multiple categories was built using Python, OpenCV, and Tensorflow (part of hand) utilised to analyse pre-harvesting variables (disease identification, plant illness, and pest detection). They also used to analyse harvesting methods (real-time fruit identification within tree, fruit categorization, outdoor fruit detection, date fruit classification, and fruit harvesting robots) (Meshram et al., 2021).

f) ML IN PERSONALISED MEDICINES

Using cutting-edge ML methods to conduct research on immune-mediated inflammatory illnesses clinically could aid in the development of personalised medicine strategies and better infection control. Recurrent neural networks (RNN) and convolutional neural networks (CNN), two common deep learning algorithms, are effective tools in the area of computer vision, whereby diagnostic imaging detection is extensively explored for illness assessment, prediction, and subtypes identification. Precision oncology has benefited significantly from deep learning approaches due to rapid recognition, subtype identification, early metastatic prediction, and clinical decision assistance. Obtaining a much more precise subpopulation characterisation based on their individual pathogenetic patterns is a useful method of providing individualised care. Genomes, metabolomics, immunophenotyping, as well as other kinds of information can all be used to derive patterns. Even though both supervised and unsupervised ML's clustering algorithms are available for dividing up complicated, high-dimensional data and supervised ML is the optimal tool for recognizing distinctive patterns. A growing number of studies have used ML techniques to separate patient subgroups and have demonstrated encouraging progress toward more individualised care. Applications of ML could be utilised to precisely assist the therapy of pathogenic infections and forecast the efficacy of drugs (Peng et al., 2021).

G) DRUG DEVELOPMENT

It is challenging to assess new drug targets using conventional techniques due to the abundance of molecular and biological data produced by contemporary high-throughput techniques. A platform to speed up drug discovery and direct clinical application, BANDIT (Bayesian machine learning technique), can also find targets for orphan drugs (Peng et al., 2021).

H) ACUTE ILLNESS DETECTION

Early diagnosis of acute disease is essential for starting therapy on time. Nevertheless, patients with serious illnesses like sepsis who are hospitalised to the emergency department (ED) or intensive care unit (ICU) frequently exhibit a variety of indications, rendering diagnosis difficult. There are various essential signs of severe illness that can be successfully imitated, such as "a fatigued appearance," "pale complexion and/or lips," "swollen cheeks," and "hanging eyelids." Using a neural transfer convolutional neural network (NT-CNN) for data augmentation, facial features that were altered to match an acute illness condition were employed to identify illness-related aspects and create a synthetic dataset of images of people who appeared to be in those state molecules. Then, a final, stacked CNN that identified people as healthy or acutely unwell was created by concatenating four different CNNs that were trained on various aspects of the facial photos. An algorithm using deep learning skills on synthetic dataset including the clinical totality of critical infection was capable of distinguishing passably between healthy and keenly ill individuals. The sensitivity and specificity revealed for every design comprise to the best designs in the task of binary classification and are based on the extracted features (Forte et al., 2021).

2.3   ML TECHNIQUES EMPLOYED IN BIOINFORMATICS (FIGURE 7.6.)

  a.   Dimensionality reduction
  b.   Clustering
  c.   Deep learning
  d   Classification/regression

A. DIMENSIONALITY REDUCTION

Automation, visual imaging, face detection, handwritten digit categorization, remote sensing, health research, genomics, and electronics are just a few of the fields where dimensionality reduction (DR) has been successfully applied to solve issues. Dimensionality reduction approaches have been used to maintain the same number of variables in the reaction environment while reducing the dimensionality of the reaction environment to the greatest extent feasible.

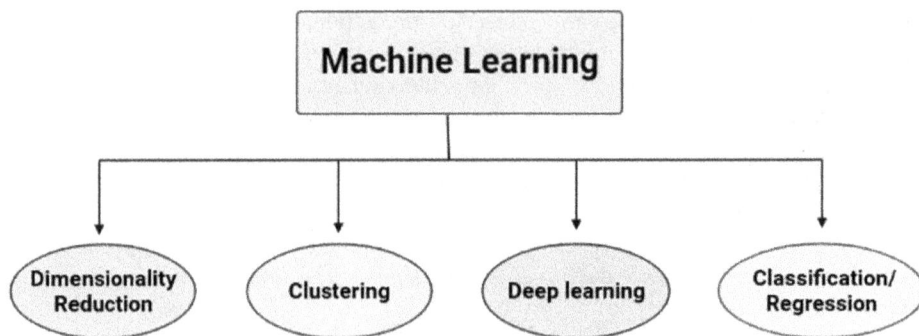

FIGURE 7.6   Techniques of machine leaning.

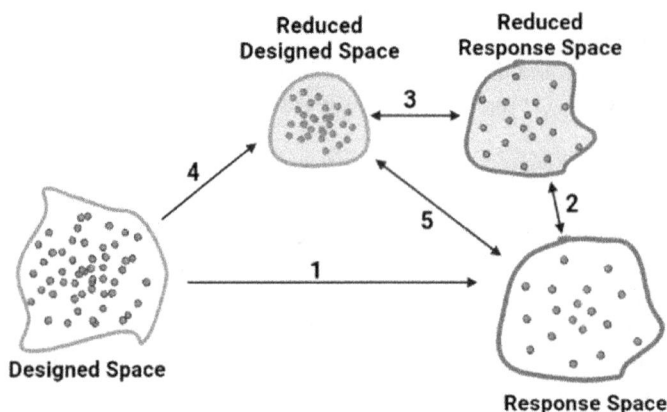

FIGURE 7.7   Applying the DR technique to response and design spaces.

By lowering the size of the two spaces, the technique not only resolves the non-uniqueness problem but also significantly lowers the computing cost. Our DR-based technique has a special quality that makes calculation simple while considering the issue's several more aspect. However, compared to previous approaches, ours requires less processing because the issue's dimensions are reduced. The key benefit of the DR technique is that it avoids employing adaptive or brute-force methods that are more commonly used. Even if it is obvious how important DR approaches are in lessening the severity of the non-uniqueness issue and the complexity of computing, it is still possible that the relationship between the reaction environment and the smaller design space will still be only somewhat several more (although much more manageable than that between the original design space and the response space) shown in Figure 7.7.

Once the best dimension for the condensed design space is chosen, the efficiency of this technique approaches one-to-one. While currently implemented through trial and error, additional stringent methods for enhancing this characteristic of the DR procedures should be taken into consideration in the future. Development of more sophisticated sampling approaches for minimising the number of simulations needed to collect training data would be beneficial to push the boundaries of this technique's usefulness with acceptable computing (Kiarashinejad et al., 2020).

B.   CLUSTERING

Applications for data analysis and data mining revolve around clustering. As sets of data expand and their attributes and interrelationships with other data change, it is very important to be able to

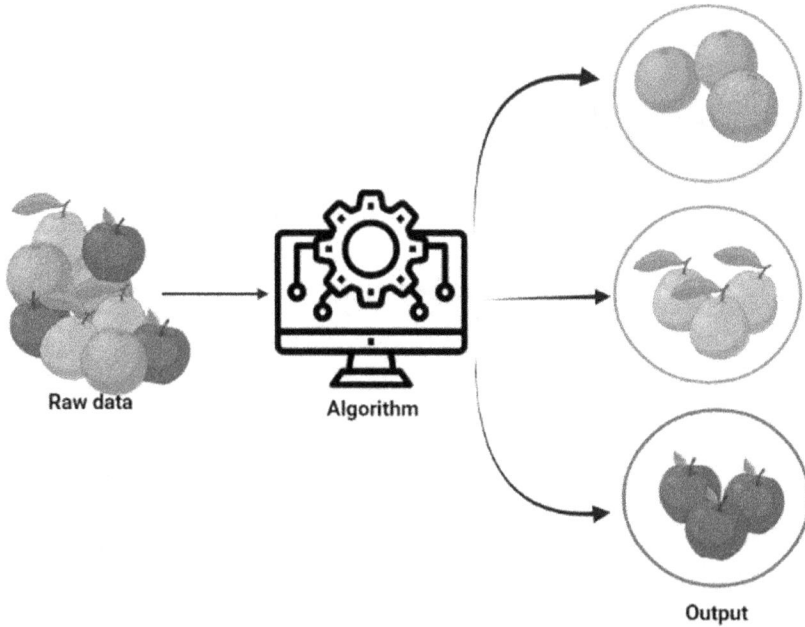

FIGURE 7.8   Working of the clustering algorithm.

identify highly associated regions of items when their number gets very vast. The distinction that any clustering "is a classification of the objects into groups based on a set of rules–it is neither true nor false" is noteworthy (Rai & Shubha, 2010).

Observing the differences and documenting the variations using several clustering programmes on the same dataset. The clustering approach inevitably produces a large number of false positives. Therefore, using various clustering algorithms and then voting to analyse the results can increase the likelihood that a data point belongs to a particular class. A data instance falls into the suspicious category if several clustering methods disagree about whether it belongs in the same class. Unsupervised ML uses the clustering approach to separate data into unlabeled yet separable points. Similar data instances are placed together into one cluster, and a metric of dissimilarity is used to distinguish the clusters. diversified strategy, to limit false positives, several clustering and classification methods, Voting process for the ultimate labelling. (*Clustering Based Semi-Supervised Machine Learning for DDoS Attack Classification | Elsevier Enhanced Reader*, n.d.). from the Figure 7.8, we can understand clustering.

c)   DEEP LEARNING

Deep learning models outperform standard data analysis methods and shallow ML techniques in several applications. Among these developments an important one was the development of artificial neural networks (ANNs) into deeper and deeper neural network topologies with improved learning capabilities, or deep learning. Deep neural networks frequently have multiple hidden layers that are arranged in layered network topologies. In addition, they frequently include more complex neurons than simple ANNs; rather than employing a straightforward activation function, they may instead involve sophisticated processes (such as convolutions) or numerous activations in a single neuron. Due to these features, deep neural networks may be fed with unprocessed data and automatically find the representations required for a given learning task. This is the essential capability of the networks, which is also referred to as deep learning (Janiesch et al., 2021).

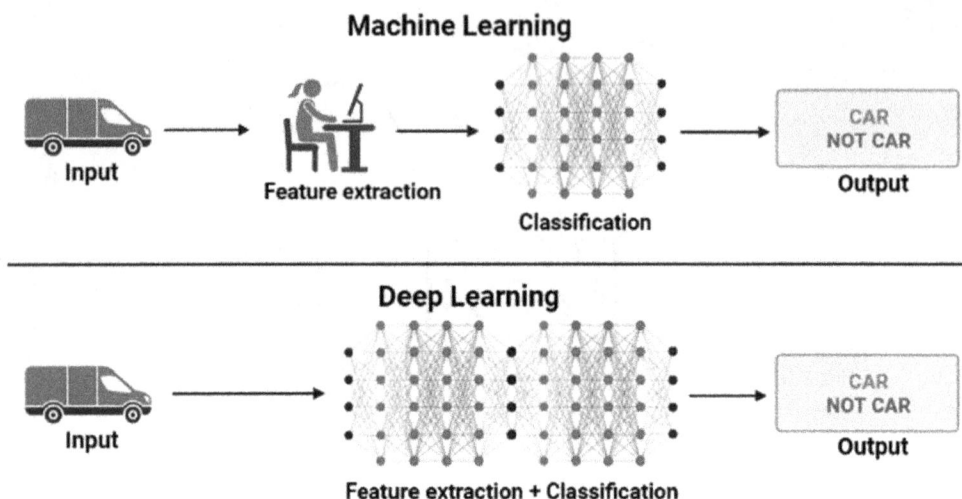

FIGURE 7.9 Pictorial representation of machine learning and deep learning.

Machine learning, which may be thought of as neural networks with many hidden layers, includes deep learning as a subfield. Deep learning techniques need a lot more training data than applications based on shallow learning. The performance of deep learning models is also significantly influenced by the network's structure.

The four primary fields of deep learning techniques under development can be divided into natural language processing, object detection, action recognition, picture segmentation, and in accordance with the data types mentioned in the gathered literature (Xu et al., 2021). A clear explanation of deep learning and machine learning shown in Figure 7.9.

d) Classification/Regression

Machine learning advancement depends on benchmarking since it enables an unbiased comparison of available techniques. The category of supervised learning activities in which the targeted class is linked with the characteristics is known and includes the classification and pattern recognition. The approximating of a function with continuous attributes is another topic of regression. Utilising interpolation, extrapolation, regression analysis, or curve fitting, the function is roughly approximated. An important part of validating a model is residual analysis, which investigates the distribution of residuals. It is closely tied to the chosen loss function because of how the residual is evaluated. Residual analysis might take the form of a statistical hypothesis test, aggregated numeric results, or a graphical visualisation. A classifier's performance on a certain classification task is determined by the statistical test (Hoffmann et al., 2019). In Figure 7.10, we can see clearly classification and regression.

A variety of encoder and regression techniques, or supervised learning methods, are used to meet expectations in continuous or categorical data elements. Unsupervised methods are also used to build a model that strengthens the grouping of data (Dara et al., 2022). Data is accurately classified and divided into several groups using an algorithm. It can identify specific entities in the provided dataset and makes assumptions about how to label those entities. Regression is used to comprehend the relationship between dependent and independent variables (Manne, 2021).

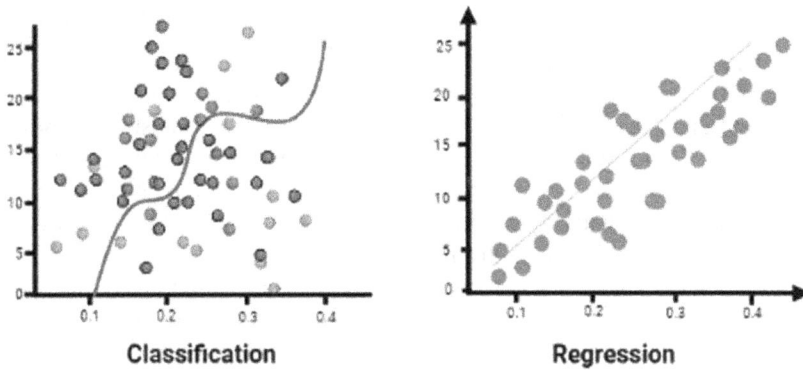

FIGURE 7.10   Pictorial representation of classification and regression.

3.1   DATA MINING

Data mining is a futuristic concept that was introduced as a fresh method of data analysis and knowledge finding in the middle of the 1990s. Although interdisciplinary work in the disciplines of statistics and ML were the two parts of the development of data mining, it also includes artificial intelligence, pattern recognition, visualisation, and database design.

Data mining is described as "the study of (sometimes enormous) observational data sets to identify unforeseen associations and to summarise the data in unique ways that are both understandable and helpful to the data owner" (Smith, 2002). According to this description, data mining's goal is to unearth novel, profound insights and a never-before-seen understanding of enormous datasets, which are usually acquired for operational purposes (Yoo & Song, 2008). These insights can then be used to support decision-making. Scientific hypotheses can also be generated via data mining from biomedical literature and huge experimental data sets. This database containing clinical records and death certificates of diabetes individuals, which helps to identify patterns that explain relationships between early mortality and observations made of patients during initial hospital visits (Richards et al., 2001).

Data mining has developed into a method for closing the gap between the use of information gained from digital data and its increasing availability (Fayyad et al., 1996). Due to the similarities in their methods and results, it is necessary to look at the differences between data mining and knowledge discovery in databases (KDD). According to the Cross-Industry Standard Process for Data Mining, the six steps in the data mining process are: business understanding, data understanding, data preparation, modeling, evaluation, and deployment (CRISP-DM). CRISP-DM was developed by four companies in late 1996 with the aim of developing "a comprehensive data mining methodology and process model" (Berger & Berger, 2004).

Knowledge discovery in databases is divided into data selection, data pre-processing, interpretation, transformation of data, evaluation, and data mining. The procedures of data mining and KDD are similar enough that the two words have been used interchangeably. However, according to several experts, data mining is one of the KDD process's key components (Fayyad et al., 1996; Jiawei Han et al., 2006). Data mining is therefore seen as a component of KDD, and in the KDD process, data mining refers to a set of applications of certain algorithms for identifying patterns in pre-processed or "ready-to-data-mine" data.

The efficient use of data mining produces new biomedical and healthcare knowledge that can effectively support both clinical and administrative decision-making in the provision of healthcare (Iavindrasana et al., 2009). Healthcare practitioners can directly benefit from the knowledge gained through data mining by giving patients better care (Bellazzi & Zupan, 2008).

3.2 DIFFERENTIATION BETWEEN DATA MINING AND STATISTICS

Statistics have been the primary data analysis approach employed in the majority of scientific domains in recent history under the auspices of scientific methodology. Currently, it is thought that data mining is best understood as secondary data analysis, whereas statistics is considered to be primary data analysis (Hand, 1998). This viewpoint is supported, in part, by the fact of data mining in the CRISP-DM (described in "What is data mining?") is data interpretation, which necessitates the application of complex statistical procedures. Data mining, in contrast to statistics, "represents a distinction of kind rather than degree" (Oliveira et al., 2012).

Statistics begins with conservative analysis procedures when mining data, but data mining is more flexible about which methods should be utilised in which order. Ad hoc methods are avoided in statistics in favour of a more rigorous mathematical approach. Although statistics and data mining are basically based on mathematics, many data mining methodologies also use heuristics to partially address problems in the real world, particularly when discrete (categorical) data are included. Data mining can work with a variety of data kinds, in contrast to statistics, which only works with numeric data because it is a completely mathematical technique. Examples include CT/MRI pictures, audio, text, and other types of data. However, because most databases contain categorical properties certain particular data mining methods might only work with categorical or numerical data (Hand, 1999).

3.3 DATA MINING TECHNIQUES

Researchers must be aware of the many types of data mining algorithms and how they work before using them on medical data. Usually, data mining techniques are classified into two categories: descriptive (or unsupervised learning) and predictive (or supervised learning) (Solomatine, 2000; Patil-Sonawane, 2006). (Fayyad et al., 1996). By examining how similar objects (or records) are to one another, descriptive data mining organises data to make it easier for users to quickly understand a vast volume of data. Additionally, it finds in the data patterns relationships that weren't previously apparent. Data mining that is exploratory in nature and descriptive includes clustering, association, summarization, and sequence discovery, among other things (Dunham, 2006).

Prediction data mining applies the rules to unpredicted/unclassified data by first inferring the rules from training data (also referred to as classification/prediction models). Predictive data mining encompasses prediction, classification, time series analysis, and regression. Three of the most well-known data mining algorithms (classification, clustering, and association) will be discussed here, along with recommendations for how to apply each one, in order to help academics understand the significance of data mining and application of data mining techniques (Dunham, 2006).

3.4 CLASSIFICATION OF DATA MINING

Typically, data points are categorised into numerous classes using the usual data mining classification techniques. It can be used to organise a wide range of data sets, from tiny, straightforward datasets to complex, enormous ones. In order to increase the quality of the data, it mostly requires employing easily adaptable algorithms. This is a key factor in the widespread use of supervised learning with classification in data mining approaches. Linking a variable of interest with the necessary variables is the primary objective of classification. The crucial factor must be of the qualitative kind (*Classification and Predication in Data Mining–Javatpoint*, 2019).

A) REGRESSION

In terms of statistics, regression analysis is the act of figuring out and looking at the relationship between variables. This shows that one variable does not depend on another, but the opposite is true. Normally, forecasting and prediction are done using it. All of these data mining techniques

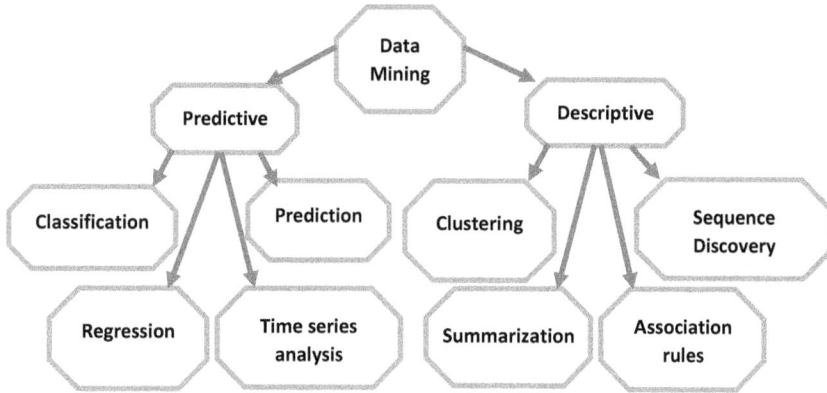

FIGURE 7.11 Representation of classification of data mining.

can be used to examine various data from different perspectives. Using this information, the best way to transform data into information that can be used to solve a variety of company problems and increase revenue, please clients, or save expenses (Dasgupta et al., 2011).

B) Time Series Analysis

The information is stored in this kind of sequence at a regular level and is of the numeric data type. They result from economic processes like stock market research and clinical examinations. They are useful for studying natural phenomena. Forecasting time series: Making predictions based on historical and current facts in order to foresee the future is known as forecasting. Trend analysis is a method for forecasting time data. It is a function that develops historical patterns in time series data that are then used in both short- and long-term projections. Popular methods include extended memory time series modeling, SARIMA, and ARIMA ("Data Mining–Time-Series, Symbolic and Biological Sequences Data," 2022).

C) Clustering

With the aim of achieving extremely high similarity between things inside a cluster and extremely low similarity between objects from different clusters, the process of clustering divides objects (records) into a preset number of clusters. Two objects are contrasted using the characteristics' values. Early in its history, biology used clustering to classify similar plants and animals into taxonomies based on their traits (such as the number of petals and the number of legs). Over the past few decades, numerous clustering techniques have been developed and put to use. Most of these algorithms fall under the hierarchical and partitional categories (Kaufman & Rousseeuw, 2009).

A detailed cluster analysis approach for genome-wide expression data from DNA microarray hybridization that groups genes based on how similar their patterns of gene expression are. The results are shown visually, presenting both the grouping and the underlying expression data in a way that is understandable to biologists (Eisen et al., 1998).

D) Sequence Discovery Analysis

It is all about knowing what to do and in what order. For instance, a customer might regularly buy shaving foam in a store before a razor. The owner of the store can then arrange the items suitably based on the user's sequence of product purchases.

E) Summarization Analysis

It includes techniques for getting a clear description of a dataset. Sales and marketing managers can get a basic summary of the data, for instance, by summarising a huge number of things linked to holiday season sales.

F) Association Rules Analysis

This technique assists in identifying intriguing connections between distinct variables in sizable databases. The retail sector is the best illustration. For example, data mining is used to stock up on chocolates as the holiday season approaches, resulting in an increase in sales before the holidays.

G) Predictive Data Mining

Data mining has a short history and was once incorrectly believed to comprise only data analysis techniques from domains other than statistics. These methods were shown to have the following characteristics: the handling of enormous volumes of data; the utilisation of diverse data types (various types of characteristics, text mining); extremely flexible modeling (e.g., the inclusion of non-linearity); and the automation of the majority of the analytical process. Due to the initial success of methods in domains like market basket analysis, as well as the overemphasis of ML and pattern recognition methods in creating data mining suites, several statisticians urged their community to become engaged and contribute to the field (Hand, 1998).

Clinical medicine predictive models are "instruments to support decision-making that combine two or more patient data items to forecast clinical outcomes" (Wyatt & Altman, 1995). These models can be utilised by doctors in a variety of clinical settings and may enable a quick response to adverse circumstances (Kattan et al., 2000). At least three interconnected components of data mining may successfully contribute to the creation of therapeutically valuable predictive models: (a) a thorough and purposeful approach to data analysis that applies techniques and methodologies from several scientific fields; (b) the ability of such models to provide explanations; and (c) the capacity to incorporate data into the data analysis process. **Predictive data mining**

SL no	Name of Data Mining Techniques	Description	References
1.	Decision Trees	Use repatriating partitioning, create transparent classifiers, as their performance may be affected by data segmentation. Decision trees' leaves may contain insufficient data to produce accurate predictions.	(Breiman, 2017)
2.	Decision Rules	The formula "IF condition-based-on-attribute-values THEN outcome-value" as in the cases of the AQ and CN2 algorithms can be extracted from the data. While these algorithms function similarly to decision trees in most respects, they may be more expensive computationally.	(Clark & Niblett, 1989; Michalski & Kaufman, 2001)
3.	Logistic Regression	A potent and well-known statistical technique. It is a development of conventional regression, and it may simulate a two-valued outcome that typically denotes the existence or absence of some event.	(Jr et al., 2013)

SL no	Name of Data Mining Techniques	Description	References
4.	Artificial Neural Networks (ANNs)	The method of artificial intelligence-based data modeling that is most frequently applied in healthcare. They may have a number of disadvantages, such as high sensitivity to the method's parameters, including those that determine the network's architecture, high computational cost during training, and model induction that may, at best, be difficult to interpret by domain experts, but this is probably due to their strong predictive abilities.	(Schwarzer et al., 2000)
5.	Support Vector Machines (SVMs)	Most effective classification algorithm for accuracy of prediction. They are based on statistical learning theory and solid mathematical foundations. The method's key step is a process that separates examples of various outcomes along a hyperplane.	(Cristianini et al., 2000; Vapnik, 1999)
6.	The Naïve Bayesian Classifier	A strategy whose effectiveness, despite its simplicity, is frequently at least on par with that of other, more advanced strategies. A classifier may be used as a benchmark algorithm in comparison studies due to its quick induction.	(Andreassen et al., 1991; Lavrač et al., 1998)
7.	Bayesian Networks	Models that use conditional probability distributions to conveniently represent a joint probability distribution across a number of variables in probabilistic graphics. Each node in a Bayesian network represents a stochastic variable, and the arcs between nodes and their parents indicate their probabilistic dependencies.	(Andreassen et al., 1991; Galán et al., 2002; Luciani et al., 2003)
8.	The K-Nearest Neighbors	Algorithm is inspired by the strategy frequently used by domain experts that base their choices on previously encountered similar circumstances. The k-nearest neighbours classifier finds the k most comparable training cases for a given data instance and classifies based on their dominant class. It may take some time to find the examples that are the most comparable, thus retrieving the entire training set at the time of classification is necessary.	(Hastie et al., 2001)

4 ROLE OF MACHINE LEARNING AND DATA MINING IN BIOINFORMATICS

Recent advancements in medical science and health informatics have paved the way for in-depth analytics, which are now necessary due to the production, gathering, and storage of vast amounts of data. In the meantime, a new era is beginning in which unique technologies are beginning to evaluate and traverse information from enormous data, opening up endless possibilities for information expansion. One unavoidable fact is that uses of ML and deep learning techniques from the perspective of biological data play a more substantial role in the achievement of bioinformatics investigation, and a link is found in both industry and academia. It is key to connect these two data analytics methods with bioinformatics. Therefore, it is desirable to develop more potent theoretical techniques and useful tools for analysing and collecting valuable information from the complicated bio-data indicated above. Prediction, classification, clustering, outlier identification, sequential processing, frequent item query, spatial/temporal data analysis, deep network architecture creation, and visualisation are just a few of the important concepts underlying the many effective and scalable approaches and tools. Building models, finding associated rules, doing classifications and clustering, making predictions, and then spotting desired designs have proven to be successful methods for passing on knowledge from the raw parent data. This method of data mining is known as ML, and it is the core methodology used in many successful data mining applications.

There are numerous significant datasets behind the incremental datasets, circumstances that call for high-capacity data storage systems and high-performance analytic tools. In terms of technicality, the data mining or ML method is also highly important. Characteristics of abstract data delineation ensure that data mining can accomplish the objective of accuracy and dependability performance. However, abstraction and human analysis are not appropriate for huge data that has a large number of instances as well as high-dimensional features. Additionally, the data expansion rate is far higher than the conventional manual techniques for analysis. Whenever we lack the skills to provide consumers with accurate data in an intelligible format, the meaning of existence is likewise lost. In order to properly utilise this kind of data to support clinical diagnostics and examine the clinical effects of drugs on experimental data, it is crucial to provide an automatic data analysis approach for evaluating the high-level data. Bioinformatics has undergone extensive study and development, and as a result, a wide variety of ML techniques are now accessible for data exploration and analysis (Han, 2002).

Bio-electronics and the nervous system, computational biology and bioinformatics, biomedical imaging and processing and visualisation, gene engineering, biomedical modeling, nano-biological analysis, medical cell biology, nuclear magnetic resonance/ECG/CT, physiological signal processing, etc., are just a few of the biomedical study areas where data mining is used. The majority of data mining procedures used in the previous decades were utilising conventional knowledge discovery in a well-organized way using a relational database primarily composed of numerical data by fulfilling the criteria of business in industries. The techniques of data analysis applications were dominated by tools like straightforward logic, statistics, etc. The field of data mining involves investigating diverse data in addition to homogenous data (among the kinds of structured data, using key ideas, write in both semi-structured and unstructured formats), pattern recognition, ML, and processing tools, to perform more intelligent learning via artificial intelligence, etc.

It is also important to consider deep in addition to huge through advanced technologies like deep neural networks (DNN), cloud computing, metaheuristics, parallel and distributed computing, fuzzy logic, etc. In various scientific and academic domains the inclination toward data mining not only leads to the efficient processing of massive data, but also offers more in-depth knowledge extraction. It has been developed and amassed for data in the biomedical field in a way that is unique in scope and depth these days, and that results in a lot of machinery discovering techniques for extracting information from bioinformatics. A wide range of real-world scenarios are covered through practical applications, in which huge data scale also presents difficulties, such as greater process capacity for diverse nature of data, higher learning adaptability for huge volume and more dimensions, etc.

More recently, ML has evolved from CNN approaches based on the fundamental idea of neuron processors in order to transfer nonlinear relationships through reactions of every level (Hinton & Salakhutdinov, 2006). With its advantages in higher-level abstractions of characteristics of large raw data, it provides fundamental and effective applications in the majority of fields, including speech and image recognition, biomedical research, and natural language processing.

5   CHALLENGES IN THE HANDLING OF BIG DATA

While ML and deep learning have demonstrated strong aptitudes and strengths in processing of bioinformatics data in different fields, they continue to face a number of inherent difficulties and new problems that need to be taken into account. Currently, massive amounts of data are being produced continuously in all stages of life with the daily use of various systems and technologies.

In a methodical manner, data is collected, saved, compiled, and examined without losing its volume, velocity, variety, and veracity (i.e., 4V) attributes.

Currently, industry endeavours and research activities are divergent disciplines are a variety of effective methods that deliver the most accurate analysis of this massive amount of data originated from various sources. Technology advancements have produced enormous amounts of both organized and unorganized data. A number of data sources can not be analysed by traditional analytics tools used previously. The "4V" is highlighted by the International Data Cooperation (IDC), which holds a similar perspective. Big data encompasses a variety of data types (numerical, visual, with many representational formats (video, text, etc.), and can be created by linked devices and social media platforms and stored in relational databases in an organised fashion. Devices, sensors, and information systems are being developed by many vendors, each of which is establishing its own algorithms and data visualisations.

The analysis and processing of data are hampered by this heterogeneity. Additionally, data volume is expanding quickly. Data streaming frequently requires the use of a variety of applications, including e-commerce, healthcare, and using statistical and machine learning methods. These algorithms must provide the ability to find data from specific sources by combining data from many sources, specialised and complicated analytics are produced. Technically speaking, the key obstacles in information technology for mining of data and processing are heterogeneity and the vast volume of data (Banaee et al., 2013).

Conventional methods include Relational Database Management Systems, will not support unorganised data using big data technologies. Innovative methods and procedures that handle massive and complicated data will store and analyse the capabilities of modern or conventional systems (Kwon et al., 2014). RDBMS operations get more sophisticated, which slows down performance and the speed at which applications respond, particularly when real-time data processing is needed (e.g., to block or foresee occurrences).

Small data for prediction and business intelligence were once the main focus. However, there is a flood of data everywhere now. Greater data correlation capacity enables us to find new and improved data. We may forecast the future from the vast quantity of diverse types of data, uncover important masked information, and determine preventative measures that might boost productivity. Manufacturers use sensor data directly from their manufacturing lines to create acts that are self-tuning to prevent time-consuming human involvement. High-performance computer clusters examine the data, and the outcomes are used in time-sensitive information, environmental factors can be used to boost productivity conditions or other external factors that have an impact on the manufacturing process.

Likewise, big data is used effectively to lower service costs and provide users with value. Large retail businesses can locate thousands of people's digital footprints. Customers can mimic their behaviour online by looking at their site and social media activity (Manyika et al., 2011). The value of data in the healthcare industry could exceed $300 billion annually. Additionally, they predicted that big data in the industrialised world may save more than 100 billion.

The field of big data needs to solve a number of obstacles and problems especially in data management. These issues should be resolved holistically by enlisting experts from several fields of computer science. Some of the difficulties in processing huge amounts of data are data capture, cleansing, acquisition, storage, analysis, and transfer as well as ethical concerns that result from these processes.

5.1 CLEANSING, OBTAINING, AND CAPTURING DATA

Big data must meet a few basic quality requirements, just like all other data. These "quality" characteristics imply that they must be accessible, easily recognisable, and easily readable in a consistent format for processing. Scrubbing of data, also known as data cleansing, is the process of getting rid of data that is inaccurate, outdated, or has lost its validity over time. It is cured after cleaning and is then ready for contemporary use and also discovery and reuse. When data are acquired, there is a lot of "noise" if it is not cleaned. In some cases big data attained the quality criteria (clean and consistent data), but may generate a huge amount of data. Thus, the next challenge is to develop filtering mechanisms capable of keeping targeted information (Agrawal et al., 2011). In order to facilitate its replication, the "robust" portion of the data that remains after filtering needs to be described using metadata schemes (Buneman et al., 2000).

5.2 STORAGE, DISTRIBUTION, AND TRANSFER

The storage of huge data is a significant problem as well. Elastic web services that provide peta byte scale data warehouses have been the key new storage models to emerge. Additionally, improved RDBMS with Hadoop Map/Reduce integration schemes and NoSQL DBMS must maintain low input/output latency from huge data repositories in addition to the storage issue. Because its properties are not bound to a particular data model but can instead be stored in any suitable structure or format, alternative DBMS schemes (like NoSQL) have the advantage of enabling modifications to be made without expensive reorganisation at the storage layer (Price & Flach, 2013).

As a result, the difficulty in storage of big data is to manage high data-volume acquisitions and support a range of heterogeneously structured data without experiencing delay. In terms of sharing and transferring both big data and the alleged "Internet plumping problem" are factors in this. This is due to the expansion of wired/optical and wireless infrastructures. Moreover, a sizable volume of machine-to-machine data is produced and thus networking infrastructures are more loaded. All of the aforementioned problems lead to key difficulty in data "flow" through the network.

5.3 ANALYSIS AND RESULT GATHERING

The wide distribution of errors that occur in each individual study across heterogeneous data structures, on the other hand, is a significant problem that needs to be addressed in the future. Errors may come at each level of analysis, as numerous operational functions (such Hadoop Map/Reduce integration schemes) are conducted. This reduces the effectiveness of standard algorithms, which are best used for small data sets, in drawing conclusions. Additionally, the precision and accuracy of outcomes from big data analysis depend on the target audience using these results, while data analysis and the conclusions obtained may also be time-dependent. A target audience, for instance, might be more interested in obtaining detailed trend statistics (Wu et al., 2013).

5.4 ETHICS-RELATED MATTERS

The handling of large data raises a number of ethical difficulties. Identity, privacy, ownership, and reputational concerns are among the ethical considerations (Anagnostopoulos et al., 2016). During the gathering, storing, sharing, and processing of massive data, problems also crop up (Mann, 2012). Veracity and trust concerns are intertwined. When a user is identified based on their online behaviour, they no longer have that anonymity (privacy loss). When a person records, stores, or shares an event, web programmes that allow the upload of audio-visual content intrude into their personal space. The "digital DNA" left behind by an individual may disclose distinctive characteristics about them that would otherwise go unnoticed (Michael & Miller, 2013).

The privacy of patients may be in danger from big data cleansing to analysis because of the disclosure of medical information to unauthorised beneficieries. For instance, it has been demonstrated that even if important descriptive fields from electronic health records (EHRs), such as name and date of birth, are masked, the chances to identify the patient with a high degree of accuracy is there when the remaining data are properly combined with other data sources and their characteristics (Patil & Seshadri, 2014).

6 CONCLUSION

A big opportunity has been created for biomedical research, knowledge discovery, data analytics, and creative application due to the rapid expansion and development of bioinformatics. A highly promising strategy to accomplish such aims in both explicit and implicit IT methods is data mining with basic machine learning algorithm techniques. But some problems including the big data challenge lead to severe bottlenecks. Deep learning, however, expands the capabilities of traditional machine learning and gives it a wider range of problems to solve.

The significance of machine learning and data mining in bioinformatics and the problems associated with handling of big data have been discussed in this chapter, along with the fundamental uses of data mining and machine learning methodologies in bioinformatics. Specifically, multiple data preprocessing, classification, and clustering algorithms of machine learning and data mining were discussed, and various deep frameworks were evaluated in the context of bioinformatics applications. Because of this, it is firmly believed that data mining and machine learning will collaborate to significantly enhance possibilities in the fields of medical bioinformatics.

REFERENCES

2001. (n.d.). MIT Technology Review. Retrieved October 29, 2022, from www.technologyreview.com/10-breakthrough-technologies/2001/

Agrawal, D., Bernstein, P., Bertino, E., Davidson, S., Dayal, U., Franklin, M., Gehrke, J., Haas, L., Halevy, A., & Han, J. (2011). *Challenges and opportunities with Big Data 2011-1*.

Anagnostopoulos, I., Zeadally, S., & Exposito, E. (2016). Handling big data: Research challenges and future directions. *The Journal of Supercomputing*, *72*(4), 1494–1516.

Andreassen, S., Jensen, F. V., & Olesen, K. G. (1991). Medical expert systems based on causal probabilistic networks. *International Journal of Bio-Medical Computing*, *28*(1), 1–30. doi.org/10.1016/0020-7101(91)90023-8

Aromolaran, O., Aromolaran, D., Isewon, I., & Oyelade, J. (2021). Machine learning approach to gene essentiality prediction: A review. *Briefings in Bioinformatics*, *22*(5), bbab128. doi.org/10.1093/bib/bbab128

Banaee, H., Ahmed, M. U., & Loutfi, A. (2013). Data mining for wearable sensors in health monitoring systems: A review of recent trends and challenges. *Sensors*, *13*(12), 17472–17500.

Breiman, L. (2017). *Classification and Regression Trees*. Routledge. doi.org/10.1201/9781315139470

Buneman, P., Khanna, S., & Tan, W.-C. (2000). Data provenance: Some basic issues. *International Conference on Foundations of Software Technology and Theoretical Computer Science*, 87–93.

Clark, P., & Niblett, T. (1989). The CN2 Induction Algorithm. *Machine Learning*, *3*(4), 261–283. doi.org/10.1023/A:1022641700528

Classification and Predication in Data Mining—Javatpoint. (n.d.). Www.Javatpoint.Com. Retrieved October 30, 2022, from www.javatpoint.com/classification-and-predication-in-data-mining

Clustering based semi-supervised machine learning for DDoS attack classification | Elsevier Enhanced Reader. (n.d.). doi.org/10.1016/j.jksuci.2019.02.003

Cristianini, N., Shawe-Taylor, J., & Shawe-Taylor, D. of C. S. R. H. J. (2000). *An Introduction to Support Vector Machines and Other Kernel-based Learning Methods*. Cambridge University Press.

Dara, S., Dhamercherla, S., Jadav, S. S., Babu, C. M., & Ahsan, M. J. (2022). Machine Learning in Drug Discovery: A Review. *Artificial Intelligence Review*, *55*(3), 1947–1999. doi.org/10.1007/s10462-021-10058-4

Data Mining—Time-Series, Symbolic and Biological Sequences Data. (2022, February 18). *GeeksforGeeks*. www.geeksforgeeks.org/data-mining-time-series-symbolic-and-biological-sequences-data/

Davenport, T., & Kalakota, R. (2019). The potential for artificial intelligence in healthcare. *Future Healthcare Journal*, *6*(2), 94.

Dunham, M. H. (2006). *Data Mining: Introductory And Advanced Topics*. Pearson Education India.

Eisen, M. B., Spellman, P. T., Brown, P. O., & Botstein, D. (1998). Cluster analysis and display of genome-wide expression patterns. *Proceedings of the National Academy of Sciences*, *95*(25), 14863–14868. doi.org/10.1073/pnas.95.25.14863

Fayyad, U., Piatetsky-Shapiro, G., & Smyth, P. (1996). From Data Mining to Knowledge Discovery in Databases. *AI Magazine*, *17*(3), Article 3. doi.org/10.1609/aimag.v17i3.1230

Forte, C., Voinea, A., Chichirau, M., Yeshmagambetova, G., Albrecht, L. M., Erfurt, C., Freundt, L. A., Carmo, L. O. e, Henning, R. H., Horst, I. C. C. van der, Sundelin, T., Wiering, M. A., Axelsson, J., & Epema, A. H. (2021). Deep Learning for Identification of Acute Illness and Facial Cues of Illness. *Frontiers in Medicine*, *8*. www.frontiersin.org/articles/10.3389/fmed.2021.661309

Galán, S. F., Aguado, F., Díez, F. J., & Mira, J. (2002). NasoNet, modeling the spread of nasopharyngeal cancer with networks of probabilistic events in discrete time. *Artificial Intelligence in Medicine*, *25*(3), 247–264. doi.org/10.1016/S0933-3657(02)00027-1

Han, J. (2002). How can data mining help bio-data analysis? *Proceedings of the 2nd International Conference on Data Mining in Bioinformatics*, 1–2.

Hand, D. J. (1998). Data Mining: Statistics and More? *The American Statistician*, *52*(2), 112–118. doi.org/10.1080/00031305.1998.10480549

Hand, D. J. (1999). Statistics and data mining: Intersecting disciplines. *ACM SIGKDD Explorations Newsletter*, *1*(1), 16–19. doi.org/10.1145/846170.846171

Hastie, T., Friedman, J., & Tibshirani, R. (2001). *The Elements of Statistical Learning*. Springer New York. doi.org/10.1007/978-0-387-21606-5

Hinton, G. E., & Salakhutdinov, R. R. (2006). Reducing the dimensionality of data with neural networks. *Science*, *313*(5786), 504–507.

Hoffmann, F., Bertram, T., Mikut, R., Reischl, M., & Nelles, O. (2019). Benchmarking in classification and regression. *WIREs Data Mining and Knowledge Discovery*, *9*(5), e1318. doi.org/10.1002/widm.1318

Hsieh, S.-L., Hsieh, S.-H., Cheng, P.-H., Chen, C.-H., Hsu, K.-P., Lee, I.-S., Wang, Z., & Lai, F. (2012). Design ensemble machine learning model for breast cancer diagnosis. *Journal of Medical Systems*, *36*(5), 2841–2847.

Hu, J., Niu, H., Carrasco, J., Lennox, B., & Arvin, F. (2020). Voronoi-based multi-robot autonomous exploration in unknown environments via deep reinforcement learning. *IEEE Transactions on Vehicular Technology*, *69*(12), 14413–14423.

Janiesch, C., Zschech, P., & Heinrich, K. (2021). Machine learning and deep learning. *Electronic Markets*, *31*(3), 685–695. doi.org/10.1007/s12525-021-00475-2

Jr, D. W. H., Lemeshow, S., & Sturdivant, R. X. (2013). *Applied Logistic Regression*. John Wiley & Sons.

Kattan, M. W., Zelefsky, M. J., Kupelian, P. A., Scardino, P. T., Fuks, Z., & Leibel, S. A. (2000). Pretreatment nomogram for predicting the outcome of three-dimensional conformal radiotherapy in prostate cancer. *Journal of Clinical Oncology: Official Journal of the American Society of Clinical Oncology*, *18*(19), 3352–3359. doi.org/10.1200/JCO.2000.18.19.3352

Kaufman, L., & Rousseeuw, P. J. (2009). *Finding Groups in Data: An Introduction to Cluster Analysis*. John Wiley & Sons.

Kharya, S. (2012). Using data mining techniques for diagnosis and prognosis of cancer disease. *ArXiv Preprint ArXiv:1205.1923*.

Kiarashinejad, Y., Abdollahramezani, S., & Adibi, A. (2020). Deep learning approach based on dimensionality reduction for designing electromagnetic nanostructures. *Npj Computational Materials*, *6*(1), 1–12.

Kwon, O., Lee, N., & Shin, B. (2014). Data quality management, data usage experience and acquisition intention of big data analytics. *International Journal of Information Management*, *34*(3), 387–394.

Lavrač, N., Kononenko, I., Keravnou, E., Kukar, M., & Zupan, B. (1998). Intelligent data analysis for medical diagnosis: Using machine learning and temporal abstraction. *AI Communications*, *11*(3–4), 191–218.

Li, J., Wong, L., & Yang, Q. (2005). Guest Editors' Introduction: Data Mining in Bioinformatics. *IEEE Intelligent Systems*, *20*(6), 16–18.

Luciani, D., Marchesi, M., & Bertolini, G. (2003). The role of Bayesian Networks in the diagnosis of pulmonary embolism. *Journal of Thrombosis and Haemostasis*, *1*(4), 698–707. doi.org/10.1046/j.1538-7836.2003.00139.x

Mahesh, B. (2019). *Machine Learning Algorithms–A Review*. doi.org/10.21275/ART20203995

Mann, S. (2012). Through the glass, lightly. *IEEE Technology and Society Magazine*, *31*(3), 10–14.

Manne, R. (2021). Machine Learning Techniques in Drug Discovery and Development. *International Journal of Applied Research, 7*(4), 21–28.

Manyika, J., Chui, M., Brown, B., Bughin, J., Dobbs, R., Roxburgh, C., & Hung Byers, A. (2011). *Big data: The next frontier for innovation, competition, and productivity.* McKinsey Global Institute.

Meshram, V., Patil, K., Meshram, V., Hanchate, D., & Ramkteke, S. D. (2021). Machine learning in agriculture domain: A state-of-art survey. *Artificial Intelligence in the Life Sciences, 1,* 100010.

Michael, K., & Miller, K. W. (2013). Big data: New opportunities and new challenges [guest editors' introduction]. *Computer, 46*(6), 22–24.

Michalski, R. S., & Kaufman, K. A. (2001). Learning Patterns in Noisy Data: The AQ Approach. In G. Paliouras, V. Karkaletsis, & C. D. Spyropoulos (Eds.), *Machine Learning and Its Applications: Advanced Lectures* (pp. 22–38). Springer. doi.org/10.1007/3-540-44673-7_2

Oliveira, T., A., O., & Bonilla, A. (2012). *Data Mining and Quality in Service Industry: Review and Some Applications* (p. 25).

Patil, H. K., & Seshadri, R. (2014). Big data security and privacy issues in healthcare. *2014 IEEE International Congress on Big Data,* 762–765.

Peng, J., Jury, E. C., Dönnes, P., & Ciurtin, C. (2021). Machine Learning Techniques for Personalised Medicine Approaches in Immune-Mediated Chronic Inflammatory Diseases: Applications and Challenges. *Frontiers in Pharmacology, 12.* www.frontiersin.org/articles/10.3389/fphar.2021.720694

Price, S., & Flach, P. A. (2013). A Higher-order data flow model for heterogeneous Big Data. *2013 IEEE International Conference on Big Data,* 569–574.

Rai, P., & Shubha, S. (2010). A Survey of Clustering Techniques. *International Journal of Computer Applications, 7.* doi.org/10.5120/1326-1808

Ramchandra, N. R., & Rajabhushanam, C. (2021). Traffic Prediction System Using Machine Learning Algorithms. *I3CAC 2021: Proceedings of the First International Conference on Computing, Communication and Control System, I3CAC 2021, 7-8 June 2021, Bharath University, Chennai, India,* 424.

Richards, G., Rayward-Smith, V. J., Sönksen, P. H., Carey, S., & Weng, C. (2001). Data mining for indicators of early mortality in a database of clinical records. *Artificial Intelligence in Medicine, 22*(3), 215–231. doi.org/10.1016/s0933-3657(00)00110-x

Sah, S. (2020). *Machine Learning: A Review of Learning Types.* doi.org/10.20944/preprints202007.0230.v1

Santosh, K. C., Vajda, S., Antani, S., & Thoma, G. R. (2016). Edge map analysis in chest X-rays for automatic pulmonary abnormality screening. *International Journal of Computer Assisted Radiology and Surgery, 11*(9), 1637–1646.

Sarker, I. H. (2021). Machine Learning: Algorithms, Real-World Applications and Research Directions. *SN Computer Science, 2*(3), 160. doi.org/10.1007/s42979-021-00592-x

Sarker, I. H., Hoque, M. M., Uddin, M., & Alsanoosy, T. (2021). Mobile data science and intelligent apps: Concepts, AI-based modeling and research directions. *Mobile Networks and Applications, 26*(1), 285–303.

Sarker, I. H., Kayes, A. S. M., Badsha, S., Alqahtani, H., Watters, P., & Ng, A. (2020). Cybersecurity data science: An overview from machine learning perspective. *Journal of Big Data, 7*(1), 1–29.

Schwarzer, G., Vach, W., & Schumacher, M. (2000). On the misuses of artificial neural networks for prognostic and diagnostic classification in oncology. *Statistics in Medicine, 19*(4), 541–561. doi.org/10.1002/(SICI)1097-0258(20000229)19:4<541::AID-SIM355>3.0.CO;2-V

Scowen, M., Athanasiadis, I. N., Bullock, J. M., Eigenbrod, F., & Willcock, S. (2021). The current and future uses of machine learning in ecosystem service research. *Science of the Total Environment, 799,* 149263.

Singh, Y., Bhatia, P. K., & Sangwan, O. (n.d.). *A REVIEW OF STUDIES ON MACHINE LEARNING TECHNIQUES.* 15.

Smith, A. (2002). Principles of Data Mining–D. Hand, H. Mannila, P. Smyth (Eds.), MIT Press, Cambridge, MA, 2001, 546 pp. + xxxii, ISBN: 0-262-08290-X. *Artificial Intelligence in Medicine, 26,* 175–178. doi.org/10.1016/S0933-3657(02)00058-1

Team, D. (2017, July 13). *Machine Learning Tutorial—All the Essential Concepts in Single Tutorial.* DataFlair. data-flair.training/blogs/machine-learning-tutorial/

Vapnik, V. N. (1999). An overview of statistical learning theory. *IEEE Transactions on Neural Networks, 10*(5), 988–999. doi.org/10.1109/72.788640

Wang, H., Ma, C., & Zhou, L. (2009). A Brief Review of Machine Learning and Its Application. *2009 International Conference on Information Engineering and Computer Science*, 1–4. doi.org/10.1109/ICIECS.2009.5362936

Weissler, E. H., Naumann, T., Andersson, T., Ranganath, R., Elemento, O., Luo, Y., Freitag, D. F., Benoit, J., Hughes, M. C., & Khan, F. (2021). The role of machine learning in clinical research: Transforming the future of evidence generation. *Trials*, *22*(1), 1–15.

Wu, X., Zhu, X., Wu, G.-Q., & Ding, W. (2013). Data mining with big data. *IEEE Transactions on Knowledge and Data Engineering*, *26*(1), 97–107.

Wyatt, J. C., & Altman, D. G. (1995). Commentary: Prognostic models: clinically useful or quickly forgotten? *BMJ*, *311*(7019), 1539–1541. doi.org/10.1136/bmj.311.7019.1539

Xu, Y., Zhou, Y., Sekula, P., & Ding, L. (2021). Machine learning in construction: From shallow to deep learning. *Developments in the Built Environment*, *6*, 100045. doi.org/10.1016/j.dibe.2021.100045

Yoo, I.-H., & Song, M. (2008). Biomedical Ontologies and Text Mining for Biomedicine and Healthcare: A Survey. *Journal of Computing Science and Engineering*, *2*(2), 109–136. doi.org/10.5626/JCSE.2008.2.2.109

8 Application of Deep Learning in Chemistry and Biology

Priya Patel, Vaishali Thakkar, Saloni Dalwadi, Prexita Patel,
Kevinkumar Garala, Bhupendra Gopalbhai Prajapati

LIST OF ABBREVIATIONS

DL   Deep Learning
AI   Artificial Intelligence
ML   Machine Learning
ANN   Artificial neural networks
CNN   convolutional neural networks
DNN   Deep neural networks

1   INTRODUCTION TO DEEP LEARNING

Deep learning is a form of mimic of the human brain, much like artificial neural networks are. A branch of machine learning called "deep learning" is solely dependent on neural networks. We do not have to explicitly programme everything in deep learning since deep learning is not a brand-new idea [1].

This is a picture of a single neuron, which makes up one of the approximately 100 billion neurons in the human brain. Each neuron is connected to thousands of its neighbours.

How do we simulate these neurons on a computer is the question at hand. To begin, we build a synthetic network of nodes and neurons, known as an artificial neural network. Between the neurons utilised for input and output values, there may be a significant number of connected neurons in the hidden layer [2].

1.1   HISTORY OF DEEP LEARNING

Deep learning networks, as we now know, are essentially deep neural networks with multiple layers. In 1943, Walter Pitts and Warren McCulloch developed the first mathematical model of a neural network that illustrated how the human brain thinks. The voyage of deep learning and neural networks started from this point.

- Frank Rosenblatt published a paper in 1957 titled "The Perceptron: A Perceiving and Recognizing Automaton," which described an algorithm or a technique for pattern recognition utilising a two-layer neural network.
- The first functional neural network was created in 1965 by Alexey Ivakhnenko and V.G. Lapa. In 1971, Alexey Ivakhnenko created an 8-layer deep neural network, which was used in the computer identification system Alpha. This served as deep learning's actual introduction.
- In the year 1980, Kunihiko Fukushima created the "Neocognitron," a synthetic deep neural network with several convolutional layers for recognising visual patterns.

DOI: 10.1201/9781003353768-8

- In 1985, Terry Sejnowski developed NETtalk, a programme that taught itself how to speak words in English.
- In 1989, Yann LeCun created a system that could read handwritten numbers using convolution deep neural networks.
- In 2009, Fei-Fei Li introduced ImageNet, a sizable database of labelled images, because deep learning models need a lot of labelled data to train themselves in supervised learning.
- 2012–Google Brain's "The Cat Experiment" findings were made public. The deep neural network in this experiment used unsupervised learning to identify patterns and features in the photos of cats from unlabeled data. However, it could only accurately identify 15% of the photos.
- In 2014, Facebook created DeepFace, a deep learning system that can recognise and tag people' faces in photos [3].

1.2   How and When to Use Deep Learning

Machine intelligence is helpful in many circumstances and is comparable to or superior to human specialists in several areas [4–7]. Therefore, DL could be a solution in the following cases:

- Situations in which there are no human experts on hand.
- Situations in which people are unable to explain choices they made using their knowledge.
- Cases when the solution to the problem changes over time.
- Cases where the solutions need to be modified based on particular cases.
- Situations in which a problem's size is so big that it defies our meagre capacity for reasoning about it [8].

The differences between Artificial Intelligence, Machine Learning, and Deep Learning are listed in Table 8.1 [9].

1.3   The Position of Deep Learning in AI

These days, the phrases AI, ML, and DL are all frequently used to refer to systems or software that exhibit intelligent behaviour. ML and the larger field of AI both include DL as a component. ML is a technology that automates the development of analytical models by learning from data or experience, whereas AI, in general, adds human behaviour and intelligence to computers or systems [10,11]. Data-driven learning methods that employ multi-layer neural networks and processing for computation are also referred to as DL [12].

2   DEEP LEARNING AND ITS APPLICATION

Digital data is expanding fast in all forms and sizes, and it is expected that by 2025, the quantity of data produced per day will reach 463 exabytes worldwide [13]. The healthcare industry is currently responsible for around 30% of all data volume creation. The compound annual growth rate of healthcare data will be 36% by 2025 [14]. Big dataset exploration and analysis are of considerable interest, and has prompted the introduction of techniques for machine learning like DL. Computer games, semantic analysis, naturalistic speech interpretation, voice identification, computer vision, financial modelling, autonomous vehicles, and prediction fields using machine learning are just a few of the many applications where DL has experienced great success [15-16].

Since the volume of biological information available in open sources is growing quickly, integrative assessment is becoming more challenging due to the diverse structure of this information. Machine learning techniques are substantially and steadily being employed in a wide range of

TABLE 8.1
Differences between artificial intelligence, machine learning, and deep learning

Artificial Intelligence (AI)	Machine Learning (ML)	Deep Learning (DL)
AI allows machines to emulate human behaviour using a specific algorithm.	ML uses statistical techniques to allow machines to get better over time.	DL is a field of study that uses neural networks to simulate human brain function. Neural networks are analogous to the neurons found in the human brain.
The larger family that includes ML and DL as its constituents is known as AI.	ML is the subset of AI.	DL is the subset of ML.
AI is a type of computer algorithm that demonstrates intelligence through judgement.	AI's ML method enables systems to learn from data.	DL is a ML method that analyses data using deep (more than one layer) neural networks and produces results as a result.
AI uses search trees and a lot of intricate arithmetic.	The ML component is defined if you understand the underlying logic (math) and can envision sophisticated functionalities like K–Mean, Support Vector Machines, etc.	If you understand the mathematics involved but have no notion about the features, you can break down complex functionalities into features that are linear or lower dimension by adding further layers. This defines the DL aspect.
Instead of precision, the goal is essentially to improve the likelihood of success.	The success rate is less important than accuracy in this case.	When it is trained with a lot of data, it achieves the greatest accuracy ranking.
The effectiveness of ML and DL, respectively, basically determines the effectiveness of AI.	Less effective than DL because it can not handle larger data sets or dimensions.	More effective than ML because it can readily handle bigger data sets.
Google's AI-Powered Predictions, ridesharing services like Uber and Lyft, commercial flights using an AI autopilot, etc., are a few examples of applications for AI.	Virtual personal assistants like Siri, Alexa, Google, and others, email spam filtering, and malware detection are a few examples of ML applications.	Sentiment-based news aggregation, image analysis, and caption creation are a few examples of DL applications.

biomedical sectors, from the generation of biomarkers to the development of new drugs. A large category of machine learning methods known as "deep learning" has shown significant potential in the extraction of advanced order abstractions from the unprocessed information of really big, diverse, high-dimensional data arrays. It is the kind of data biology which is currently available.

Artificial neural networks (ANNs) with numerous layers of nonlinear processing components are used in the subclass of machine learning techniques known as DL to understand data representations. ANNs with various complexities of nonlinear processing components are used to learn data representations. The very first ANN was created in 1943 by Warren McCulloch and Walter Pitts, who used mathematics and algorithms motivated by the design of the human brain to create a threshold logic-based computer model for neural networks [17].

2.1 Application of DL to Predict Compound Properties and Activities

Long-term predictions of chemical activity have been made using the machine learning approach, particularly ANN [18]. Generally, DL techniques are used to solve the original activities predictions issues. The simple approach is to utilise fully connected DNNs to generate models when substances are represented with the same number of chemical descriptors.

By means of a high number of 2D topological descriptors, Dahl et al. [19] used a DNN to the Merck Kaggle testing data, and the DNN outperformed the typical random forest (RF) technique in 13 out of the 15 objectives. Similarly, on a database with 12,000 substances comprising 12 high-throughput toxicity testing, Mayr et al. [20] revealed their multitask DNN models that took first place in the Tox21 challenge.

2.2   DE NOVO DESIGN USING DL

The method of developing new chemical constituents with desired pharmacological and physio-chemical characteristics is known as "de novo drug design." DL applications in de novo drug design have gained popularity, and numerous DL-based methods have been created for molecule synthesis problems. These methods were typically constructed using one of four frameworks: generative adversarial networks, encoder-decoder, reinforcement learning, and recurrent neural networks [21].

Gomez-Bombarelli et al. [22] proposed an innovative use of DL in chemoinformatics employing variational autoencoder (VAE) to construct new molecular compounds. After the research of Gomez-Bombarelli, Kadurin et al. [23] employed VAE as a molecular descriptor generator in conjunction with a generative adversarial network (GAN) [24], a unique NN design, to develop new molecules that were said to have potential unique anticancer activities. In order to create new molecules with expected effectiveness towards dopamine receptor type 2, Blaschke et al. [25] used VAE. Processing of natural language has seen great success with RNNs [25]. Segler et al. [26] and Yuan et al. [27] presented research work employing RNNs to build unique molecular compounds.

Segler et al. [26] investigated the application of RNNs to construct resources tailored to a target by first training a generalized preceding model followed by a concentrated and refined design using a learning algorithm on a smaller number of bioactive chemicals precise to target. Their focused algorithms were likely to provide 18% unexpected real actives for *Staphylococcus aureus* and 28% for *Plasmodium falciparum* in a retrospective study to test against two antibioactive targets.

A policy-based supervised learning strategy was put forth by Olivecrona et al. [28] to adjust the previously trained RNNs for producing compounds having specified user-defined attributes. For one testing ground for fine-tuning the model to produce active molecules expected to be effective towards the dopamine receptor type 2, the prototype produced structures of which more than 95% were projected to be effective, which includes bioactives that established by experimentation, but does not include in either the reproductive model or the action forecast prototype. The aforementioned techniques have shown promise as substitutes to conventional rule-based methodologies for *de novo* design.

2.3   USE OF DL FOR RETROSYNTHETIC ASSESSMENT AND RESPONSE PREDICTION

Since the advent of rule-based techniques in the 1960s [29], synthesis forecasts have been around for a great many years. Recently, some encouraging findings in reaction prediction utilising DL approaches were revealed. The findings demonstrate that DL can reach performances on pace against, or better than, the rule-based techniques, even though there has not been a direct comparison to further machine learning techniques. DL in reaction informatics can be used to conceptually address two different kinds of issues. Retrosynthetic predictions, where provided a finished product, predict the reactive processes that make the product, is such form. Forward reaction prediction, on the other hand, predicts the outcomes provided a collection of reagents. In one study, 3.5 million interactions were employed as the training dataset using DNN by Segler et al. [26]. For reaction predictions and retrosynthetic analyses, the best 10 accuracy rates were 97% and 95%, respectively. A further study [30] used a training dataset of 12 million reactions from the published research and coupled policy networks with Monte-Carlo tree search for retrosynthetic prediction. Their technique can resolve twice as many retrosynthesis strategies for chemicals as the rule-based method. Additionally, neural sequence-to-sequence algorithms were employed by Liu et al. [31] for retrosynthetic prediction. Researchers trained the algorithm using 50,000 reactions from US patents and achieved more accuracy than rule-based techniques.

2.4 APPLICATION OF CONVOLUTIONAL NEURAL NETWORKS TO FORECAST LIGAND–PROTEIN INTERACTIVITIES

A key component of molecular docking software is evaluating the interactions among proteins and ligands, and many scoring functions have been created primarily relying on forcefields or information through known protein–ligand complexes [32]. Applying CNNs to measure molecular interactions has lately been the subject of multiple investigations, which were motivated by the performance of CNNs in picture processing.

The experiment conducted by Ragoza et al. [33] is a perfect illustration. The discretization process used a matrix with a resolution of 0.5 Å for the protein–ligand complexes. The matrix was centred on the binding site and had 24 Å across each side. The input matrix was created by generating the atomic concentrations well over grids after every atom was specified by a function. The Caffe DL framework was used to construct and train multilayer CNN models.

2.5 INVESTIGATION OF BIOIMAGING USING DEEP LEARNING

Biomedical imaging and image processing are frequently employed at many phases of the drug discovery process, from preclinical research and development to clinical trials. With the help of imaging, researchers can see the phenotypes and behaviours of hosts (human or animal), body organs, tissues, cells, and subcellular elements. The underlying biology and pathology, in addition to the drug's mechanism of action, are made visible by digital picture processing. Fluorescently labeled or unlabeled microscopic pictures, computed tomography (CT), MRI, positron emission tomography (PET), tissue pathology scanning, and mass spectrometry imaging (MSI) are a few illustrations of imaging modalities. In biomedical image processing, DL too has found success, and various researches have shown that it performs better than traditional classifications.

CNNs were employed to divide and subtype specific fluorescently labeled cells in microscope pictures [34], in addition to unlabeled pictures from phase contrast microscopy [35]. DL can also be used to optimize additional time-consuming preclinical operations like colony counts [36] and cell tracking [35]. Due to the rich tissue morphology, pictures obtained from tissue pathology are usually more complicated than fluorescently tagged images. However, hematoxylin and eosin (H&E) staining of breast and colon tissues allowed for the segmentation and classification of individual cells at the cellular level [37,38]. At the tissue area, DL was used to identify the tumour areas in H&E-stained breast tissue, as well as to identify additional types of leukocytes and fat tissue [39]. DL has previously been employed for histological diagnosis using H&E and immunohistochemical labeled tissues, going above simple image processing [40,41].

The use of DL was also employed in the assessment of imaging studies using CT [42-44], MRI [45,46], and PET [47]. In addition to the widely used applications of image segmentation and classifications, their benefits in content-based image retrieval have additionally been demonstrated [48]. It was further revealed that DL approaches excelled the well-known ISOMAP as well as Elastic Net methodologies.

Additionally, high-quality training datasets are hard to come by in the environment of high-throughput image analysis. Hence, to execute biological imaging segmentations and classifications, image topographies trained after natural scenes and additional databases were "borrowed," and substantial performances were published [49,50].

2.6 DL FOR DRUG DEVELOPMENT

Massive amounts of data are typically required to train machine learning algorithms, and in deep learning especially; nevertheless, the human brain can learn from a small number of cases. Therefore, one of the most common problems in machine learning is how to learn through little to no sufficient database. Matching networks [51], which were introduced as a kind of one-shot learning, are an example of how to use supplementary datasets in DL to enhance a model using only a few data points. When the supplementary data were used, the outcomes were strengthened. Drug discovery, since medicinal chemists frequently act on novel objectives using minimal available datasets, is a field where techniques similar one-shot learning are applicable. In order to develop algorithms using a relatively limited training dataset, Altae-Tran et al. [52] applied the LSTM approach to chemoinformatics datasets. They obtained encouraging results. Memory-enhanced neural networks are a different sort of structure that have only recently been applied to deep learning. The neural Turing machine was the earliest version. A differentiable neural computer (DNC) drastically enhanced such architecture [53]. DNCs have been used to solve a variety of issues, including determining the shortest path in graphs and question-answering systems. But as of yet, drug discovery has not made use of these further sophisticated frameworks.

3 APPLICATION OF DEEP LEARNING IN CHEMISTRY

There are several chemistry fields that deep learning has so far affected, with examples in each that highlight specific accomplishments (Figure 8.1). This discussion will adhere to a hypothetical chemical pathway in order to construct a logical narrative. Creating techniques to precisely correlate any given structure to the property would be necessary before creating a molecule with a specific property. With the help of them, a molecule can then be cleverly designed to maximise the desired attribute. Designing an effective synthesis from widely available starting ingredients is the last stage. As a result, a closed feedback cycle is created that allows for experimental analysis of the synthesised molecule and the subsequent improvement of the models that relate molecules to attributes. Every step of this workflow has been impacted by deep learning, starting with understanding molecules [54].

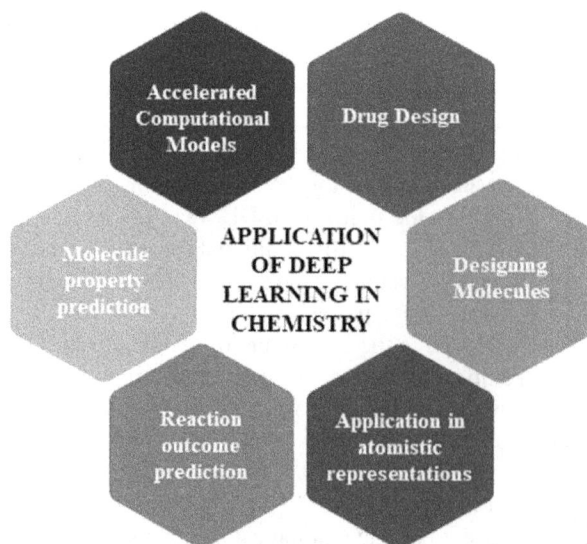

FIGURE 8.1 Applications of Deep Learning in Chemistry.

3.1   ACCELERATED COMPUTATIONAL MODELS

In chemistry, computational modeling aims to employ calculations based on physics to ascertain the characteristics and behaviour of a specific chemical system. In this context, deep learning can be applied in two different ways. To reduce computing bottlenecks, the first strategy is to combine the deep learning technique with physics-inspired techniques. The second method completely avoids physical laws by predicting attributes directly from molecular structures. To combine deep learning techniques with physics-based methods, the network must be trained to anticipate a crucial aspect of the total calculation. These involve predicting potential energy surfaces using the deep learning approach [55–57]. Behler's overview and tutorial on the use of neural networks for the forecasting of potential energy surfaces are both very effective [58–59]. The energy of the system is determined by adding the energetic contributions of all the atoms, a technique first proposed by Behler and Parrinello [60] in 2007 and used by many of these techniques. This technique uses radial symmetry functions that capture the details of each atom's immediate surroundings to change the Cartesian coordinates of a molecule. The contribution of this atom to the total energy is then predicted by a neural network using the modified representation. This fundamental technique of employing functions to represent an atom's immediate surroundings, then estimating its energy through a network, and lastly aggregating these contributions has been improved in a number of different contexts. The research by Schutt et al. has been improved on and turned into an open source software programme (SchNetPack) that can be used to forecast attributes. This method has the advantages of being more flexible than mapping a structure to a property and being easier to understand because of its physical foundation. The problem is that these approaches can not perform as quickly as those that directly map from a structure to a specific property because they frequently still need physics-based calculations. It is significant to highlight that employing kernel ridge regression as the ML approach is supported by a substantial body of literature [61–62].

3.2   DRUG DESIGN

Possibly one of chemistry's most significant applications is drug design. Fundamentally, it entails locating molecules that perform a specific biological task with the greatest efficiency. These can be created from scratch or derived from natural resources. In each scenario, the task is to optimise the attributes of the initial molecule or compounds to increase potency, specificity, reduce side effects, and lower production costs. Autoencoders [63], GANs [64], and reinforcement learning [65] are all employed in drug design generative models in a manner similar to that of generic molecular design in an effort to produce effective therapeutic compounds. Predicting anti-cancer drug synergy [66] and creating a benchmarking for generative models in drug design [67] are two further unique ways to drug development rather than molecular design. Due to the high cost of acquiring data, drug design methodologies may struggle with it more than any other discipline. One-shot learning was used in work by Altae-tran et al. [68] to address this shortcoming and create knowledgeable predictions about medication candidates with scant data. Finally, although it is not a generative system that optimises molecules, work by Segler et al. [69] developed methods to develop focused libraries of drug candidates for screening using RNNs.

3.3   DESIGNING MOLECULES

A different use of artificial intelligence in molecular design has led to ground-breaking discoveries in the discipline, while the detection of molecular characteristics is very helpful in the field of chemistry.

Scientists have been able to collect historical information and create chemical relationships by constructing molecules. Researchers have advanced in finding compounds by incorporating AI algorithms, and this has undoubtedly assisted them in making ground-breaking discoveries in AI chemical synthesis.

Additionally, creating molecules opens up a variety of valuable applications that have advanced chemistry significantly [70].

3.4 APPLICATIONS IN ATOMISTIC REPRESENTATIONS

In this section, we give a few illustrations of how to use deep learning techniques trained on atomistic data to solve materials science challenges. Atomic coordinates and data about the material's atomic composition often make up the atomic structure of a substance. Applying conventional ML algorithms to make atomistic predictions in systems with arbitrary numbers of atoms and different sorts of elements is difficult. DL-based approaches are a clear way to solve this issue. The representation of crystals and molecules using fixed-size descriptors, such as the Coulomb matrix [71–73], has been attempted previously on a number of occasions. Direct property predictions, materials screening, and force-field development are DL uses for atomistic materials. Along with the aforementioned ideas, we also explain various contemporary generative adversarial networks and alternatives to atomistic approaches [74].

3.5 REACTION OUTCOME PREDICTION

To forecast the outcomes of reactions, machine learning techniques have been trained on databases containing millions of chemical reactions. These applications are built using recurrent neural networks or graph-convolutional neural networks. They operate in the following manner: after receiving the reactants, the software predicts both the primary reaction product and a number of additional potential products, many of which are reaction byproducts. One drawback of these programmes is that they do not consider account the process conditions, including temperature, for their predictions [75].

3.6 MOLECULE PROPERTY PREDICTION

When creating novel molecules for a particular use, scientists must first create them in order to do an experimental verification that they have the desired qualities. If not, the researchers create brand-new compounds (which might be analogues of previously synthesized molecules, for instance) and repeat the process until they find molecules that meet their needs (properties, performance, price, toxicity, environmental impact, etc.). It costs both time and money to complete this iterative process. Researchers would be able to synthesise only the most promising chemicals and avoid synthesising and testing several molecules that do not have the desired qualities if they could precisely forecast the attributes of hypothetical molecules. Long-established techniques for predicting molecular characteristics are known as quantitative structure-activity relationships (QSAR) or quantitative structure-property relationships (QSPR). These techniques typically rely on physical principles or empirical correlations that link molecules' structures to their properties (often inadvertently, via a set of predetermined descriptors). Machine learning methods can also be used to predict chemical characteristics. Numerous other types of attributes, including bioactivity, toxicity, solubility, melting temperatures, atomization energies, HOMO/LUMO molecular orbital energies, and many others, have also been predicted using these algorithms. They are wholly data-driven and not dependent on physical principles or carefully crafted empirical relationships. In essence, these AI systems are trained by being fed numerous samples of molecules together with their corresponding attributes (supervised learning). Different techniques for regression or classification can be utilized, including neural networks, support vector machines, random forests, and linear regression.

Artificial intelligence (AI)-based algorithms are especially well-suited to problems where the physical laws that determine the molecular properties to be predicted are not precisely known, or when establishing empirical relationships would be too difficult, for example, because of significant nonlinearities or correlations between parameters. It is interesting to note that using AI in conjunction with other prediction techniques, such as mathematical formulas or empirical relationships, can produce forecasts that are even more precise. A method known as stacking allows AI to, for instance, use the outcomes of predictions provided by physical or empirical relationships as input data [75].

4 TOOLS AND METHODS USED IN CHEMISTRY

Drug–target interaction is the term used to describe the interactions between chemical compounds and bimolecular drug targets in the human body. This interaction is crucial for the discovery and development of new drugs since it leads to the therapeutic effect. There is a vast gap between known and unknown drug-target pairs as a result of the extremely restricted and incomplete knowledge about drug–target interactions based on wet-lab studies, which has prompted attention in the quest for effective approaches to DTI prediction [76]. Various tools used in chemistry are shown in Figure 8.2.

4.1 DTI-CNN

Convolutional neural networks (CNNs) are a type of NN that are frequently employed to analyze visual data. DTI–CNN [77-79] is a straightforward DL-based tool for predicting drug–target interactions that is claimed to outperform current state-of-the-art approaches through the intelligent interaction of three components: (1) a feature extractor based on heterogeneous networks, (2) a feature selector based on denoising-auto encoders, and (3) a CNN-based interaction predictor [80–81].

4.2 DeepCPI

A new framework-based tool called DeepCPI uses unsupervised representation learning and DL approaches to predict DTI [82]. It is claimed that in the beginning, latent semantic analysis and Word2vec techniques are used to unsupervisedly learn the low-dimensional feature representations of both chemicals and proteins.

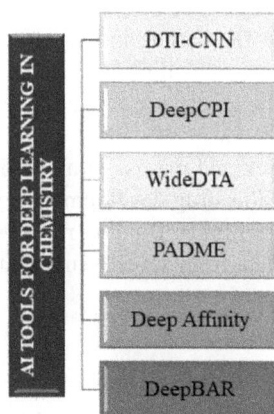

FIGURE 8.2   AI Tools used in Chemistry.

The drug–protein interaction is successfully predicted by the DL network using the feature embedding of the chemicals and protein from the first step. Two key components of the DeepCPI framework are (1) representation learning for both chemicals and proteins, and (2) CPI (or DTI) prediction using a multimodal DNN [83].

4.3 WideDTA

A CNN DL model called WideDTA uses information from four text-based sources, including ligand SMILES (LS), protein sequences (PS), ligand maximum common substructure (LMCS), and protein domains and motifs (PDM). In contrast to DeepDTA, WideDTA portrays LS and PS as a group of words rather than as complete sequences. A word has three residues in the PS sequences while having eight residues in the LS sequences. Because of the poor signal-to-noise ratio in full-length sequences, the creators claim that the WideDTA is a word-based model rather than a character-based one [84–85]. In order to predict binding affinity, WideDTA uses four text-based data sources: the protein sequence, ligandSMILES, protein domains and motifs, and maximum common substructure words. For the prediction of drug–target binding affinity, WideDTA outperformed one of the cutting-edge deep learning techniques [86].

4.4 PADME

Protein and drug molecule interaction prediction is the name of the DL-based DTBA forecasting technique that applies fingerprints and drug–target properties to several DNNs (PADME). The tool is referred to as PADME–ECFP when extended-connectivity fingerprint is used as the input for the representation of medicines. The other PADME version is known as PADME-GraphConv [87,88].

4.5 DeepAffinity

DeepAffinity depends on the structural property sequence representation to represent proteins. This representation annotates the sequence with structural information, is shorter than previous representations, rapidly supplies structural data, and gives superior resolution of the sequences. The full model trained from beginning to end, including data representation, embedding learning, and joint supervised learning. Basically, it has been discovered that the RNN-CNN pipeline produces results with great accuracy when compared to other ML-based techniques used on the same dataset [89,90].

4.6 DeepBAR

MIT researchers created DeepBAR, a DL-based binding affinity prediction tool, by fusing chemistry and ML [91]. The drug with the lowest binding free energy value should be picked from a set of potential medications in order to get the greatest results because it will be able to effectively disrupt the protein's normal function.

Thus, it is safe to say that the quick and precise computation of standard binding free energy has several significant applications in this drug development process. The Bennett acceptance ratio approach is denoted by the BAR in DeepBAR. Binding ree power is determined by the dated BAR algorithm. Bennet acceptance ratio, data from several endpoints, and other intermediate situations are all used by DeepBAR [92,93].

4.7 Methodology

The main AI techniques utilized in power systems include artificial neural networks, fuzzy logic, particle swarm optimization, simulated annealing, and evolutionary computing. Some of the numerous techniques include the following.

Expert Systems in Medicine

Expert systems (ES) are the most prevalent class of AI system used in everyday clinical practise. ES are capable of handling facts to draw logical judgments. Among their many utilities, ES are used for picture interpretation, diagnosis support, and alert creation.

Key features of an ES are:

A system for gathering information and the rules the ES will employ to address the problems put forth is known as a knowledge acquisition system (KAS). The expert or knowledge engineer can directly input data into this procedure, or they can use a database of previous case studies and their findings.

b. A knowledge base: It contains information and guidelines regarding the particular issue the ES is supposed to tackle.

c. A control system called an inference engine performs the reasoning process by applying the information and rules stored in the knowledge base to the data.

The most prevalent ES utilized in the diabetes area are fuzzy systems, rule-based reasoning (RBR), and case-based reasoning (CBR).

FL

Fuzzy ES are used to describe expert knowledge that use ambiguous terminology in a form that is computer intelligible. According to traditional wisdom, blood glucose levels >180 mg/dl are high and those 80 mg/dl are low. Making decisions using this classification is not particularly helpful. In real life, a blood glucose level of 181 mg/dl usually justifies a different course of action than one of 281 mg/dl. In other words, 281 mg/dl is extremely high and far from acceptable, while 181 mg/dl is high but almost normal. FL expresses this uncertainty by giving several categories varying degrees of membership. In the previous illustration, 181 mg/dl pertains 70% to the group of "high," but only 30% to the category of "high" but only 30% to the category of "very high" [94].

AI-ONE

Nathan ICE, our in-house, biologically inspired language core technology, powers the Analyst Toolbox. The BrainDocs application, which offers a platform for processing document libraries, creating agents, and analysing findings, serves as the heart of the Analyst Toolbox. The BrainDocs API is a component of our cloud service housed on MS Azure and is accessible to enterprise developers for the building of applications.

With the help of this technology, programmers can incorporate intelligent assistants into virtually any software programmes. API, construction agents, and document repository are included in ai-Analyst one's Toolbox.

This tool's capacity to transform data into universal sets of rules that support intricate ML and AI frameworks is its main advantage [95].

AI in Designing Drug Molecule

Yu et al. used two RF models to predict potential drug–protein interactions by combining pharmacological and chemical data and validating them with excellent sensitivity and specificity against well-known platforms, such as SVM. Additionally, these modes could forecast drug–target relationships, which could then be expanded to anticipate associations between target–disease and target–targets, accelerating the drug development process [96].

Li et al. demonstrated the usage of KinomeX, an online AI platform that uses DNNs to recognize polypharmacological kinases based on their chemical structures. The DNN used by this platform was trained using over 14,000 bioactivity data points produced by more than 300 kinases. Thus, it has practical use in determining a drug's overall selectivity towards certain kinase subfamilies and kinases, which aids in the development of new chemical modifiers. This investigation used the model compound NVP-BHG712 to reasonably anticipate both its major targets and its off-targets [97,98].

4.8 Deep Learning Algorithms

Deep learning algorithms are known for their self-learning representations, but they also use artificial neural networks (ANNs) to model how the brain processes information. Throughout the training phase, algorithms take advantage of unknown aspects in the input distribution to extract features, categorize objects, and find pertinent data patterns. This occurs on numerous levels, using algorithms to build models in a manner similar to teaching machines to learn on their own.

Deep learning models employ a number of different algorithms. No network is perceived as perfect, yet some algorithms are better suited to complete specific jobs. To make the best decisions, it is advantageous to gain a complete understanding of all important algorithms [99].

Types of Algorithms Used in Deep Learning

Deep Belief Networks (DBNs), Self Organizing Maps (SOMs), Restricted Boltzmann Machines (RBMs), Long Short Term Memory Networks (LSTMs), Convolutional Neural Networks (CNNs), Generative Adversarial Networks (GANs), Recurrent Neural Networks (RNNs), Multilayer Perceptrons (MLPs), Radial Basis Function Networks (RBFNs), etc., have been used in DL.

5 BASICS OF DEEP LEARNING IN BIOLOGY

AI refers to a machine's capability to mimic intelligent human behavior. ML refers to the use of AI that enables a system to automatically learn and enhance an experiment. It is a cutting-edge technique for problem resolution and work automation. ML, in particular, is concerned with the creation and deployment of algorithms that can discover patterns in data and utilize them for predictive modeling, compared to domain specialists manually constructing rules for prediction tasks. ANNs are a type of ML method and model that have grown and are now considered DL. Figure 8.3 depicts how AI, ML, and DL are related with each other.

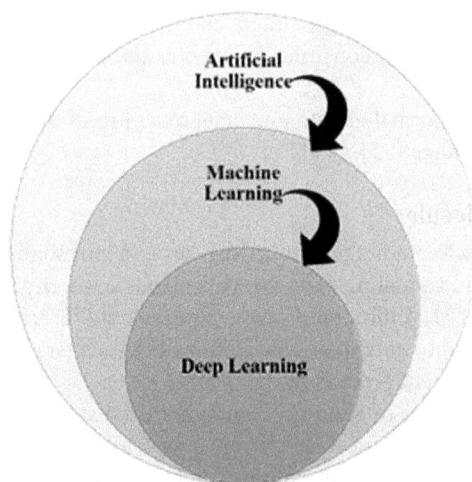

FIGURE 8.3 Hierarchy of theories within the field of artificial intelligence.

DL comprises neural networks including several layers in addition to methods that allow them to perform successfully. These neural networks are made up of artificial neurons stacked in layers and are designed to mimic the human brain, despite the fact that the building blocks and learning algorithms might differ [100]. Each layer receives input from the network's initial levels (the first of which reflects the input data) and then communicates an updated version of their own weighted output to the network's subsequent layers. As a result, "training" a neural network refers to the process of altering the weights of the network's layers in order to minimize a cost or loss function that acts as a surrogate for the prediction error. Because the loss function is differentiable, the weights may be modified automatically to minimize the loss. DL employs ANN with many layers (thus the word "deep"). Given recent computer breakthroughs, DL can now be used to large data sets and in an infinite number of settings. DL can understand more complicated relationships and generate more precise predictions than previous approaches in many situations. As a result, deep learning has evolved into its own branch of ML. It is increasingly being utilized in biological research to generate unique insights from high-dimensional biological data [101].

Rapid advancements in hardware-based techniques over the last several eons have enabled life scientists to collect multimodal data in a variety of application domains (e.g., Omics, Medical Imaging, Bioimaging, and [Brain/Body]-Machine Interfaces), thereby creating novel possibilities for the creation of data-intensive ML techniques. Recent studies in DL, Reinforcement learning (RL), and their combination (Deep RL) have the potential to transform AI. Scientists in a variety of areas are now able to apply these approaches to datasets that were previously intractable due to their size and intricacy because of the increase in processing power, quicker and expanded data storage, and lowering computing costs. A hypothetical illustration of the DL, RL, and deep RL frameworks for biological applications is illustrated in Figure 8.4 A to F. G is the schematic diagram of the learning framework as a part of AI. Broadly ES and ML may be viewed as the two major parallel evolutions of AI. ES makes expert judgements utilizing rule-based inferences from provided factual facts. When applied to various datasets, ML extracts characteristics from data mostly through statistical modeling and produces prediction output. As a subfield of ML, DL extracts more abstract characteristics from a bigger collection of training data, often in a systematic pattern replicating the brain's operation. The other branch, RL, offers a software agent that collects approach that relies on the environment's interactions through a series of activities and attempts to maximize cumulative performance. H is the potential AI uses for biological data [102].

5.1 APPLICATION OF DEEP LEARNING IN BIOLOGY WITH EXAMPLES

• Omics

The term omics describes biological sciences disciplines that ending in -omics, such as proteomics, genomics, and metabolomics. The suffix -omics is used to refer to the subjects of research in these disciplines, such as genomics, metabolomics, and proteomics. The overarching goal of omics sciences is to describe, discover, and characterize all biological components that contribute to the structure, function, and dynamics of a cell, tissue, or organism [103].

• Genomics

Specifically, genomics is the area of biology that studies the structure, function, evolution, and mapping of genomes, as well as the characterization and quantification of genes, which regulate the production of proteins through enzymes and messenger molecules [103]. Several basic genomic Bayesian (or non-Bayesian) prediction techniques, such as the basic additive gene effect model, have been established, with variable elements generated using mixed-model equations. DL techniques have recently been examined in the realm of genomic prediction. The DL methods are nonparametric models that can adapt to complicated interactions between input and output as well as highly complex patterns [104]. Table 8.2 shows examples of DL in genomics.

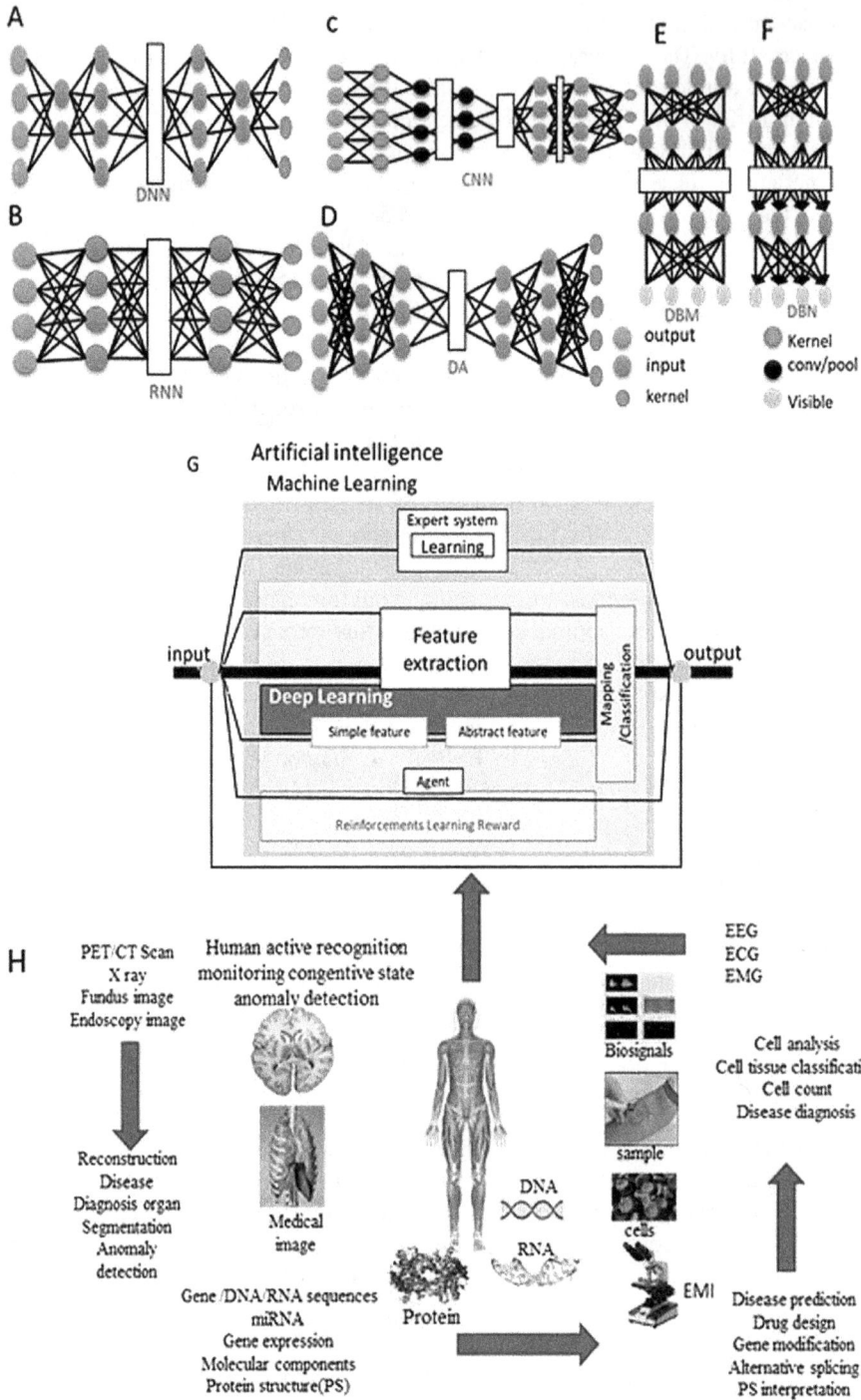

FIGURE 8.4 An illustration of DL frameworks that are used for biological applications.

TABLE 8.2
Application of deep learning in genomics

Name of Author	Purpose	Model	References
Zhang et al.	Using the main sequence and secondary and tertiary structural profiles to predict the binding position of RNA-binding proteins (RBPs) yielded an AUC of 0.98 for certain proteins.	DBN	[105]
Alipanahi et al.	DeepBind, a CNN-based model used even when directed and tested with in vitro and in vivo data, to determine DNA–and RNA-binding protein molecule-binding sites.	CNN	[106]
Lanchantin et al.	CNN was employed in both investigations to predict transcription factor-binding sites (TFBSs), and its performance was superior to that of DeepBind (AUC of 0.894).	CNN	[107]
Yoon at al.	Devised a unique deep belief network (DBN)-based technique to predict splice junctions at the DNA level that was trained on RBMs. In addition, their method uncovered subtle non-canonical splicing patterns to achieve greater precision and resilience.	DBN	[108]
Frey et al.	Provided a splicing code as a statistical inference problem. This model predicted splicing designs in individual tissues and changes in splicing designs between tissues using genomic characteristics and tissue context as inputs.	Bayesian neural network (BNN)	[109]
Koumakis et al.	The purpose of the model development was target prediction, identification/reconstruction of biological signals and pattern, gene expression inference, identification of key genes and miRNAs, classify gene expression predict drug response, predicts missing methylation states and detects sequence motifs, predicting the function of DNA directly from sequence alone, predicting transcription factor target, predict tissue-of-origin, normal or disease state and cancer type. The data types included were gene expression, miRNA-mRNA pairing, expression of landmark genes, histone modifications, DNA methylation and miRNA expression, cell-line with drug response, mRNA and miRNA, Single cell methylation, scRNA-seq, DNA-seq	DBN, Deep neural networks (DNN), CNN, RNN, ANN	[110]

- **Proteomics**

Proteome refers to the entire protein content of a cell, tissue, or organism. Proteomics is the study of proteins in relation to their biochemical characteristics and functional activities, as well as how their amounts, alterations, and structures vary throughout development and in response to both internal and external stimuli [103]. Proteomics presents numerous difficult computational challenges to resolve. Estimating entire protein architectures in 3D space from biological sequences is a difficult NP-hard task. On the other hand, the protein constructions can be divided into independent sub-problems (e.g., torsion angle, dihedral angles, access surface area, etc.) and managed to solve in similar and estimate the secondary protein architectures (2-PS). Envisaging compounds–protein collaboration (CPI) is intriguing and difficult to address from a drug discovery standpoint [111]. Examples of DL in protiomics are listed in Table 8.3.

TABLE 8.3
Application of deep learning in proteomics

Name of Author	Purpose	Model	References
Wang et al.	DeepCNF (Deep Convolutional Neural Fields) is enhanced to predict protein SS. Conditional Neural Fields (CNF) are the outcome of combining Conditional Random Fields (CRF) with shallow neural networks. DeepCNF is a Deep Learning implementation of CNF. DeepCNF, which uses a deep hierarchical structure to capture not just intricate sequence–structure interactions, but also the interdependence of surrounding SS labels, is significantly more successful than CNF.	DeepCNF	[112]
Li et al.	Suggested a methodology based on DA learning to recreate protein architecture from a template for prediction of CPI.	DNN	[113]
Bocicor et al.	RL with the help of a system called by using an RL-based framework unraveled the problematic of DNA portion assembly.		[114]
Spencer et al.	Projected Domein name system (DNS), a deep learning network architecture-based ab initio method for determining the secondary structure of proteins. DNS established a site-specific marking matrix of the protein sequence and Atchley's factors of residues, and the GPU and computed unified device architecture were tuned to accelerate computation Compute Unified Device Architecture (CUDA).	DNS	[115]
Snderby et al.	Convolutional neural network designs for predicting DNA–protein interaction employed a bidirectional RNN (BRNN) with short, and long-term memory cells to enhance the secondary structure prediction, obtaining a better degree of accuracy (0.671) than the current standard (0.664).	RNN	[116]

- **Metabolomics**

A biological cell, tissue, organ, or organism's whole set of metabolites is referred to as its metabolome. Metabolites are the end products of cellular activity and may be found in any biological cell, tissue, organ, or organism. The study of any chemical processes that include metabolites is referred to as metabolomics. The analysis of all small-molecule metabolite profiles is what this refers to [103].

The distribution of raw data collected directly from mass spectrometry (MS)-based metabolomics (e.g., gas chromatography [GC] coupled with MS, or GC–MS liquid chromatography [LC] coupled with MS, LC–MS) implies multiple phases of imbrication, such as deconvolution, metabolite identification, and feature detection. Common open foundation platforms for MS data processing include XCMS [117], MS-DIAL [118], and MZmine 2 [119], as well as specific vendor-specific software. When comparing the techniques used for GC–MS and LC–MS, care must be taken to notice their differences. Potential metabolites are identified based on their mass-to-charge ratio and retention length, followed by tandem MS (fragment spectra). While traditional GC–MS is coupled with hard ionization techniques such as electron impact (EI) that create in-source fragmentation of metabolites, hence producing a spectrum that is difficult to deconvolution. Chemical Ionization (CI) and Atmospheric Pressure Chemical Ionization (APCI) may be used in GC-based methods. Nonetheless, even more delicate ionization techniques typically result in a loss in sensitivity. Recent research [120] created a multi-algorithm, ML-based approach for peak identification in GC–MS data for a Waste Isolation Pilot Plant [WiPP]. WiPP can evaluate the quality of discovered peaks and provide feature detection for metabolite identification and subsequent analysis using a classification method and seven peak classifiers.

A CNN technique was developed in order to find missing peaks in LC–MS metabolomics data [121]. The peak can only recommend the elimination of peaks with a low intensity, or noise, and the insertion of peaks that were accurately anticipated. The method, although having an impact on the number of real positives, was successful in eliminating 90% of false-positive peaks (noise). In general, the system has a high degree of confidence (between 84 and 96%) in its ability to recognize true positives. It is possible that up to 30% of the peaks in breath samples will be identified as false positives [122].

- **Bioimaging**

Biomedical imaging is a vast, multidisciplinary area concerned with acquiring, analyzing, visualizing, and interpreting structural and functional pictures of living organisms for therapeutic or research purposes. In bioimage analysis, DL is used for three types of tasks: 1) image restoration, in which an input image is transformed into an enhanced output image; 2) image partitioning, in which an input image is partitioned into regions and/or objects of interest; and 3) image quantification, in which objects are classified, recorded, or totaled. Applications of DL in bioimaging are listed in Table 8.4.

- **Image Restoration**

While dealing with developing systems, obtaining a high signal-to-noise ratio (SNR) while imaging an object of concern is a constant problem. Noise in microscopy may originate from a wide range of sources, including the microscope's optics and/or its related detectors or camera. Furthermore, in growing organisms, areas of interest are usually situated within the body, distant from the microscope objective. As a result of dispersion, light emitted by fluorescent markers may be twisted and less intense once it reaches the goal. Photobleaching and phototoxicity also grow more troublesome at greater depth into the tissue, resulting in poor SNR as the impact is mitigated by decreasing laser intensity and raising camera exposure or detector voltage [123]. When utilized from the perspective of image restoration algorithms, which turn input pictures into output images with increased SNR, DL has proven effective in overcoming these obstacles. The high-SNR pictures are used to train a DL model that is based on the U-net architecture [124]. After that, the trained network may be used to restore noiseless, higher-resolution pictures from previously unnoticed noisier datasets.

- **Image Partitioning**

In most cases, analyzing particular items in a biological picture necessitates image partitioning, which is the extraction of the subjects of interest from the image backgrounds. Image partitioning may be accomplished by recognizing a line segment around objects (object identification) or by identifying the collection of pixels that comprise each item (segmentation). While images with a few elements may be partitioned by hand, larger datasets need automation. DL methods derived from computer vision have substantially accelerated and improved the speed and accuracy of object recognition and segmentation in biological pictures. DL models have evolved to recognize bioimage-specific objects [125]. Many of these algorithms draw on partitioning problems in real pictures, therefore there is still a direct relationship to computer vision.

- **Image Quantification**

After detecting objects in individual images, the next step is to quantify them. Quantification may be used to describe the number of objects (totaling), their classification, their morphometry, or their tracking, along with other things. Classification may be done for a full thing (e.g., wild-type vs. mutant) or for a single element of an entity (e.g., the shape of internal components). Even when performed by specialists, manual item classification is time-consuming and prone to bias. DL-powered image classification may decrease annotation variability in conjunction to expediting the

TABLE 8.4
Application of deep learning in bioimaging

Application	Purpose	Data Type	Model Used	Reference
Image segmentation	Pixel-by-pixel image segmentation of the nucleus, cell, cytoplasm, and nuclear membranes	Electron Microscope Image (EMI)	CNN	[127]
	Identify mitosis in breast image	Histology images	DNN	[102]
	To locate nuclei in breast cancer	Histopathology images	Stacked Sparse DA	[128]
	Categorization of colon cancer images for cell and tissue-level analysis	Images of colon cancer	Multiple Instance Learning (MIL) using DNN-learned features	[128]
	For label-free cell analysis in cancer diagnostics	Microscopy images of yeast	DNN	[129]
	Bacterial Colony Counting and classification	Digital Microbiology Imaging	CNN	[130]
	Brain tumor localization	Magnetic resonance imaging (MRI)	Cascade Convolutional Neural Network (C-ConvNet/C-CNN)	[131]
	Breast cancer	Digital Database Mammography (DDSM)	CNN	[132]
Brain Decoding	To study the affective state recognition of brain	Electroencephalogram (EEG)	Deep Belief Networks (DBN)	[133]
	Affective state classification	EEG	DBN + Generative Restricted Boltzmann Machine (RBM)	[134]
	Emotion Recognition (Positive, neutral and negative)	EEG dataset	DBNs	[135,136]
	Classifying students based on cognitive state	EEG dataset and its EEG signal	Siamese neural network (SNN)	[137]
	To evaluate the cognitive states of the driver	Eye movement and driving performance measures	DBN	[138]
	In the detection of clinical depression	Electroencephalography (EEG) signals	CNN + Long Short-Term Memory (LSTM)	[139]
	To Classify Alzheimer's Disease	MRI scan brain images	CNN	[140]
	Automatic seizure detection	Clinical EEG data	Three-dimensional (3D) CNN	[141]
Diagnosis of disease	To distinguish bacterial and viral pneumonia	Chest X-rays	Transfer learning techniques	[142]
	Classification of Skin Disease	Input image	CNN, Fine-Tuned Neural Networks (FTNN)	[143]
	Liver disease	Real-time data collected from patients	ANN model	[144]
	Kidney illness	Kidney ultrasound images	ANN	[145]
	COVID-19 disease	Chest X-ray dataset	DenseNet121, ResNet50	[146]
	Diabetic disease	Real-time data collected from diabetic and non-diabetic people	SVM, CNN, Long Short Term Memory	[147]
	Alzheimer's disease	ADNI database	SVM, KNN	[148]
	Squamous Cell Carcinoma	Scalp cSCC patients data	Artificial Neural Network	[149]
	Tuberculosis	CT TB images	ResNet	[150]
	Staging of Bone Lesions	Computed tomography scans (CTs)	ResNet	[151]

process. We have studied DL in the context of large-scale biological imaging data processing from three perspectives: segmentation, disease detection, and brain decoding. Throughout biology, DL designs taught the NN on the high-resolution level of a biological picture. By layering biological pictures, pixel noise may be decreased while abstract qualities are accentuated.

- **Deep Learning for Segmentation**

The technique of splitting a picture based on specified patterns is known as segmentation. Image segmentation is the technique of automatically recognizing areas of interest in an image, and it is a critical component of quantification of microscope images in biological studies. Segmentation refers to the segmentation of tissues or organs in medical pictures. Although extremely effective, it has limits in that it requires enormous datasets of labeled data to train the models, which may be time-consuming and expensive to set up [126].

- **Deep Learning in (Body/Brain) Interfaces**

DL techniques have been used to measure BMI signals (e.g., electroencephalogram, EEG, electrocardiogram, ECG, electromyogram, EMG) primarily from the perspectives of (brain) function decoding and anomaly identification.

- **Deep Learning in the Diagnosis of Disease**

Understanding complicated illnesses may be aided by using information from biological networks. Traditional techniques to studying illness-related biomolecules concentrate mainly on identifying and characterizing certain elements of the disease and are unable to completely appreciate the pathophysiology of the disease and the associated biomolecule connections. Analyzing biological network data structures may reveal more detailed etiology and disease-related biomolecules and can be used to examine the link between illnesses utilizing molecular network data.

6 TOOLS, METHODS, AND ALGORITHM USED IN BIOLOGY

- **RNA-PROTEIN–BINDING SITES PREDICTION WITH CNN**

RNA-binding proteins (RBPs) are crucial in regulating biological processes, including gene regulation. Understanding their activities, such as their binding location, can help treat RBP-associated disorders. With the development of high-throughput methods such as CLIP-seq, the RBP-binding sites may be verified in bulk [152]. Despite their effectiveness, these high-throughput technologies can be costly and time-consuming. The prediction of the RBP-binding site [153] may thus be facilitated by computational methods based on machine learning that is rapid and inexpensive. In reality, DL techniques are ideally suited for these challenges. As is known, RBP can exhibit sequence preference, identifying certain motifs and local structures and binding selectively to those locations [154]. CNN is very adept at identifying unique patterns at various scales. Previous research has demonstrated CNNs' effectiveness in discovering themes [155]. This example demonstrates how the RBP-binding location estimation is done via CNN and data [153].

The aim is to predict if a particular RBP, definite for one model, may bind to a specific RNA sequence (i.e., a binary arrangement problem). The conversion of "AUCG" RNA sequence strings using one-hot encoding into 2D tensors. For example, the letter "A" is represented by the vector [1, 0, 0, 0], while the letter "U" is represented by the vector [0, 1, 0, 0]. By concatenating these 1D vectors into a 2D tensor in the same sequence as the initial sequence, the one-hot encoding for a particular RNA sequence is obtained. They employ the same settings as the preceding instances in relation to the activation function, optimizer, loss function, etc. Observe that the one-hot encoding can be viewed as either a 2D map with a single network or a 1D vector with four channels.

- ## CNN AND RNN FOR DNA SEQUENCE FUNCTION ESTIMATION

Understanding the characteristics and activities of DNA orders is a difficult and demanding issue for bioinformatics [156]. Amongst these sequences, determining the functionality of noncoding DNA is particularly difficult, as over 98% of the human genome consists of noncoding DNA. It is extremely tough to conduct biological experiments to investigate the functionality of each non-coding DNA sequence fragment [157]. Low-cost and highly parallelizable computational methods can greatly assist in resolving the issue. Previous research [130] has demonstrated the effectiveness of employing DL to estimate the working of noncoding DNA sequences. In this example, RNN and CNN are utilized to estimate the function of noncoding DNA sequences. These sequences' labels were produced by gathering profiles from the Roadmap and ENCODE Epigenomics data releases, resulting in a 919-bit binary vector for each sequence (690 transcription factor-binding profiles, 125 DNase I-hypersensitive profiles, and 104 histone-mark profiles). To encode the DNA sequence string into a mathematical form that can be fed to the model using one-hot encoding. The model combines RNN and CNN; on top of 1D convolutional layers, there is a bi-directional LSTM layer because, for DNA sequences, the interaction between the upstream and downstream motifs plays essential roles in defining the sequence functioning, but the specific motifs also matter. The original implementation [158] depends on Theano, which has been deprecated. They reimplemented the concept using only Keras. This example can accomplish other critical predictions related to this challenge for DNA and RNA sequences, like extended noncoding RNA function predictions [159] and DNA methylation state predictions [160].

- ## TRANSFER LEARNING AND RESNET BY USING BIOMEDICAL IMAGE CLASSIFICATION

In biomedical diagnosis, the categorization of biomedical data has a variety of applications. For instance, the classification of biological images has been utilized to aid in identifying skin cancer [161] and retinal diseases [162], with performance reaching expert levels. The renowned ImageNet collection has more than 10 million photos [163]. Due to technological limitations and privacy concerns, this quantity of biomedical data is typically unavailable. However, transfer learning is typically performed for DL models as follows. When fine-tuning the model, the weights of convo-lutional layers are frozen, and the last fully connected layer in the standard model is replaced with a new fully connected layer with the same number of nodes as the real application's classes. Then, the weights of this new layer are trained. These frozen convolutional layers are the feature extractor, while the freshly added fully connected layer is the classifier.

- ## GRAPH EMBEDDING FOR NOVEL PROTEIN INTERACTION PREDICTION USING GCN

To predict novel protein–protein interactions in the yeast PPI network. Understanding protein–protein interactions can help predict the activity of uncharacterized proteins and design medications. Due to the recent advent of huge PPI screening approaches, including the yeast two-hybrid test [164], the quantity of PPI data has grown substantially. Despite the creation of these tools, the present PPI network is incomplete and noisy due to a large number of proteins and complex interactions. Using computational approaches to predict protein–protein interactions can be a convenient way to uncover novel and significant interactions [165]. As the dataset, they employ the yeast protein–protein interaction network. Model-wise, they employ the graph neural network [166].

- ## BIOLOGY IMAGE SUPER-RESOLUTION USING GAN

For image processing and structure reconstruction will demonstrate using GAN to do a simple yet useful task: image super-resolution. Cryo-EM images [167] and fluorescence images [168] are typically noisy and of low resolution due to the limitations of the imaging acquiring technology in biology, necessitating postprocessing to obtain high-resolution, noise-free images. Additionally, the super-resolution image must include more accurate details and fewer false ones. This image super-resolution approach has had remarkable success in rapid fMRI image acquisition [169] and fluorescent microscopy [170]. Two DL models are trained: a generator and discriminator networks. The generator network produces high-resolution images from low-resolution photos. The discriminator attempts to differentiate between the super-resolution images created by the model and the actual high-resolution photos. The discriminator network interacts with the generator network to encourage the generator network to provide the most realistic super-resolution images possible. During training, they simultaneously train both networks to improve both them.

- **VAE Used for High-Dimensional Biological Data Embedding and Generation**

VAE, like GAN, is a generative model. Hence, VAE may theoretically perform the same tasks as GAN. It is possible to sample new data from the distribution approximation. They can build more effective medications by linking the concepts of VAE with optimization [171]. Moreover, as the measurement of the bottleneck layer is typically significantly lower than the original input, this approach is utilized to reduce dimensionality and extract the significant features from the original input. They utilize the database from, which has been preprocessed from the TCGA database, which contains gene expression records for over 10,000 distinct cancers [172]. The RNA-seq data within the database describe the multidimensional state of each tumor. In the dataset used, the dimension for each tumor is 5,000. They can decrease the dimension to 100 using VAE. In this region, it is straightforward to recognize the shared patterns and signs of various cancers. In contrast to gene expression data [173], this example can be used in other uses, such as protein fingerprinting [174], that face the problem of high dimensionality.

7 GOALS AND ADVANCEMENTS

Chemists are increasingly using AI to carry out a variety of jobs. Initially, the need to speed up drug development, lower its astronomical costs, and shorten the period it requires for new drugs to become available in the marketplace were the major driving forces behind development in AI extended to chemistry. In terms of accelerating drug discovery research and development, AI has so far achieved tremendous strides. Although drug discovery is one of the applications of AI in chemistry that is covered in this chapter, it is not the only one. As we outline the broad picture of how AI may make chemists' research quicker and much more innovative. The major objective was to use the ML algorithm to analyze the atomic interactions inside a molecules in order to find the underlying patterns in molecular behaviour and to anticipate novel molecular events. Deep learning is a subclass of machine learning, which is only a neural network having three or more layers. Such ANNs attempt to imitate how well the human brain works, but they are unable to replicate it. This allows the network to "learn" from enormous amounts of data. Furthermore, DL is used to get better with every new piece of information. This involves having the ability to change its fundamental structure in order to properly evaluate data. Once that network has been completely constructed utilizing the test data, it enables more individualized use of customer analytics. Additionally, machine learning methods are unable to access the electrical structure of molecules yet may precisely determine atomistic chemical characteristics. It is more crucial than ever to code quickly. Upcoming DL programmers are expected to employee integrated, open, cloud-based development environments that provide them approaches to a variety of commercially accessible and customizable algorithm packages.

8 CHALLENGES AND OUTLOOK

Although AI offers many opportunities, there are also risks. Although the area of AI has recently made enormous strides, really intelligent and broadly usable AI solutions for medicine are still years away. Technology, data curation/sharing and security, workflow and integration, clinical evaluation and installation, as well as social, economic, political, ethical, and legal issues are just a few of the hurdles. These are highlighted [175].

8.1 Technical Challenges

As powerful tools, DL and AI excel at finding patterns in data and creating predictions for challenging scenarios. The main areas of current AIM research include automation, categorization, regression, and detection. But as it stands, AI technology is far from perfect and does not even come close to acting like a human being. AI needs to be less artificial and smarter than it is now if technology is to have a real impact and be widely adopted in clinical practice [175].

8.2 Data Security and Storage

The bulk of AI systems use a significant amount of data to learn and come to good conclusions. Utilizing a significant amount of data has the drawback of potentially causing storage issues for businesses. The use of data to automate organizational activities could also lead to issues with data security.

In order to apply AI, enterprises must adopt the best and most appropriate data management infrastructure. In addition to providing sensitive data with increased protection, such a data management platform will also make it straightforward for enterprises to access siloed data for AI and ML initiatives [176–177].

8.3 Legal Issues

The development and usage of AI applications raise a number of legal concerns for businesses. The algorithms' collection of user data is highly delicate. AI applications that use faulty data governance mechanisms and algorithms will always result in inaccurate forecasts and lower corporate revenues. Additionally, it can violate laws or guidelines, putting businesses in danger of legal issues [177].

8.4 Economical, Political, Social, Ethical, and Legal Aspects

At 18% of GDP, healthcare expenses in the United States are the highest per capita in the world. They have risen faster than the rate of our economy's expansion and are not entirely unsustainable. Prior research has demonstrated that between 1960 and 2007, medical technology increased healthcare expenses by 27% to 48%. 166 Cost-effectiveness studies must receive a lot of consideration when developing AI for medical applications. The expense of developing and using this technology is one issue that needs to be addressed. Will it be utilized in place of or in addition to a more expensive present technology? What is a reasonable estimate of the benefits that patients will experience as a result of AI use (e.g., fewer problems, improved survival, quicker onset of treatment or shorter duration of treatment)?

AI's explosive growth and ongoing innovations should be evaluated in conjunction with their effects on society. In fact, AI has the potential to have a huge impact on society and our daily lives. Numerous reports have been put together after relevant sources were analyzed by subject-matter specialists [175, 178–191].

9 CONCLUSION

Future human development will be influenced by deep learning in almost every sector. Deep learning will permeate a variety of businesses more deeply in the near future. Currently, it is found in our cars, smart devices, healthcare systems, and favorite apps. One should consider what higher types of understanding computers can help us reach in order to advance the field of AI for chemistry and materials research. The interpretability of the models we create is one of the tasks we set forth for theoretical chemistry in the twenty-first century. By accelerating scientific research, this will aid the irreducible duo of self-driving laboratory/human in resolving a number of the world's present challenges.

REFERENCES

1. www.geeksforgeeks.org/difference-between-artificial-intelligence-vs-machine-learning-vs-deep-learning/?ref=rp
2. www.geeksforgeeks.org/introduction-deep-learning/#:~:text
3. blog.quantinsti.com/introduction-deep-learning-neural-network/
4. Carvelli L, Olesen AN, Brink-Kjær A, Leary EB, Peppard PE, Mignot E, Sørensen HB, Jennum P. Design of a deep learning model for automatic scoring of periodic and non-periodic leg movements during sleep validated against multiple human experts. Sleep Med. 2020;69:109–19.
5. De Fauw J, Ledsam JR, Romera-Paredes B, Nikolov S, Tomasev N, Blackwell S, Askham H, Glorot X, O'Donoghue B, Visentin D, et al. Clinically applicable deep learning for diagnosis and referral in retinal disease. Nat Med. 2018;24(9):1342–50.
6. Topol EJ. High-performance medicine: the convergence of human and artificial intelligence. Nat Med. 2019;25(1):44–58.
7. Kermany DS, Goldbaum M, Cai W, Valentim CC, Liang H, Baxter SL, McKeown A, Yang G, Wu X, Yan F, et al. Identifying medical diagnoses and treatable diseases by image-based deep learning. Cell. 2018;172(5):1122–31.
8. Alzubaidi, L., Zhang, J., Humaidi, A.J. et al. Review of deep learning: concepts, CNN architectures, challenges, applications, future directions. J Big Data .2021;8:53
9. www.geeksforgeeks.org/difference-between-artificial-intelligence-vs-machine-learning-vs-deep-learning/
10. Sarker IH. Machine learning: Algorithms, real-world applications and research directions. SN Computer. Science. 2021;2(3):1–21
11. Sarker IH, Furhad MH, Nowrozy R. Ai-driven cybersecurity: an overview, security intelligence modeling and research directions. SN Computer. Science. 2021;2(3):1–18.
12. Sarker, I.H. Deep Learning: A Comprehensive Overview on Techniques, Taxonomy, Applications and Research Directions. Sn comput. Sci. 2021;2: 420.
13. seedscientific.com/how-much-data-is-created-every-day/J. Wise, "How Much Data Is Created Every Day in 2022? [New Stats]," Earthweb, 2022, Accessed: Aug. 24, 2022.
14. www.rbccm.com/en/gib/healthcare/episode/the_healthcare_data_explosion "RBC Capital Markets | The healthcare data explosion." (accessed Aug. 24, 2022).
15. N. Sharma, R. Sharma, and N . Jindal. Machine Learning and Deep Learning Applications-A Vision. Glob. Transitions Proc.2021;2(1): 24–28.
16. H. Chen, O. Engkvist, Y. Wang, M. Olivecrona, and T. Blaschke, The rise of deep learning in drug discovery. Drug Discov. Today.2018; 23(6):1241–1250.
17. H. Chen, O. Engkvist, Y. Wang, M. Olivecrona, and T. Blaschke. The rise of deep learning in drug discovery. Drug Discov. Today.2018;23(6): 1241–1250.
18. T. Tanebe and T. Ishida, End-to-end learning for compound activity prediction based on binding pocket information. BMC Bioinformatics.2021;22(3):1–11.
19. J. Ma, R. P. Sheridan, A. Liaw, G. E. Dahl, and V. Svetnik. Deep neural nets as a method for quantitative structure-activity relationships. J. Chem. Inf. Model.2015;55(2): 263–274.
20. A. Mayr, G. Klambauer, T. Unterthiner, and S. Hochreiter. DeepTox: Toxicity prediction using deep learning. Front. Environ. Sci. 2016; 3:80.

21. M. Wang et al. Deep learning approaches for de novo drug design: An overview. Curr. Opin. Struct. Biol. 2022;72: 135–144.

22. R. Gómez-Bombarelli et al. Automatic chemical design using a data-driven continuous representation of molecules. ACS Cent. Sci. 2016; 4(2):268–278.

23. A. Kadurin, S. Nikolenko, K. Khrabrov, A. Aliper, and A. Zhavoronkov. DruGAN: An Advanced Generative Adversarial Autoencoder Model for de Novo Generation of New Molecules with Desired Molecular Properties in Silico. Mol. Pharm. 2017; 14(9): 3098–3104.

24. I. J. Goodfellow et al. Generative Adversarial Nets. Adv. Neural Inf. Process. Syst. 2014; 27 : 1–10.

25. S. Fernández, A. Graves, and J. Schmidhuber. An application of recurrent neural networks to discriminative keyword spotting. Lect. Notes Comput. Sci. (including Subser. Lect. Notes Artif. Intell. Lect. Notes Bioinformatics). 2007; 4669 : 220–229.

26. M. H. S. Segler, T. Kogej, C. Tyrchan, and M. P. Waller. Generating focused molecule libraries for drug discovery with recurrent neural networks. ACS Cent. Sci.2018; 4(1): 120–131

27. W. Yuan et al. Chemical Space Mimicry for Drug Discovery. J. Chem. Inf. Model. 2017; 57(4): 875–882.

28. M. Olivecrona, T. Blaschke, O. Engkvist, and H. Chen. Molecular de-novo design through deep reinforcement learning. J. Cheminform. 2017; 9(1): 1–14.

29. E. J. Corey and W. Todd Wipke. Computer-assisted design of complex organic syntheses,. Science. 1969; 166 (3902): 178–192.

30. M. H. S. Segler and M. P. Waller. Neural-Symbolic Machine Learning for Retrosynthesis and Reaction Prediction. Chem.–A Eur. J. 2017;23(25): 5966–5971.

31. J. Tudela, M. Martínez, R. Valdivia, J. Romo, M. Portillo, and R. Rangel. Enhanced Reader.pdf. Nature. 2010; 388:539–547, 2010.

32. B. Liu et al., Retrosynthetic Reaction Prediction Using Neural Sequence-to-Sequence Models. ACS Cent. Sci. 2017; 3(10):1103–1113.

33. N. S. Pagadala, K. Syed, and J. Tuszynski. Software for molecular docking: a review. Biophys. Rev. 2017; 9(2):91–102.

34. M. Ragoza, J. Hochuli, E. Idrobo, J. Sunseri, and D. R. Koes. Protein-Ligand Scoring with Convolutional Neural Networks. J. Chem. Inf. Model. 2017; 57(4): 942–957.

35. C. Angermueller, T. Pärnamaa, L. Parts, and O. Stegle. Deep learning for computational biology. Mol. Syst. Biol. 2016; 12(7): 878.

36. F. Ning, D. Delhomme, Y. LeCun, F. Piano, L. Bottou, and P. E. Barbano. Toward automatic phenotyping of developing embryos from videos. IEEE Trans. Image Process. 2005; 14(9):1360–1371.

37. A. Ferrari, S. Lombardi, and A. Signoroni. Bacterial colony counting by Convolutional Neural Networks. Proc. Annu. Int. Conf. IEEE Eng. Med. Biol. Soc. EMBS.2015; 7458–7461.

38. K. Sirinukunwattana, S. E. A. Raza, Y. W. Tsang, D. R. J. Snead, I. A. Cree, and N. M. Rajpoot. Locality Sensitive Deep Learning for Detection and Classification of Nuclei in Routine Colon Cancer Histology Images. IEEE Trans. Med. Imaging. 2016; 35(5):1196–1208.

39. D. C. Cireşan, A. Giusti, L. M. Gambardella, and J. Schmidhuber. Mitosis detection in breast cancer histology images with deep neural networks. Lect. Notes Comput. Sci. (including Subser. Lect. Notes Artif. Intell. Lect. Notes Bioinformatics). 2013; 8150:411–418.

40. R. Turkki, N. Linder, P. E. Kovanen, T. Pellinen, and J. Lundin. Antibody-supervised deep learning for quantification of tumor-infiltrating immune cells in hematoxylin and eosin stained breast cancer samples. J. Pathol. Inform.2016;7(1): 1–11.

41. G. Litjens et al. Deep learning as a tool for increased accuracy and efficiency of histopathological diagnosis. Sci. Rep.2016;6:21–30.

42. M. E. Vandenberghe, M. L. J. Scott, P. W. Scorer, M. Söderberg, D. Balcerzak, and C. Barker. Relevance of deep learning to facilitate the diagnosis of HER2 status in breast cancer. OPEN. 2017;21–30.

43. K. H. Cha, L. Hadjiiski, R. K. Samala, H. P. Chan, E. M. Caoili, and R. H. Cohan. Urinary bladder segmentation in CT urography using deep-learning convolutional neural network and level sets. Med. Phys.2016; 43(4):1882–1898.

44. J. Z. Cheng et al. Computer-Aided Diagnosis with Deep Learning Architecture: Applications to Breast Lesions in US Images and Pulmonary Nodules in CT Scans. Sci. Rep. 2016;6:32–38.

45. Singh RK, Pandey R, Babu RN. COVIDScreen: explainable deep learning framework for differential diagnosis of COVID-19 using chest X-rays. Neural Comput Appl. 2021;33(14):8871–8892.

46. R. Li et al. Deep learning based imaging data completion for improved brain disease diagnosis. Lect. Notes Comput. Sci. (including Subser. Lect. Notes Artif. Intell. Lect. Notes Bioinformatics). 2014; 8675: 305–312.

47. M. R. Avendi, A. Kheradvar, and H. Jafarkhani. A combined deep-learning and deformable-model approach to fully automatic segmentation of the left ventricle in cardiac MRI. Med. Image Anal. 2016; 30:108–119.

48. S. Liu, H. Che, K. Smith, and L. Chen.Contamination event detection using multiple types of conventional water quality sensors in source water. Environ. Sci. Process. Impacts. 2014; 16(8):2028–2038.

49. R. Turkki, N. Linder, P. E. Kovanen, T. Pellinen, and J. Lundin. Antibody-supervised deep learning for quantification of tumor-infiltrating immune cells in hematoxylin and eosin stained breast cancer samples. J. Pathol. Inform. 2016; 7(1): 38.

50. W. Zhang et al. Deep model based transfer and multi-task learning for biological image analysis. Proc. ACM SIGKDD Int. Conf. Knowl. Discov. Data Min. 2015; 2015:1475–1484.

51. O. Vinyals, G. Deepmind, C. Blundell, T. Lillicrap, K. Kavukcuoglu, and D. Wierstra, Matching Networks for One Shot Learning. Adv. Neural Inf. Process. Syst. 2016; 29.

52. H. Altae-Tran, B. Ramsundar, A. S. Pappu, and V. Pande. Low Data Drug Discovery with One-Shot Learning. ACS Cent. Sci.2017; 3(4): 283–293.

53. A. Graves et al. Hybrid computing using a neural network with dynamic external memory. Nature. 2016; 538(7626): 471–478.

54. Adam C. Mater and Michelle L. Coote, Journal of Chemical Information and Modeling 2019; 59 (6):2545–2559.

55. Smith, J. S.; Isayev, O.; Roitberg, A. E. ANI-1: an extensible neural network potential with DFT accuracy at force field computational cost. Chemical Science. 2017; 8: 3192–3203.

56. Behler J. Atom-centered symmetry functions for constructing high-dimensional neural network potentials. J. Chem. Phys. 2011; 134: 074108.

57. Behler J. Neural network potential-energy surfaces in chemistry: a tool for large-scale simulations. Phys. Chem. Chem. Phys. 2011; 13: 17930–17955

58. Behler J. First Principles Neural Network Potentials for Reactive Simulations of Large Molecular and Condensed Systems. Angew. Chem., Int. Ed. 2017; 56: 12828–12840.

59. Behler J. Constructing high-dimensional neural network potentials: A tutorial review. Int. J. Quantum Chem. 2015; 115 :1032–1050

60. Behler J, Parrinello M. Generalized Neural-Network Representation of High-Dimensional Potential-Energy Surfaces. Phys. Rev. Lett. 2007; 98: 146401

61. Schütt, KT, Kessel P, Gastegger M, Nicoli KA, Tkatchenko A, Müller KR. A Deep Learning Toolbox For Atomistic Systems. J. Chem. Theory Comput. 2019; 15: 448–455.

62. von Lilienfeld, O. A. Quantum Machine Learning in Chemical Compound Space. Angew. Chem., Int. Ed. 2018; 57 : 4164–4169

63. Polykovskiy D, Zhebrak A, Vetrov D, Ivanenkov Y, Aladinskiy V, Mamoshina P, Bozdaganyan M, Aliper A, Zhavoronkov A, Kadurin A. Entangled Conditional Adversarial Autoencoder for de Novo Drug Discovery. Mol. Pharmaceutics. 2018; 15 : 4398–4405.

64. Kadurin A, Nikolenko S, Khrabrov K, Aliper A, Zhavoronkov A. druGAN: An Advanced Generative Adversarial Autoencoder Model for de Novo Generation of New Molecules with Desired Molecular Properties in Silico. Mol. Pharmaceutics. 2017; 14 : 3098–3104.

65. Popova M, Isayev O, Tropsha, A. Deep reinforcement learning for de-novo drug design.2017; 1711.10907.

66. Preuer K, Lewis RPI, Hochreiter S, Bender A, Bulusu KC, Klambauer G. DeepSynergy: predicting anti-cancer drug synergy with Deep Learning. Bioinformatics. 2018;34(9):1538–1548.

67. Preuer K, Renz P, Unterthiner T, Hochreiter S, Klambauer G. Fréchet ChemNet Distance: A Metric for Generative Models for Molecules in Drug Discovery. J Chem Inf Model. 2018;58(9):1736–1741.

68. Han Altae-Tran, Bharath Ramsundar, Aneesh S. Pappu, and Vijay Pande. Low Data Drug Discovery with One-Shot Learning. ACS Central Science 2017 3 (4), 283–293

69. Marwin H. S. Segler, Thierry Kogej, Christian Tyrchan, and Mark P. Waller. Generating Focused Molecule Libraries for Drug Discovery with Recurrent Neural Networks. ACS Cent. Sci. 2018; 4: 120–131

70. www.analyticssteps.com/blogs/5-ai-applications-chemistry

71. Matthias Rupp, Alexandre Tkatchenko, Klaus-Robert Müller, and O. Anatole von Lilienfeld. Fast and accurate modeling of molecular atomization energies with machine learning. Phys. Rev. Lett. 2012; 108: 058301.

72. Bartók AP, Kondor R, & Csányi G.On representing chemical environments. Phys. Rev. B. 2013; 87: 184115.

73. Faber FA.et al. Prediction errors of molecular machine learning models lower than hybrid dft error. J. Chem. Theory Comput. 2017; 13: 5255–5264.

74. Choudhary K, DeCost B, Chen C, et al. Recent advances and applications of deep learning methods in materials science. npj Comput Mater. 2022; 8(59).

75. chemintelligence.com/ai-for-chemistry

76. Fangping Wan, Yue Zhu, Hailin Hu, Antao Dai, Xiaoqing Cai, Ligong Chen, Haipeng Gong, Tian Xia, Dehua Yang, Ming-Wei Wang, Jianyang Zeng, DeepCPI: A Deep Learning-based Framework for Large-scale in silico Drug Screening. Genomics, Proteomics & Bioinformatics. 2019; 17 (3):478–495.

77. Peng J, Li J, Shang X. A learning-based method for drug-target interaction prediction based on feature representation learning and deep neural network. BMC Bioinform.2020; 21(13):1–13.

78. Li Y, Qiao G, Wang K, Wang G.Drug–target interaction predication via multi-channel graph neural networks. Brief Bioinform.2022; 23(1): 348.

79. Ding X, Zhang B .DeepBAR: a fast and exact method for binding free energy computation. J Phys Chem Lett. 2021; 12(10):2509–2515.

80. Li Y, Qiao G, Wang K, Wang G.Drug–target interaction predication via multi-channel graph neural networks. Brief Bioinform.2022; 23(1): 346

81. Peng J, Li J, Shang X. A learning-based method for drug-target interaction prediction based on feature representation learning and deep neural network. BMC Bioinform.2020;21(13):1–13

82. Wan, F., Hong, L., Xiao, A., Jiang, T., and Zeng, J. NeoDTI: neural integration of neighbor information from a heterogeneous network for discovering new drug-target interactions. Bioinformatics.2019; 35: 104–111.

83. Fangping Wan, Yue Zhu, Hailin Hu, Antao Dai, Xiaoqing Cai, Ligong Chen, Haipeng Gong, Tian Xia, Dehua Yang, Ming-Wei Wang, Jianyang Zeng, DeepCPI: A Deep Learning-based Framework for Large-scale in silico Drug Screening, Genomics, Proteomics & Bioinformatics.2019; 17 (5),: 478–495.

84. Lim S, Lu Y, Cho CY, et al. A review on compound-protein interaction prediction methods: Data, format, representation and model. Computational and Structural Biotechnology Journal. 2021;19:1541–1558.

85. Thafar M, Raies AB, Albaradei S , Essack M, Bajic VB . Comparison study of computational prediction tools for drug-target binding affinities. Front Chem. 2019a; 7:782.

86. Lin XLi XLin XA. review on applications of computational methods in drug screening and design.Mo lecules.2020;25(6);1375.

87. Feng Q, Dueva E, Cherkasov A, Ester M. Padme: A deep learning-based framework for drug-target interaction prediction. arXiv preprint arXiv. 2018;180709741

88. Nag S, Baidya ATK, Mandal A, Mathew AT, Das B, Devi B, Kumar R. Deep learning tools for advancing drug discovery and development. 3 Biotech. 2022;12(5):110.

89. Karimi M., Wu D., Wang Z., Shen Y. DeepAffinity: interpretable deep learning of compound-protein affinity through unified recurrent and convolutional neural networks. Bioinformatics. 2019; 35:3329–3338.

90. Thafar M, Raies AB, Albaradei S, Essack M, Bajic VB. Comparison Study of Computational Prediction Tools for Drug-Target Binding Affinities. Front Chem. 2019;7:782.

91. Ding X, Zhang B. DeepBAR: a fast and exact method for binding free energy computation. J Phys Chem Lett. 2021; 12(10):2509–2515

92. Nag S, Baidya ATK, Mandal A, Mathew AT, Das B, Devi B, Kumar R. Deep learning tools for advancing drug discovery and development. 3 Biotech 2022;12:110

93. Miller BR, McGee TD, Swails JM, Homeyer N, Gohlke H. An Efficient Program for End-State Free Energy Calculations. J. Chem. Theory Comput. 2012; 8: 3314–3321.

94. Rigla M, García-Sáez G, Pons B, Hernando ME. Artificial Intelligence Methodologies and Their Application to Diabetes. J Diabetes Sci Technol. 2018;12(2):303–310.

95. digi117.com/blog/top-5-tools-for-artificial-intelligence-ai-and-machine-learning-ml-development. html

96. Yu H. A systematic prediction of multiple drug–target interactions from chemical, genomic, and pharmacological data. PLoS One. 2012;7:e37608

97. Li Z. KinomeX: a web application for predicting kinome-wide polypharmacology effect of small molecules. Bioinformatics. 2019;35:5354–5356

98. Paul D, Sanap G, Shenoy S, Kalyane D, Kalia K, Tekade RK. Artificial intelligence in drug discovery and development. Drug Discov Today. 2021;26(1):80–93.

99. www.simplilearn.com/tutorials/deep-learning-tutorial/deep-learning-algorithm#how_deep_learning_algorithms_work

100. Macpherson, T., Churchland, A., Sejnowski, T., DiCarlo, J., Kamitani, Y., Takahashi, H., & Hikida, T. Natural and Artificial Intelligence: A brief introduction to the interplay between AI and neuroscience research. Neural Networks. 2021; 144: 603–613

101. Ching T, Himmelstein DS, Beaulieu-Jones BK, Kalinin AA, Do BT, Way GP, et al. Opportunities and obstacles for deep learning in biology and medicine. Journal of the Royal Society Interface. 2018;15.

102. Mahmud M, Kaiser MS, Hussain A, Vassanelli S. Applications of Deep Learning and Reinforcement Learning to Biological Data. IEEE Trans Neural Networks Learn Syst. 2018;29(6):2063–79.

103. Vailati-Riboni M, Palombo V, Loor JJ. What are omics sciences? Periparturient Dis Dairy Cows A Syst Biol Approach. 2017;1–7.

104. Montesinos-López OA, Montesinos-López A, Pérez-Rodríguez P, Barrón-López JA, Martini JWR, Fajardo-Flores SB, et al. A review of deep learning applications for genomic selection. BMC Genomics. 2021;22(1):1–23.

105. Zhang S, Zhou J, Hu H, Gong H, Chen L, Cheng C, et al. A deep learning framework for modeling structural features of RNA-binding protein targets. Nucleic Acids Res. 2015;44(4).

106. Alipanahi B, Delong A, Weirauch MT, Frey BJ. Predicting the sequence specificities of DNA–and RNA-binding proteins by deep learning. Nat Biotechnol. 2015;33(8):831–8.

107. Lanchantin J, Singh R, Lin Z, Qi Y. Deep Motif: Visualizing Genomic Sequence Classifications. Machine Learning. 2016;1–5.

108. Lee T, Yoon S. Boosted Categorical Restricted Boltzmann Machine for Computational Prediction of Splice Junctions. In: International conference on machine learning. 2015.

109. 1Xiong HY, Barash Y, Frey BJ. Bayesian prediction of tissue-regulated splicing using RNA sequence and cellular context. bioinformatics. 2011;27(18):2554–62.

110. Koumakis L. Deep learning models in genomics; are we there yet? Comput Struct Biotechnol J. 2020;18:1466–73.

111. Chandramouli K, Qian P-Y. Proteomics: Challenges, Techniques and Possibilities to Overcome Biological Sample Complexity. Hum Genomics Proteomics. 2009;1(1).

112. Wang S, Peng J, Ma J, Xu J. Protein Secondary Structure Prediction Using Deep Convolutional Neural Fields. Scientific Reports. Nature Publishing Group. 2016;4.

113. Mahmud M, Kaiser MS, Mcginnity TM, Hussain A. Deep Learning in Mining Biological Data DL4J. Cognitive Computation. Springer US. 2021; 1–33.

114. Huang KW, Chen J Le, Yang CS, Tsai CW. A memetic particle swarm optimization algorithm for solving the DNA fragment assembly problem. Neural Comput Appl. 2015;26(3):495–508.

115. Price GJ. A deep learning network approach to ab initio protein secondary structure prediction. IEEE/ACM Trans Comput Biol Bioinform. 2015;48(1):193–7.

116. Sønderby SK, Winther O. Protein Secondary Structure Prediction with Long Short Term Memory Networks. 2014;

117. Smith CA, Want EJ, O'Maille G, Abagyan R, Siuzdak G. XCMS: Processing mass spectrometry data for metabolite profiling using nonlinear peak alignment, matching, and identification. Anal Chem. 2006;78(3):779–87.

118. Tsugawa H, Cajka T, Kind T, Ma Y, Higgins B, Ikeda K, et al. MS-DIAL: data-independent MS/MS deconvolution for comprehensive metabolome analysis. Nat Methods 2015 128. 2015;12(6):523–8.

119. Pluskal T, Castillo S, Villar-Briones A, Orešič M. MZmine: Modular framework for processing, visualizing, and analyzing mass spectrometry-based molecular profile data. BMC Bioinformatics. 2010;11.

120. Borgsmüller N, Gloaguen Y, Opialla T, Blanc E, Sicard E, Royer AL, et al. WiPP: Workflow for Improved Peak Picking for Gas Chromatography-Mass Spectrometry [GC-MS] Data. Metabolites. 2019;9(9).

121. Melnikov AD, Tsentalovich YP, Yanshole V V. Deep Learning for the Precise Peak Detection in High-Resolution LC-MS Data. Anal Chem. 2020;92(1):588–92.

122. Risum AB, Bro R. Using deep learning to evaluate peaks in chromatographic data. Talanta. 2019;204:255–60.

123. Boka AP, Mukherjee A, Mir M. Single-molecule tracking technologies for quantifying the dynamics of gene regulation in cells, tissue and embryos. 2021; 4:21–29.

124. Falk T, Mai D, Bensch R, Çiçek Ö, Abdulkadir A, Marrakchi Y, et al. U-Net–Deep Learning for Cell Counting, Detection, and Morphometry. 2019;70:67–70.

125. Waithe D, Brown JM, Reglinski K, Diez-sevilla I, Roberts D, Eggeling C. Object detection networks and augmented reality for cellular detection in fluorescence microscopy. 2020;219(10).

126. Villoutreix P. What machine learning can do for developmental biology. 2021;15.

127. F. Ning, D. Delhomme, Y. Lecun, F. Piano, L. Bottou, et al. Toward automatic phenotyping of developing embryos from videos. IEEE Transactions on Image Processing, Institute of Electrical and Electronics Engineers. 2005; 14 (9).1360–137129.

128. Quazi S. Artificial intelligence and machine learning in precision and genomic medicine. Med Oncol 2022;39:120.

129. Chen CL, Mahjoubfar A, Tai LC, Blaby IK, Huang A, Niazi KR, et al. Deep Learning in Label-free Cell Classification. Sci Rep. 2016;6:1–18.

130. Shen J, Valagolam D. A Systematic Implementation of Machine Learning Algorithms for Multifaceted Antimicrobial Screening of Lead Compounds. ECA 2022 2022.

131. Ranjbarzadeh R, Kasgari AB, Ghoushchi SJ. Brain tumor segmentation based on deep learning and an attention mechanism using MRI multi - modalities brain images. Sci Rep. 2021;1–17.

132. Stages E. applied sciences TTCNN: A Breast Cancer Detection and Classification towards Computer-Aided Diagnosis Using Digital Mammography in. 2022;1–27.

133. Li K, Li X, Zhang Y, Zhang A. Affective state recognition from EEG with deep belief networks. Proc–2013 IEEE Int Conf Bioinforma Biomed. 2013;305–10.

134. Jia X, Li K, Li X, Zhang A. A novel semi-supervised deep learning framework for affective state recognition on EEG signals. In: IEEE 14th International Conference on Bioinformatics and Bioengineering, BIBE 2014; 2014. p. 30–7.

135. Zheng WL, Lu BL. Investigating Critical Frequency Bands and Channels for EEG-Based Emotion Recognition with Deep Neural Networks. IEEE Trans Auton Ment Dev. 2015;7(3):162–75.

136. Jirayucharoensak S, Pan-Ngum S, Israsena P. EEG-Based Emotion Recognition Using Deep Learning Network with Principal Component Based Covariate Shift Adaptation. Sci World J. 2014;1–10.

137. Shaw R, Patra BK. Classifying students based on cognitive state in flipped learning pedagogy. Futur Gener Comput Syst. 2022;126:305–17.

138. Liang Y, Lee JD. A hybrid Bayesian Network approach to detect driver cognitive distraction q. Transp Res Part C. 2014;38 :146–55.

139. Paul P, Anna T, Surekha D, Varghese M. EEG - based deep learning model for the automatic detection of clinical depression. Phys Eng Sci Med. 2020;[0123456789].

140. Salehi AW. A CNN Model: Earlier Diagnosis and Classification of Alzheimer Disease using MRI. 2020;156–61.

141. Access O. Automatic seizure detection using three–dimensional CNN based on multi-channel EEG. 2018;18(5).

142. Kermany DS, Goldbaum M, Cai W, Lewis MA. Identifying Medical Diagnoses and Treatable Diseases by Image-Based Deep Learning Resource Identifying Medical Diagnoses and Treatable Diseases by Image-Based Deep Learning. Cell. 2018;172(5):1122–1131.

143. 4Srinivasu PN, Sivasai JG, Ijaz MF, Bhoi AK, Kim W, Kang JJ. Networks with MobileNet V2 and LSTM. 2021;1–27.

144. Iot U, Dogiwal SR, Dadheech P, Making ED, Verma S. Machine Learning in liver disease diagnosis: Current progress and future opportunities Machine Learning in liver disease diagnosis: Current progress and future opportunities. 2022;18(2):1–10.

145. Nithya A, Appathurai A, Venkatadri N, Ramji DR, Palagan CA. Kidney disease detection and segmentation using artificial neural network and multi-kernel k-means clustering for ultrasound images. Measurement. 2020;149:106952.

146. Vasal S. COVID-AI: An Artificial Intelligence System to Diagnose COVID-19 Disease. 2020;9(8):62–7.

147. Swapna G, Vinayakumar R, Kp S. Diabetes detection using deep learning algorithms. ICT Express. 2018;4(4).

148. Jo T, Nho K, Saykin AJ. Deep Learning in Alzheimer ' s Disease: Diagnostic Classification and Prognostic Prediction Using Neuroimaging Data. 2019;11.

149. Damiani G. Arti fi cial neural networks allow response prediction in squamous cell carcinoma of the scalp treated with radiotherapy. 2020;1369–73.

150. Gao XW, James-reynolds C, Currie E. Analysis of tuberculosis severity levels from CT pulmonary images based on enhanced residual deep learning architecture. 2020;6(12):131.

151. Masoudi S, Mehralivand S, Harmon SA, Lay N, Lindenberg L, Mena E, et al. Deep Learning Based Staging of Bone Lesions From Computed Tomography Scans. 2021;87531–42.

152. Li JH, Liu S, Zhou H, Qu LH, Yang JH. StarBase v2.0: Decoding miRNA-ceRNA, miRNA-ncRNA and protein-RNA interaction networks from large-scale CLIP-Seq data. Nucleic Acids Res. 2014;42:92–7.

153. Pan X, Shen H Bin. Predicting RNA-protein binding sites and motifs through combining local and global deep convolutional neural networks. Bioinformatics. 2018;34(20):3427–38.

154. Ray D, Kazan H, Chan ET, Castillo LP, Chaudhry S, Talukder S, et al. Rapid and systematic analysis of the RNA recognition specificities of RNA-binding proteins. Nat Biotechnol. 2009;27(7):667–70.

155. Dai H, Umarov R, Kuwahara H, Li Y, Song L, Gao X. Sequence2Vec: A novel embedding approach for modeling transcription factor binding affinity landscape. Bioinformatics. 2017;33(22):3575–83.

156. Venter JC, Smith HO, Adams MD. The Sequence of the Human Genome. Clin Chem. 2015;61(9):1207–8

157. Pennacchio LA, Ahituv N, Moses AM, Prabhakar S, Nobrega MA, Shoukry M, et al. In vivo enhancer analysis of human conserved non-coding sequences. Nature. 2006;444(7118):499–502.

158. Quang D, Xie X. DanQ: A hybrid convolutional and recurrent deep neural network for quantifying the function of DNA sequences. Nucleic Acids Res. 2016;44(11):1–8.

159. Angermueller C, Lee HJ, Reik W, Stegle O. DeepCpG: Accurate prediction of single-cell DNA methylation states using deep learning. Genome Biol. 2017;18(1).

160. Mercer TR, Dinger ME, Mattick JS. Long non-coding RNAs: Insights into functions. Nat Rev Genet. 2009;10(3):155–9.

161. Esteva A, Kuprel B, Novoa RA, Ko J, Swetter SM, Blau HM, et al. Dermatologist-level classification of skin cancer with deep neural networks. Nature. 2017;542(7639):115–8.

162. Kermany DS, Goldbaum M, Cai W, Valentim CCS, Liang H, Baxter SL, et al. Identifying Medical Diagnoses and Treatable Diseases by Image-Based Deep Learning. Cell. 2018;172(5):1122–1131.e9.

163. Fei-Fei L, Deng J, Li K. ImageNet: Constructing a large-scale image database. J Vis. 2010;9(8):1037–1037.

164. Ito T, Chiba T, Ozawa R, Yoshida M, Hattori M, Sakaki Y. A comprehensive two-hybrid analysis to explore the yeast protein interactome. Proc Natl Acad Sci U S A . 2001;98(8):4569–74.

165. Grover A, Leskovec J. Node2Vec: Scalable Feature Learning for Networks. In: 22nd ACM SIGKDD International Conference on Knowledge Discovery and Data Mining. 2018. p. 855–64.

166. Kipf TN, Welling M. Semi-superwised classification with graph convolutional network. Astron Astrophys. 2017;587:1–14.

167. Merk A, Bartesaghi A, Banerjee S, Falconieri V, Rao P, Davis MI, et al. Breaking Cryo-EM Resolution Barriers to Facilitate Drug Discovery. Cell. 2016;165(7):1698–707.

168. Ruhlman CD, Bartizek RD, Biesbrock AR. Comparative Efficacy of Two Battery-Powered Toothbrushes on Dental Plaque Removal. J Clin Dent. 2002;13(3):95–9.

169. Chen Y, Xie Y, Zhou Z, Shi F, Christodoulou AG, Li D. Brain MRI super resolution using 3D deep densely connected neural networks. In: IEEE 15th International Symposium on Biomedical Imaging. 2018. p. 1–11.

170. Li Y, Xu F, Zhang F, Xu P, Zhang M, Fan M, et al. DLBI: Deep learning guided Bayesian inference for structure reconstruction of super-resolution fluorescence microscopy. Bioinformatics. 2018;34(13):284–294.

171. Rampášek L, Hidru D, Smirnov P, Haibe-Kains B, Goldenberg A. Dr.VAE: Improving drug response prediction via modeling of drug perturbation effects. Bioinformatics. 2019;35(19):3743–51.

172. Way GP, Greene CS. Extracting a biologically relevant latent space from cancer transcriptomes with variational autoencoders. Pacific Symp Biocomput. 2018;23:80–91.

173. Pierson E, Yau C. ZIFA: Dimensionality reduction for zero-inflated single-cell gene expression analysis. Genome Biol. 2015;16(1):1–10.

174. Das P, Moll M, Stamati H, Kavraki LE, Clementi C. Low-dimensional, free-energy landscapes of protein-folding reactions by nonlinear dimensionality reduction. In: Proceedings of the National Academy of Sciences of the United States of America. 2008. p. 9885–90.

175. Lang T, Pelaseyed T. Discovery of a MUC3B gene reconstructs the membrane mucin gene cluster on human chromosome 7. PLoS One. 2022 Oct 18;17(10):e0275671

176. 10xds.com/blog/challenges-implementing-artificial-intelligence/

177. www.spiceworks.com/tech/artificial-intelligence/guest-article/top-10-ai-development-and-implementation-challenges/

178. Margetts H, Dorobantu C. Rethink government with AI. Nature 2019;568:1635.

179. Magnus D, Batten JN. Building a trustworthy precision health research enterprise. Am J Bioeth 2018;18:12.

180. Char DS, Shah NH, Magnus D. Implementing machine learning in health care addressing ethical challenges. N Engl J Med 2018;378:9813.

181. Kreitmair KV, Cho MK, Magnus DC. Consent and engagement, security, and authentic living using wearable and mobile health technology. Nat Biotechnol 2017;35:61720.

182. Grigorovich A, Kontos P. Towards Responsible Implementation of Monitoring Technologies in Institutional Care. Gerontologist. 2020 Sep 15;60(7):1194-1201

183. Gordon JS. Building moral robots: ethical pitfalls and challenges. Sci Eng Ethics 2020;26:14157.

184. Racine E, Boehlen W, Sample M. Healthcare uses of artificial intelligence: challenges and opportunities for growth. Healthc Manage Forum 2019;32:2725.

185. Nebeker C, Torous J, Bartlett Ellis RJ. Building the case for actionable ethics in digital health research supported by artificial intelligence. BMC Med 2019;17:137.

186. Cath C, Wachter S, Mittelstadt B, Taddeo M, Floridi L. Artificial intelligence and the 'Good Society': the US, EU, and UK approach. Sci Eng Ethics 2018;24:50528.

187. Horvitz E, Mulligan D. Policy forum. Data, privacy, and the greater good. Science 2015;349:253–5.

188. Price, W. Nicholson, II. "Health Care AI: Law, Regulation, and Policy." Douglas Mcnair, co-author. In Artificial Intelligence in Health Care: The Hope, the Hype, the Promise, the Peril, edited by Michael Matheny et al., 181–213. Washington, D.C.: The National Academy of Medicine, 2019.

189. builtin.com/artificial-intelligence/artificial-intelligence-future

190. A. Aspuru-Guzik, R. Lindh, and M. Reiher. Exploration of Reaction Pathways and Chemical Transformation Networks. ACS Cent Sci. 2018; 4: 144–152.

191. Schlick T, Portillo-Ledesma S. Biomolecular modeling thrives in the age of technology. Nat Comput Sci. 2021; 1: 321–331.

9 Text Mining in Chemistry for Organizing Chemistry Data

Unnati Patel, Prajesh Prajapati, Dharmendrasinh Bariya, Jigna Prajapati

1 INTRODUCTION

Humans are unable to process the volume of data produced each day. It is difficult for us to derive inferences from the vast bulk of this data because it is unorganized and unlabeled. The explosion of big data, a large portion of text, has prompted researchers to begin developing computer algorithms to deal with textual data [1]. Hence, text mining is helpful here because it enables researchers to extract meaningful information from massive, unrestricted texts. Hans Peter Luhn first suggested the concept of text mining in the *IBM Journal* in 1958, and it is also acknowledged as one of the earliest forms of artificial intelligence [2]. In his paper, Luhn described a technique for extracting words from documents and categorizing them based on frequency statistics [2]. Much productive research followed the idea of text mining [3].

Scientific literature has switched from print publications to digital electronic publishing as a result of the expansion of the internet. This migration has significantly impacted research in academic and industrial settings [4]. Since the 1940s, the emergence of machine-readable encoding techniques in the chemistry sector has made faster movement [4]. Similarly, the patents in the field of chemistry have been multiplied five times annually (from 1000 per year to around 5000 per year) since the early 1990s [5]. It is interesting to note that, aside from life scientists, chemistry experts read more scholarly articles per person than their counterparts in other fields [6]. Compared to other researchers, they study scientific material the most frequently on chemistry due to its difficulties in literature [4].

The identification of exciting compounds in chemistry, which are based on the compound's structure that, frequently is accessed [5, 6]. This can be used for the preliminary stages of chemical prediction modeling or quantitative structure–activity relationships (QSAR) modelling in medicinal chemistry [5, 6]. Chemical study depends on understanding the structures of chemical compounds; hence while investigating the chemical realm, most chemists concentrate on chemical structures or substructures [4].

Chemical-related information can be included in chemistry publications in several ways, whether they are journal articles or patents [7]. Various chemical IDs can be used to store the data in the textual portion of the document (naming conventions) [7]. Additionally, the data can be kept in tabular form or in 1D, 2D, or 3D chemical diagrams. Manually locating and extracting pertinent information from texts relating to chemicals in the form of scientific publications and patents is becoming more and more challenging due to the volume of such texts constantly growing [8]. Processing large amounts of unstructured data manually from numerous sources is difficult.

Different strategies can be considered in order to get beyond this challenge, amongst these techniques, mining is available as commercial or open-source chemical databases, which use

DOI: 10.1201/9781003353768-9

chemical text-mining to extract information from the published documents. Chemical entities extracted from a text can aid information retrieval systems by pointing to documents that mention the compound. Another approach that can be used to assess the data is customized search engines in conjunction with other relevant information (e.g., biological activity extracted from text) [9]. It will be more helpful if it can differentiate the appropriate compounds from a large number of extracted compounds within a document [7, 10]. This chapter focuses on resources for extracting chemical information, followed by comprehensive details of the text mining process. Cheminformatics is also discussed in this chapter. Finally, the challenges and future trends in text mining of chemistry data are discussed.

2   WHAT ARE CHEMICAL INFORMATION RESEARCHERS LOOKING FOR?

Researchers frequently look for chemical information such as chemical names, structures, molecular formulas, physicochemical characteristics, toxicity, biological activities, and usage of chemicals. Chemicals can be classified according to their chemical properties (hydrocarbon), their physical properties (Volatile Organic Solvents), their usage (Sanitizers, Pesticides), radioactive materials, etc. Chemicals are typically categorized by generic name (Aspirin), brand name (Disprin®), systematic (acetylsalicylic acid), or by abbreviation/acronyms (ASA). Most of the above chemical names are inaccessible to Information Retrieval (IR) users as they lack knowledge of the elemental chemical structure (i.e., the relationship between atoms and bonds). These structural details are captured by SMILES (such as "C1CCCCC1") and InChi/InChIKey codes (such as "InChI = 1S/C6H12/c1-2-4-6-5-3-1/h1-6H2"), which make them suitable for this use.

Initially, chemicals were given a trivial name. In order to standardize chemical compound notation and conduct a more thorough analysis of chemical information representation, the International Union of Pure and Applied Chemistry (IUPAC) was established in 1919 [11]. Since 1921, the IUPAC has formed committees to study chemical nomenclature with the goal of codifying standards for systematically naming compounds [11]. A molecule with intricate stereochemistry can not be effectively described by its IUPAC designations alone. Other initiatives to offer some sort of systematic chemical names exist in addition to the IUPAC names (e.g., CAS index names). In addition to commercial codes (e.g., ICI204636), the chemicals were also known by their name after inventors (e.g., "Glauber's salt"; i.e., sodium sulphate), "Devil's Red"; i.e., doxorubicin) and "Jim's juice"; i.e., cancel), which is also famous as chemical nicknames [6, 12, 13].

Markush structures and chemical reactions are additional targets of interest for mining techniques in addition to specific chemical names. The term "Markush structures" originates from Eugene Markush's first patent, issued in 1924, which included a chemical substance with generic components and maintained related compounds as per its similar structure [6]. Markush formulae are frequently used in patent applications as well as scholarly papers explaining the structure–activity relationships (SAR) of families of compounds and combinatorial libraries [6]. The generic notations used to represent the core of Markush structures are known as R-groups and they are used to enumerate atom and bond ists and homology groups (such as "heteroaryl," "a bond," and "C1–C8 alkyl") [6]. These also describe the variations at the attachment point and rates of substituents occurrence (like repeating units) [6, 14].

3   WHERE DO RESEARCHERS FIND CHEMICAL INFORMATION?

As the majority of sophisticated document retrieval systems (search engines) aggregate them, this section addresses some of the most popular sources utilized by chemists for unstructured or index repositories for information. The majority of chemical papers can be divided into three groups: patents, scientific publications, and various regulatory reports. The preparation for text mining and information retrieval is significantly impacted by each of them having different levels of

TABLE 9.1

Platforms and Portals for Document Repository Search

Data Repository	Document Type	Access	References
Global Patent Search Network	Patents	Public	[15]
SIPO Patent Search	Patents	Public	[16]
Patent Scope	Patents	Public	[17]
Free Patents Online	Patents	Public	[18]
Patent Lens	Patents	Public	[19]
Sumo Brain	Patents	Public	[20]
Total Patent	Patents	Commercial	[21]
JPDS	Patents	Commercial	[22]
Pat-Seer	Patents	Commercial	[23]
PatBase	Patents	Commercial	[24]
PriorSmart	Patents	Commercial	[25]
WIPS Global	Patents	Commercial	[26]
Google Patents	Patents	Public	[27]
OvidSP	Patents	Commercial	[28]
Google Scholar	Journal articles, Reports, Patents	Public	[29]
PubMed	Journal articles	Public	[30]
Europe PMC	Journal articles, Patents	Public	[31]
PMC Canada	Journal articles	Public	[32]
PubChem	Journal articles, Patents	Public	[33]
Springer Nature	Journal articles, books	Commercial	[34]
Science Direct (Elsevier)	Journal articles	Commercial	[35]
Scopus (Elsevier)	Abstracts	Public	[36]
SciFinder (CAS)	Journal articles, Patents	Commercial	[37]
DOAJ	Journal articles	Public	[38]
CAplus	Journal articles, Reports, Patents	Commercial	[39]
aRDi	Journal articles, Patents	Public	[40]
TOXNET	Journal articles	Public	[41]
BASE	Journal articles	Public	[42]
Sci-Hub	Journal articles, Books	Public	[43]
Reaxys	Journal articles, Patents	Commercial	[44]
GOSTAR	Journal articles, Patents	Commercial	[45]
STN	Journal articles, Patents	Commercial	[46]
Thompson Innovation	Journal articles, Patents	Commercial	[47]
Web of Science	Journal articles	Commercial	[48]
ProQuest Dialog	Journal articles, Patents, Dissertations, Books	Commercial	[49]
iScienceSearch	Journal articles, Patents	Public	[50]
Academic Search	Journal articles	Commercial	[51]

format uniformity. Some of the platforms and portals for text mining and information retrieval are listed in Table 9.1.

PMC–PubMed Central, CAS–Chemical Abstract Service, DOAJ–Directory of Open Access Journals, aRDi–Access to Research for Development and Innovation, TOXNET–The Toxicology Data Network, BASE–Bielefeld Academic Search Engine, GOSTAR–Global Online Structure Activity Relationship Database, STN–Scientific & Technical Information Network, SIPO–Slovenian Intellectual Property Office, JPDS–Japan Patent Data Service, WIPS–Worldwide Intellectual Property Service.

3.1 PATENTS

The patent format includes, bibliography with the invention title, name of all applicant(s), name of inventor(s), patent filing and publication date, classification code, and abstract, a background details with the examples, and scientific claims that supports the parameters of the invention comprise the format of patents. The largest commercial repositories listed in Table 9.1 also include index patents, including both applications and awarded patents.

National and regional patent offices make images of text documents to keep track of granted patents and patent applications in the public domain and either offer Internet access to their internal collections or provide patents in other formats, like XML [6]. The fact that patent databases are well organized by patent families rather than individual patent records makes them distinctive (e.g., Derwent World Patents Index (DWPI) [52], INPADOC [53], and PatBase [54]). The published patent documents from various nations that reveal similar technical inventions and are connected by claims of priority are known as a patent family. WIPO, national and regional offices, as well as providers of publicly available, cost-free patent databases (such as Free Patents Online [55], Patent Lens [56], and PriorSmart[57]) provide a number of search interfaces for patents alone (in Table 9.1, the largest collections patent offices, SIPO, EPO, USPTO, and WIPO are listed).

3.2 SCHOLARLY ARTICLES

Scientific journals often follow an orderly format with sections for the title page with or without author information, abstract, keywords, introduction, methodology, experimental sections, results and discussion, and conclusion. Journal literature repositories consist of some well-established and reputed references. For instance, 1) in the biomedicine field, BIOSIS Previews [58], Embase [59], and MEDLINE [60]; 2) Inspec [61] in the engineering/physics disciplines; 3) TOXLINE in the toxicology and pharmacology field; 4) CAplus for chemistry, biochemistry, and chemical-related fields [62]. Some of the platforms and portals for document repositories are listed in Table 9.1.

The majority of the time, especially for non-patent literature, the search services associated with these repositories include links to the relevant journal (e.g., SciFinder) rather than offering the option to examine publications found through searches in their full-text version [63]. On the other hand, several smaller repositories, including the Directory of Open Access Journals and PubMed Central (PMC) for the biological sciences field, offer full-text document access [64]. In addition to these gathered resources, text-based searches can be performed on open-access publishers' publications or fee-based journal databases. The ScienceDirect APIs provide access to ScienceDirect [65]. However, certain publication houses restrict complete access to the contents (e.g., TOXLINE and PubMed Central), whereas MEDLINE allows full-text mining of all of its information.

3.3 REGULATORY REPORTS

Regulatory reports provided by governmental bodies, such as DailyMed, the FDA's New Drug Application (NDA),[66] or the European Public Assessment Reports (EPAR) [67], have chemical information. Additionally, paid access to the profiles of pharmaceutical programmes, clinical research, safety analyses, and business transactions is provided by Adis Insight [68]. Many people consider the Scopus database (Elsevier) and the Web of Science (Thomson Reuters) to be comparable resources [69]. Although the Web of Science has the unique advantage that includes bibliographic resources along with chemistry and reaction databases [69].

The four primary standard tools, Web of Science [70], Reaxys [71], STN [72], and SciFinder [63], are all for-profit organizations, and highlight the need and desire to do away with freely available annotated databases. The recent debut of SureChEMBL, a patent repository with a unique access interface, further demonstrates this [6]. Furthermore, iScienceSearch is a publicly available federal search engine that permits searching over 86 accessible chemical databases, academic publications, and patents [73]. However, its information retrieval ability is inferior to that of SciFinder [73].

Google Scholar [74] and Europe PubMed Central (Europe PMC) [75] are two systems that combine scientific literature and patents in the open-access domain, but Google Scholar [74] is sometimes criticized for not outlining clear selection criteria for journal publications that are indexed. It is interesting to note that the Access to Research for Development and Innovation (aRDI) program makes it easier for patent offices worldwide and technical institutions in developing nations to access academic journals on a variety of scientific subjects (for free or for a small cost) [76].

4   TEXT MINING OF CHEMISTRY LITERATURE

It is challenging to explore the chemical domain in publications relating to chemicals, such as journal papers and patents. By employing algorithmic, statistical, and data management approaches, text mining can be used to extract pertinent data from a substantial body of literature on chemicals and from unstructured free text. Text mining shifts the burden of too much information from people to robots [77]. The intricacy of the content can affect the effectiveness of a text-mining system. The text-mining algorithms mainly focus on domains or sub-domains to attain high performance. For instance, a text-mining algorithm needs to consider the distinctions between journal articles and patents [77, 78].

A scientific text-mining pipeline typically consists of the steps shown in Figure 9.1: (1) text collection, which is document retrieval; (2) document processing converts PDFs into the plain text or from mark-up languages to plain text; (3) text representation by natural language processing; and (4) text analysis and information extraction [79]. The resulting collection may be utilized as the text mining's final product or as a resource for further mining and analysis [79].

4.1   TEXT COLLECTION

Section 3 of this chapter covered well-known repositories for mining the data in chemistry and their access options. The primary advantages of well-known databases for text mining are their usable application programming interfaces (API), uniform metadata structure, and, occasionally, analytic tools.

4.2   DOCUMENT PROCESSING

Metadata of the targeted text, like the name of the journal, its title, its authors, its keywords, and others, are often included in the content that can be retrieved. As depicted in Table 9.1, text databases provide the ordered outputs of raw text, which shows processing and analysis. Contrarily, website-extracted content normally consists of final paper files that must be converted into raw text. Most text sources now offer HTML/XML/JSON documents, while papers published earlier are mostly available as embedded or image PDFs.

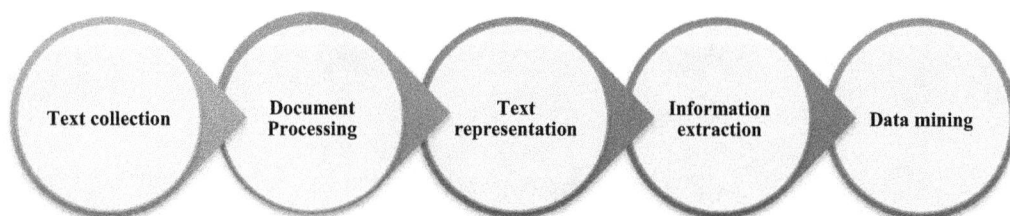

FIGURE 9.1   Text representation by Natural Language Processing (NLP) pipeline.

To create a whole record of the documents, system scanned for PDF files, HTML text documents, and XML files in order to extract the raw text, and maybe fragment data from several sources [80]. The final outcome of the document elements such as title, abstract, heading, paragraph, figure, and table forms a coherent, streamlined document structure [80]. This enables subsequent pipeline components to handle every document as same format, irrespective of its original file format.

While reading HTML/XML markups may be done with a variety of computer tools, extracting plain text from PDF files takes more effort. PDF documents pose a more significant obstacle because the format does not enable computer interpretation of the content. Available PDFs typically have a structure with columns of text interspersed with tables, figures, and equations, which has an impact on text sequencing and conversion accuracy. Scientific articles published in PDF format have attempted to recover a logical text structure using rule-based [81] and machine learning techniques [82, 83]. According to the author's experience, this can significantly affect the extraction pipeline's output [82, 83]. Many documents are only available as image PDF files, especially those that date back before the 1990s. To the best of our knowledge, when these data are translated into raw text, they still are unable to deliver high enough precision to extract chemical information consistently [84, 85].

The pragmatics margins of headers, paragraphs, figures, tables, and captions makes it simple to process content from HTML and XML sources. After isolation of the text domain, any additional available markup (such as bold and italic characters) is used for plain text for language processing naturally. The original table structures are preserved in the nest in the form of individual table cells and are regarded as distinct text domains [80].

4.3 TEXT REPRESENTATION

Raw documents are subjected to the five major stages: 1) sentence splitting, 2) tokenization, 3) part-of-speech recognition, 4) recognition of chemical substances, and 5) phrase parsing (see Figure 9.2). In this process, the text is divided into logical units (such as statements) and tokens (such as words

FIGURE 9.2   Schematic representation of steps in the text mining process.

and phrases), which are combined to form the full sentences (the text's grammatical structure) [79]. The text tokenization may be normalized by stemming or lemmatization and carried out by a part of speech tagging (POS tagging) and resolve the dependency to form the sentence structure on the final text and for the data mining purpose [79].

4.3.1 Sentence Splitting

Typically, the text is divided into sentences in this step. For sentence detection, rule-based approaches are typically used (e.g., use of exclamation mark or question mark for sentences ending with a period) [86]. It can be challenging to recognize repeated sentences in a publication related to chemicals. Systematic chemical identifiers for any chemicals like IUPAC names have punctuation marks, which makes sentence splitting more difficult.

Swain et al. published ChemDataExtractor, a toolkit to extract chemical information; in their system, a sentence splitter was trained on 3592 chemistry papers from The American Chemical Society (ACS), The Royal Society of Chemistry (RSC), and Springer's abstracts, primary texts, and captions [80]. The sentence splitter correctly differentiates between natural sentence boundaries and punctuation that can be found in abbreviations such as "equiv," "fig," "et al.," and "ref," which are commonly used in the literature. The training articles' sentence splitter can differentiate among 702,132 unique sentences [80].

4.3.2 Tokenization

Each sentence is divided into words, or tokens, that roughly match individual words and punctuation throughout the tokenization process. The use of punctuation and symbols has a notable impact on the tokenization of chemical names. Parentheses, for instance, serve as pattern separators in standard English. Parentheses can be a part of a token in chemistry, as in "$(CH_3)_2CHCH_2CH(CH_3)_2$." Splitting a sentence into its logical components is an essential step in the information extraction process because mistakes made here have a tendency to spread to other steps and affect the accuracy of the results.

Several examples demonstrate the tokenization of sentences using all-purpose tokenizers like NLTK [87] and SpaCy [88]. The random arrangement of punctuation marks in chemical formulas, and another domain-specific terminology is a significant source of errors, similar to sentence segmentation. To address the issue of over-tokenization, the chemical NLP toolkits OSCAR4 [89], ChemicalTagger [90], and ChemDataExtractor [80] each employ their own rule- and dictionary-based solutions. When it comes to chemical terminology, chemical NLP toolkits excel even when the remainder of the text has inferior tokenization accuracy.

4.3.3 Part-of-Speech Tagging

When using part-of-speech (POS) tagging, each token is given a label that indicates its grammatical function, which includes adjectives, noun, verbs, and others. This process gives linguistic and grammar-based properties to the words that are utilized as input for machine learning models instead of altering the text corpus. It can be challenging to recognize POS tags in scientific content because of the ambiguity the word's context introduces. The use of POS tags on the tokens, various tools utilized for phase resolution, crucial information extraction and for entity recognition determines the success of the tagging process as part of their inputs.

Most publicly accessible natural language recognition systems include POS taggers that were trained on news articles. As a result, they occasionally have difficulty spotting literature in the field of chemistry. Tsuruoka et al. discovered that performance of trained a POS tagger in the biomedical sector was significantly enhanced on a mixed corpus of MEDLINE abstracts (GENIA corpus) and newspaper articles (WSJ corpus) [91–93].

Table 1. Calculated Lattice Parameters of Li$_x$Ni$_{1/3}$Co$_{1/3}$Mn$_{1/3}$O$_2$

x	a, Å	c, Å	Ni–O, Å	Co–O, Å	Mn–O, Å
1[a]	2.892	14.251	(4 × 2.05, 2 × 2.07)	(4 × 1.95, 2 × 1.96)	(6 × 1.94)
5/6[b]	2.891	14.398	(2.00, 2.03, 2 × 2.05, 2.07) (1.89, 1.91, 1.99, 2 × 2.04, 2.07)	(1.89, 1.93, 1.94, 3 × 1.95) (1.91, 1.93, 2 × 1.94, 1.95, 1.98)	(1.91, 2 × 1.93, 2 × 1.95, 1.96) (1.92, 1.93, 3 × 1.94, 1.95)
2/3[a]	2.884	14.503	(2 × 1.88, 4 × 2.04)	(2 × 1.89, 4 × 1.94)	(2 × 1.90, 2 × 1.93, 2 × 1.96)
1/2[b]	2.841	14.973	(1.89, 3 × 1.91, 1.92, 1.94) (2 × 1.90, 1.91, 1.93, 2 × 2.12)	(1.89, 1.90, 1.91, 1.94, 1.95, 1.97) (2 × 1.90, 1.92, 2 × 1.93)	(1.89, 1.91, 1.93, 1.94, 2 × 1.95) (1.89, 1.91, 1.93, 1.94, 1.96)

TABLES DATA EXTRACTION

Experimental Section

Powders were synthesized by a sol-gel method. Precursors were dissolved in distilled water and mixed with citric acid. Resulting solution was stirred at 80°C for 5 h, and dried in a vacuum oven at 140 °C for 24 h. Resulted compound was calcined at 900 °C.

BUILDING SYNTHESIS ACTIONS GRAPH

INTERPRETATION OF FIGURES AND CAPTIONS

Results & Discussion

Analysis shows different particle size distributions for the two materials. The particle size of LNTMO is ~100 nm, whereas LMNO has a broad size distribution ranging from tens of nanometers to micrometers.

CHEMICAL NAMED ENTITIES RECOGNITION

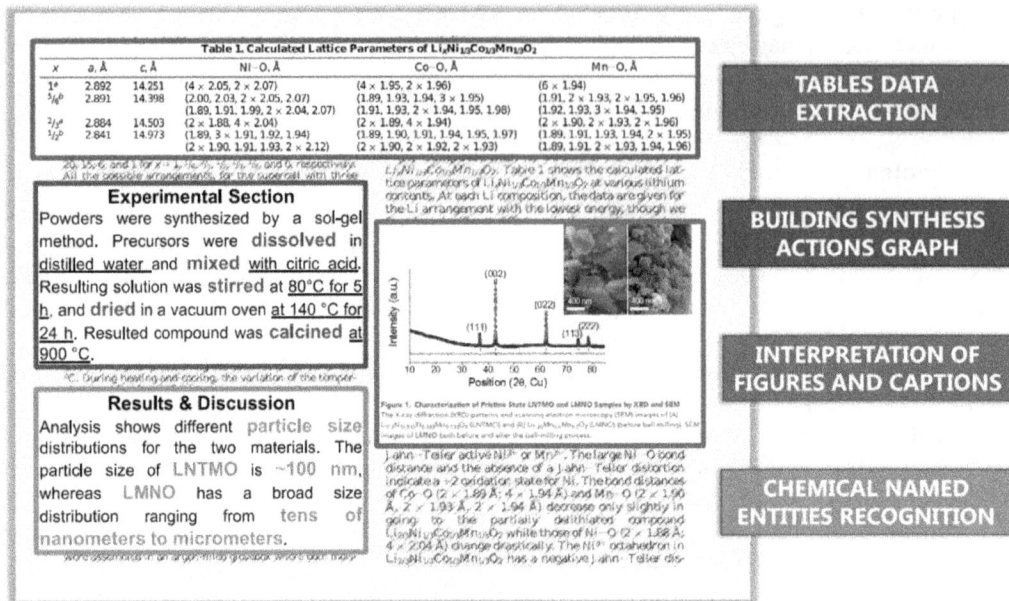

FIGURE 9.3 Information types extracted from a typical scientific paper–A Schematic. Reprinted with permission from Kononova, O.; He, T.; Huo, H.; Trewartha, A.; Olivetti, E. A.; Ceder, G. Opportunities and challenges of text mining in materials research. *Science* **2021**, *24* (3), 102155.

4.3.4 Chemical Named Entry Recognition

The technique of locating and classifying particular chemical substances within a text is known as Chemical Named Entity Recognition (NER) [78]. Chemical NER recognizes chemical compounds or their subclasses using IUPAC names, formulae, and CAS numbers, among other information. Information retrieval (IR), one of the many NLP jobs, uses pre-processed corpus to extract various types of data (see Figure 9.3) [79]. Chemical Named Entity Recognition, which assigns text tokens to one of several categories, is the most common IR task. Systems for document categorization and information retrieval can be improved by implementing NER approaches. Systems like Semedico [53], for example, improve retrieval of documents of different categories by automatically classification, like chemicals and drugs [77].

Chemical named entities are extracted from text using three distinct text-mining methods: grammar-based, dictionary-based, and statistics-based [9, 77]. The concepts that provide systematic chemical identities are utilized by grammar-based approaches to capture them [77]. Because of this, it can find the entries (like IUPAC names) that are missing from dictionaries. A set of guidelines can be used to translate a systematic name into a chemical structure. The same rules are applied by grammar-based methods to produce chemical structures for known substances. It is essential to have a solid understanding of the domain and naming conventions while developing grammar-based systems. Because naming standards change over time, these systems also need to be updated. Regular expressions can be used to locate some of these identifiers, although grammar-based algorithms are often only capable of finding a small subset of non-systematic chemical IDs [9].

Dictionary-based techniques make use of dictionaries as a foundation to find text matches for dictionary terms [9, 77]. The quality of the dictionary being utilized has a significant impact on how well these strategies work. Usually, chemical IDs from well-known chemical databases are used to generate these dictionaries. However, this strategy is restricted to words found in dictionaries. Novel chemical compounds cannot be identified using dictionary-based methods since they are not present in the databases used to create the dictionaries. As it is practically not possible to make entries in a dictionary for all available systematic chemical identifiers, dictionary-based methods are more helpful in obtaining non-systematic chemical identifiers than for retrieving systematic chemical IDs (systematic identifiers are algorithmically generated) [9]. It is critical to keep in mind that dictionary-based methods can leverage the use of chemical databases to create the vocabulary to determine the substance's structure.

Statistically based algorithms can track a classifier to recognize chemical IDs across a text by using human-supplied resources [9]. These strategies can find both systematic identifiers and non-systematic identifiers. Statistical methods are in need of a substantial, annotated corpus to prepare the algorithm, which cannot directly yield chemical structures for extracted entities [9]. As previously stated, each strategy has its own advantages and disadvantages. Some of the restrictions can be overcome by an ensemble system that integrates many methodologies. It is interesting to note that, up until recently, the majority of text-mining systems' attention was directed to the biomedical field, and relatively little study had been conducted in the area of chemical text mining [9].

Saber and group carried out an intriguing study on chemical text mining using grammar-based, dictionary-based, and statistical-based approaches [9]. They looked at NER in scholarly paper titles with structured abstracts [9]. They created an ensemble system to incorporate paper titles with abstracts, which also combines grammar and dictionaries. Only 60% of the chemicals mentioned in journals are covered by their dictionaries, according to their investigation [9]. Comparing the ensemble system to the examined individual systems, it fared better. They were able to create representations of compound structures for the known mentions using this technique [9]. The most challenging jobs concerned chemical formulas and proper tokenization. They also focused on the recognition of named entities in patent abstracts [9]. They created an ensemble system to achieve this, which combines dictionary-based and statistical-based methods.

The chemical named entities must first be recognized so that they can be used to find information about chemical properties and their relation. The most efficient solutions frequently use a hybrid approach that blends machine learning techniques with the dictionary and rule-based methods. ChemSpot, which uses a Conditional Random Field (CRF) model [94], and the OSCAR4 recognizer, are based on a well-known system of the maximum-entropy Markov model (MEMM) [95]. The CHEMDNER community challenge recently encouraged the development of various new systems and added 84,355 manually annotated chemical entities to the CHEMDNER corpus of 10,000 PubMed abstracts [96].

In order to incorporate the results from different methodologies utilizing heuristic techniques, Swain et al. devised a system that offers a modular architecture for named entity recognition. This approach takes advantage of the enormous diversity of procedures and their wide range of strengths [80]. They primarily employ a dictionary-based recognizer that excels at identifying trivial and trade names, together with a CRF-based recognizer for identifying chemical names. They also employed a recognizer working on regular expression model that recognizes database and chemical formulas [80].

A word list from the Jochem chemical dictionary [97] is used by the dictionary-based recognizer, and words that lead to false positives are automatically filtered out using a process based on Lowe and Sayle's method [98]. To eliminate the duplication between related names using a graph-like shape, directed acyclic word graph (DAWG) has kept dictionary for storage and speedy matching of string. A linear-chain CRF model, which is used by the CRF-based recognizer, was trained using the OWL-QN strategy, which is used by the CRF suite framework [80].

4.3.5   Phase Parsing

Natural language parsing is complicated in general because of ambiguities that sometimes allow for multiple interpretations of a single statement. Due to the formulaic and accurate structure of the chemical literature, this occurs less frequently in practice than it does in other fields, making it much simpler to parse to a level that can be utilized for information extraction [80]. By resolving the intrinsic grammatical relationships between the words, parsing constructs a hierarchical system from a linear succession of sentence tokens. The root token is usually at the top, and the terminal nodes are at the bottom of this dependency tree, which represents this hierarchy. Parsing grammatical dependencies establishes semantic linkages between words and phrase components and helps deal with the sentence's random word order.

The ChemicalTagger project was coined by Hawizy et al. and used rule-based grammar to parse experimental synthesis sections of chemical literature [90]. In an attempt to parse every potential input, they tried to develop a single universal language, but even within their relatively small target area, it was still very hard to use the practical bounds of a single rule-based grammar [99].

In the case of the table parsing process, as the tabular data have high-density and semi-structured data, it was an appealing target for chemical information extraction, which allows for more straightforward interpretation when compared to entirely unstructured natural language. Since it does not have the strict uniformity in a table format, many tables in the literature on chemistry adhere to broad standards that allow for reliable interpretation using rule-based procedures.

4.4 INFORMATION EXTRACTION

It is the last step of the text mining pipeline, text mining's main objective is to turn natural language into machine-processable structured data. It is necessary to maintain and label the parse trees and nodes to identify each phrase or language component within them. These data can be output in RDF, CML, and XML forms.

5 CHEMINFORMATICS

The field of cheminformatics has made significant contributions to text mining for organizing chemistry data. The goal of cheminformatics, a branch of theoretical chemistry that combines force-field molecular modelling with quantum chemistry, is to characterize molecular structures in a way that is suitable for statistical modelling (mining chemical data) [100]. It is a vast field and impossible to describe in detail in a section of a book chapter.

The potential areas of chemical study, or cheminformatics, have been greatly enlarged by silica chemistry as a result of the growing reliance on informatics in many scientific fields. Cheminformatics has a long history, despite the fact that the term did not exist until the late 1990s [101]. In 1998, Brown gave the first official definition of the term, stating that "cheminformatics is the blending of those information resources to transform data into information and information into knowledge with the intended objective of making better decisions more quickly in the field of drug lead identification and organization" [102]. Quantitative structure–property relationships (QSAR) and chemical lead identification are the main focuses of cheminformatics [102]. In order to transform molecules into characteristics and then properties, a general cheminformatics model frequently uses a "two-part method" as shown in Figure 9.4. Compounds are first encoded as feature vectors, and 2) then the feature vectors are mapped to the desired attribute using cheminformatics techniques [102, 103].

Cheminformatics differs from other subfields of computational chemistry in that it reckons with plenty of data sets that are not limited by the conventional mathematical models, and it also requires in silico mathematics, which makes data processing possible [100]. To convert chemical data into chemical information with the aid of a computer, cheminformatics relies on mathematical, statistical, and machine learning techniques [104]. Both professions substantially draw on one another's work and employ a number of similar techniques. The distinction is that while cheminformatics focuses on creating data from the description of the chemical structure, chemometrics employs multivariate data from instruments (such as spectral data), which frequently does not require any knowledge of chemical structure [104].

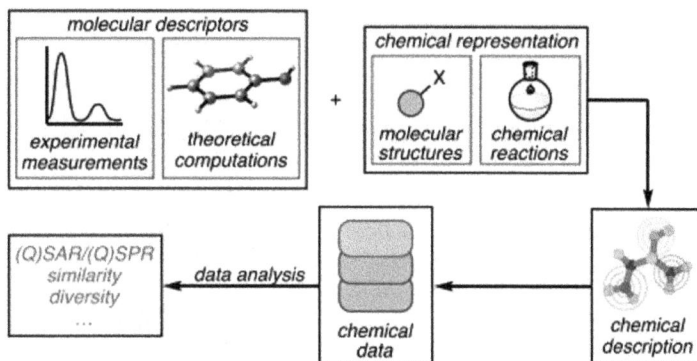

FIGURE 9.4 Workflow of Cheminformatics. Reprinted with permission from Williams, W. L.; Zeng, L.; Gensch, T.; Sigman, M. S.; Doyle, A. G.; Anslyn, E. V. The Evolution of Data-Driven Modeling in Organic Chemistry. ACS Central Science **2021**, 7 (10), 1622-1637. DOI: 10.1021/acscentsci.1c00535.

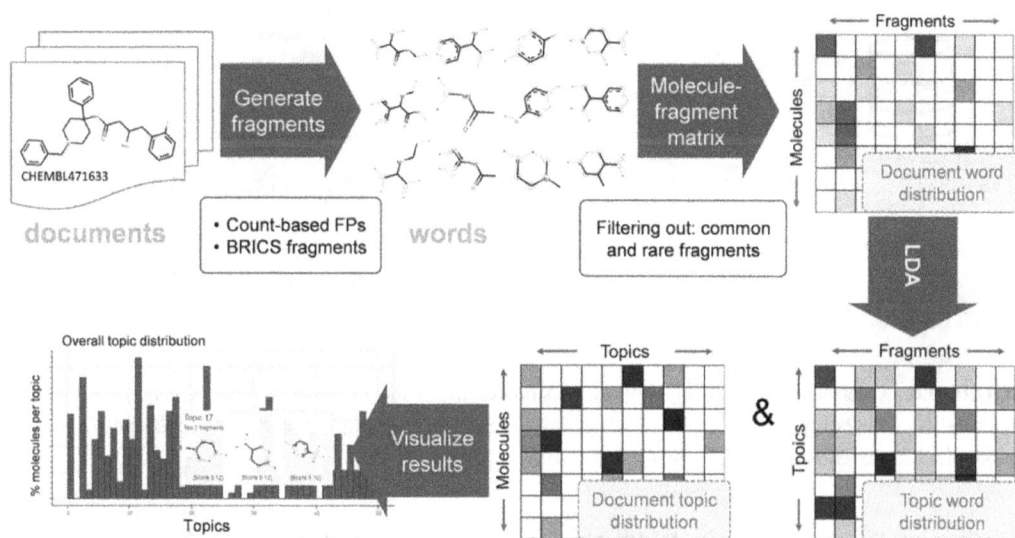

FIGURE 9.5 Chemical topic modelling workflow. Reprinted with permission from Schneider, N.; Fechner, N.; Landrum, G. A.; Stiefel, N. Chemical topic modelling: Exploring molecular data sets using a common text-mining approach. Journal of chemical information and modelling **2017**, 57 (8), 1816-1831.

6 APPLICATION

The most exciting text-mining application in the chemistry field is chemical topic modeling, which is used to organize bulk molecular data sets and study relationships between them. Schneider et al. utilized a text-mining approach for chemical topic modeling to explore molecular data sets [105]. They provided details on how they modified the topic modeling approach to work with chemical data, allowing collections of chemical compounds to be organized into chemical themes.

As shown in Figure 9.5 [105] molecules are referred to as documents in the chemical topic modeling approach, and the substructures or fragments formed from these documents are referred to as words [105]. Several fragmentation methods, such as BRICS, chemical fingerprints, or preset substructures use rule-based bond-cutting methods that may result in the formation of these fragments [106].

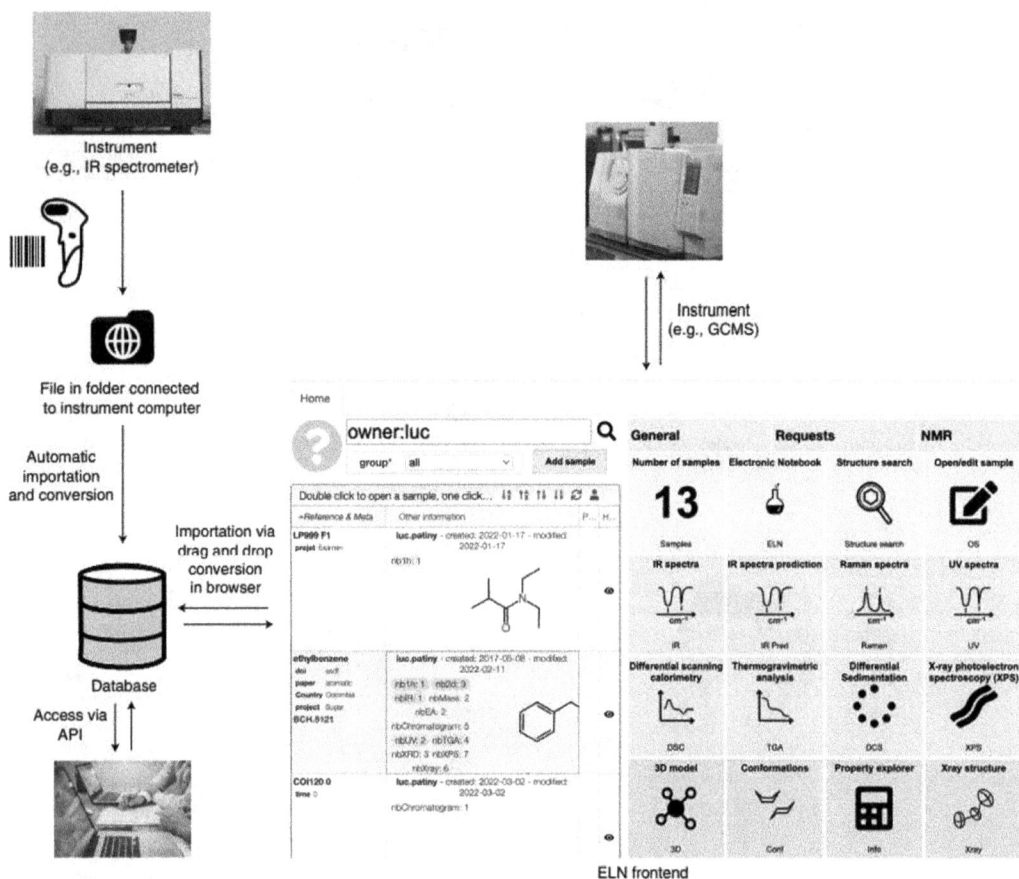

FIGURE 9.6 Overview of a possible importation procedure of the ELN. Reprinted with permission from Jablonka, K. M.; Patiny, L.; Smit, B. Making the collective knowledge of chemistry open and machine-actionable. Nature Chemistry **2022**, 14 (4), 365-376.

Following the creation of the fragments, a matrix consisting of horizontal rows denoting the molecules and vertical columns containing the counts for each fragment that occurs in each one are created. By eliminating the most prevalent and uncommon fragments, this matrix can then be further filtered [106]. Common fragments are those that appear in more than 10% of molecules, whereas uncommon fragments are those that appear in less than 1% of compounds. The topic-fragment matrix and the molecule-topic matrix are the two matrices the final Latent Dirichlet Allocation (LDA) model returns [106]. The subjects and the most likely fragments for each topic can be extracted and shown using these matrices [106].

Recently, the Smit group reported highly intriguing work that was focused on developing electronic lab notebooks and making them machine-actionable, as shown in Figure 9.6. Figure explains that barcode on the sample if scanned and if it is connected with the network, it can import all the sample information to the chemical databases. Apart from this, by using an online interface, files containing chemical information can be added by dragging and dropping. In either case, ELN is helpful in organizing data in a standard format that can be visualized on a web portal to analyze it later [107]. This data can also be accessed by using an application programming interface (API) or published repositories. Hence, it can be visualized and analyzed from different places and locations, which encourages collaboration. Their efforts are aimed at creating a modular open-science chemistry platform that will benefit not only data-mining studies but the entire chemistry community [107].

ChemicalTagger is a tool created by Hawizy et al. for organizing chemistry data [90]. They showed how ChemicalTagger might be used to extract organized scientific information from unorganized scientific publications [90]. Additionally, their team has shown that by utilizing text mining and NLP technologies, it is possible to extract chemical entities as well as the connections between those entities and turn the resulting data into a machine-processable format [90]. According to their study, experimental chemical writing may be parsed using rule-based methods and a standard language parser [90]. Using more than 10,000 patents, ChemicalTagger has successfully identified linguistic context of the solvents with an accuracy of more than 99.5% [90].

Another fascinating work on the application of text mining for chemical compounds and the prediction of their physicochemical properties was reported by the Palmbald group [108]. In semantically improved literature searches, this method was used to choose the analytical methods to utilize, collect information on structure of proteins and their ligands, or investigate their background for a particular area of chemistry [108]. The discovery of new drugs, the integration of omic data, and historical study are all possible applications of their method [108]. The TREC-CHEM project, a community challenge that looks at those patents that have chemical names and formulas to check the efficiency of the information retrieval methods, has finished evaluating chemical information retrieval engines [109].

Text mining is also used to help create and annotate drug/chemical repositories and databases. Text mining has been shown to increase curation effectiveness in the comparative toxicogenomics database (CTD), which comprises data on illnesses, genes/proteins, and cell activities associated with chemicals. Systems for information retrieval and systems that improve the reading experience are both included in this list of text-mining technologies. Some, such as SureChem, an online system for chemical information retrieval, uses chemical NER and name-to-structure translation, and even offers structure-based classification within patent records.

Several pertinent attempts have also been undertaken to identify the parameters associated with biological reactions. Hakenberg et al. identified papers that explain experimentally observed parameters for kinetic models using machine learning approaches [110]. The vocabulary utilized to anticipate parameters in the literature was regularized by Rojas et al. [111]. Heinen and colleagues developed a more accurate extraction process for kinetic information using an algorithm (rule-based) and a dictionary to recognize variables (temperatures, pH, Ki, KM, and kcat) and various chemical and biological entities (ligands, enzyme names, EC numbers, species, and localizations) essential to reproduce pathways (enzymatic) from PubMed abstracts [112]. The extraction of toxicological information is a fascinating use of text mining. In the case of the Toxicology pharmacokinetic parameter and liver toxicity data extracted can be linked to drugs or associations between drugs and cytochrome proteins are all part of the text mining in the Toxicology domain [113–115].

7 CHALLENGES

Even though text mining is an up-and-coming tool for releasing an enormous amount of information found in published studies it has a number of challenges. The creation of trustworthy, high-precision algorithms for chemical named entity recognition is significantly slowed down by the lack of a sizable dataset. The majority of the annotated sets that are now accessible were developed to support a specific objective or field of chemical study, making them challenging to apply broadly. If more work is done on creating organized databases, it might be possible that NLP techniques will be more effective and can be linked to the publications experimental data. On the basis of an established relationship between data and publications, it is possible to develop machine-annotated data.

Another issue is ambiguity and a lack of standard terminology for describing and categorizing complex chemicals. Most chemical compounds require chemical descriptions. Although IUPAC offers nomenclature recommendations for standard chemical terms, writers typically prefer to use simple chemical names or random notations for chemical names if there is no recognized standard terminology available. Chemical and material abbreviations (e.g., EDTA, PMA, LMO) are also common. This makes comparing and combining extracted data from different articles more challenging, and it calls for intensive data post-processing to standardize and converge the findings. On occasion, it results in uncertainty that cannot be resolved, or when it is, it leads to mistakes.

Authors frequently only include the most successful data in their research paper's main body and keep the less successful data to the additional information or supplementary material (mostly available as a PDF, which supports the finding in the original article and also supports meaningful automated data extraction). Machine learning models trained on these datasets experience significant issues due to this positive bias. Hence, care must be taken when choosing the queries to pose to the models. Furthermore, it is likely that some researchers investigated the precise composition, but the results were never published because they came up empty-handed.

The lack of harmful data makes defining the boundaries of machine learning outcomes impossible. The impact of human bias on the accuracy of machine learning model predictions has not been thoroughly studied and continues to be a complex problem for NLP-based data collectors.

Another obstacle in the process of text mining is the complexity of the language and the accumulation of errors. It is well recognized that a research paper's narrative has a very particular language and style. When compared to other themes, like sports or the weather, texts covering scientific subjects had the lowest readability scores, according to a corpus of newspapers on a variety of topics [116]. Since errors typically compound from step to step and reduce output quality and size, the accuracy of the text-mining pipeline is as crucial as processing [117]. As previously discussed, sentence tokenization issues have a significant impact on information extraction results, especially for chemical NER. It may be able to solve this issue by using hybrid NLP techniques that incorporate domain knowledge.

It is always advisable to use caution when interpreting findings from the application of machine learning algorithms to text-mining data and to keep the constraints of the input data in mind. In general, problems with the machine learning approach are significantly less likely to be the cause of restrictions in machine learning predictions than problems with the input data.

8   FUTURE TRENDS AND PERSPECTIVES

To manage the increasing volume of chemical data, the chemistry community must develop new methods and approaches. After a thorough examination of the literature, it seems that machine learning for text mining is a promising new area of study. Even though machine learning is a prospering field in chemical research, most published machine learning work can be produced by high-throughput computing or on relatively tiny experimental datasets by using computed data sets frequently having no more than 50–100 data items. Owing to this lack of systematic experimental data collection, text mining and natural language processing will probably be essential in fostering more data-driven chemical research. There will likely be an increase in the quantity and calibre of information retrieved from the scientific text as a result of publishers' willingness to offer access to their extensive corpus for text mining and many recent breakthroughs in the NLP field.

In recent years, the introduction of "big data" and the increased development of high-throughput computations in chemical science have moved attention to data management and curation. This has led to the design and creation of high-quality databases with adaptable graphical user interfaces that make accessing the data for mining and analysis simple and convenient. It is necessary to create a highly sophisticated infrastructure for the representation, upkeep, and dissemination of the ever-expanding volumes of data that are taken from scientific articles. For chemical science to fully profit from the new data paradigm, a significant amount of work will need to be invested in data collection.

It is clear that text mining and natural language processing are ways to make the results of a century of material inquiry available in order to actualize this paradigm.

9 CONCLUSION

There is a plethora of chemical data available in the open-access scientific literature that can be used for insightful text mining to organize chemistry data. However, retrieving this information is not always straightforward. In most circumstances, finding pertinent documents for a given chemical molecule or family is the first step in retrieving vital chemical information. The targeted retrieval of chemical documents requires the extraction of the complete list of chemicals cited in a document, together with any accompanying information by using automatic recognition of chemical entities in the text.

This chapter introduced readers to the text-mining process for organizing chemistry data. This chapter provided a comprehensive and in-depth detailed discussion of information extraction sources and their corresponding bibliographies. A strong focus was placed on the various steps of the text-mining process, which learly explained in detail. The growing interest in the integration of chemical information with biological data by utilizing automatically annotated chemical knowledge databases that integrate cheminformatics approaches was also discussed. As a roadmap for research in this developing field, the application of text mining, current challenges, and future trends were highlighted.

REFERENCES

1. A.-H. Tan, Text mining: The state of the art and the challenges, Proceedings of the pakdd 1999 workshop on knowledge discovery from advanced databases, Citeseer, 1999, pp. 6570.
2. H.P. Luhn, A business intelligence system, IBM Journal of research and development 2(4) (1958) 314–319.
3. C. Zhang, Advanced Text Mining Techniques and the Applications in Clinics, Chemistry and Manufacturing, (2022).
4. D. Young, T. Martin, R. Venkatapathy, P. Harten, Are the chemical structures in your QSAR correct?, QSAR & combinatorial science 27(11-12) (2008) 1337–1345.
5. A.J. Williams, S. Ekins, A quality alert and call for improved curation of public chemistry databases, Drug Discovery Today 16(17-18) (2011) 747–750.
6. M. Krallinger, O. Rabal, A. Lourenco, J. Oyarzabal, A. Valencia, Information retrieval and text mining technologies for chemistry, chemical reviews 117(12) (2017) 7673–7761.
7. C. Southan, P. Várkonyi, S. Muresan, Quantitative assessment of the expanding complementarity between public and commercial databases of bioactive compounds, Journal of cheminformatics 1(1) (2009) 1–17.
8. P. Stenetorp, S. Pyysalo, G. Topić, T. Ohta, S. Ananiadou, J.i. Tsujii, BRAT: a web-based tool for NLP-assisted text annotation, Proceedings of the Demonstrations at the 13th Conference of the European Chapter of the Association for Computational Linguistics, 2012, pp. 102–107.
9. S. Ahmad Akhondi, Text Mining for Chemical Compounds, (2018).
10. C. Tyrchan, J. Boström, F. Giordanetto, J. Winter, S. Muresan, Exploiting structural information in patent specifications for key compound prediction, Journal of chemical information and modeling 52(6) (2012) 1480–1489.
11. R. Fennell, R.W. Fennell, History of IUPAC, 1919-1987, Blackwell Science Oxford1994.
12. H. Gurulingappa, A. Mudi, L. Toldo, M. Hofmann-Apicius, J. Bhate, Challenges in mining the literature for chemical information, Rsc Advances 3(37) (2013) 16194–16211.
13. W.J. Wilbur, G.F. Hazard Jr, G. Divita, J.G. Mork, A.R. Aronson, A.C. Browne, Analysis of biomedical text for chemical names: a comparison of three methods, Proceedings of the AMIA Symposium, American Medical Informatics Association, 1999, p. 176.

14. W. Dethlefsen, M.F. Lynch, V.J. Gillet, G.M. Downs, J.D. Holliday, J.M. Barnard, Computer storage and retrieval of generic chemical structures in patents. 11. Theoretical aspects of the use of structure languages in a retrieval system, Journal of chemical information and computer sciences 31(2) (1991) 233–253.

15. Global Patent Search Network. www.uspto.gov/blog/systemsupdates/entry/global_patent_search_n etwork_gpsn (accessed on September 17, 2022).

16. State Intellectual Property Office (SIPO) Patent Office. english.sipo.gov.cn (accessed on September 17, 2022).

17. PATENT SCOPE. www.wipo.int/patentscope/en (accessed on September 16, 2022)

18. Free Patent Online. www.freepatentsonline.com (accessed on September 17, 2022).

19. Patent Lens. www.lens.org/lens (accessed on September 17, 2022).

20. Sumo Brain. www.sumobrain.com (accessed on September 17, 2022).

21. Total Patent. www.lexisnexis.com/totalpatent (accessed on September 17, 2022).

22. JPG. www.jpds.co.jp/eng (accessed on September 17, 2022).

23. PASSENGER. patseer.com (accessed on September 17, 2022).

24. PatBase. www.patbase.com (accessed on September, 17 2022).

25. PriorSmart. www.priorsmart.com (accessed on September 17, 2022).

26. WIPS Global. www.wipsglobal.com (accessed on September 17, 2022).

27. Google Patents. www.google.com/patents (accessed on September 17, 2022).

28. OvidSP. ovidsp.ovid.com (accessed on September 17, 2022).

29. Google Scholar. scholar.google.com (accessed on September 17, 2022).

30. PubMed. www.ncbi.nlm.nih.gov/pmc (accessed on September 17, 2022).

31. Europe PMC. europepmc.org (accessed on September 17, 2022).

32. PMC Canada. pubmedcentralcanada.ca/pmcc (accessed on September 17, 2022).

33. PubChem. pubchem.ncbi.nlm.nih.gov (accessed on September 17, 2022).

34. Springer Nature. dev.springernature.com (accessed on September 17, 2022).

35. Science Direct (Elsevier). dev.elsevier.com/api_docs.html (accessed on September 17, 2022).

36. Scopus (Elsevier). dev.elsevier.com/api_docs.html (accessed on September 17, 2022).

37. SciFinder (CAS). www.cas.org/products/scifinder (accessed on September 17, 2022).

38. DOAJ. doaj.org (accessed on September 17, 2022).

39. plus. www.cas.org/support/documentation/references (accessed on September 17, 2022).

40. aRDi. www.wipo.int/ardi/en (accessed on September 17, 2022).

41. TOXNET. toxnet.nlm.nih.gov (accessed on September 17, 2022).

42. BASE. www.base-search.net (accessed on September 17, 2022).

43. Sci-Hub. sci-hub.se (accessed on September 17, 2022).

44. Reaxys. www.elsevier.com/solutions/reaxys (accessed on September 17, 2022).

45. GOSTAR. www.gostardb.com (accessed on September 17, 2022).

46. STN. www.cas.org/products/stn (accessed on September 17, 2022).

47. Thompson Innovation. ovidsp.ovid.com (accessed on September 17, 2022).

48. Web of Science. clarivate.com/webofsciencegroup/solutions/web-of-science (accessed on September 17, 2022).

49. ProQuest Dialogue. www.proquest.com/products-services/ProQuest-Dialog.html (accessed on September 17, 2022).

50. iScienceSearch. isciencesearch.com/iss/default.aspx (accessed on September 17, 2022).

51. Academic Search. www.ebscohost.com (accessed on September 17, 2022).

52. Derwent World Patents Index. clarivate.com/products/ip-intelligence/ip-data-and-apis/derwent-world-patents-index/, (accessed on August 1, 2022).

53. INPADOC, www.epo.org/searching-for-patents/data/bulk-data-sets/inpadoc.html, (accessed on August 1, 2022).

54. PatBase. www.patbase.com (accessed August 1, 2022).

55. Free Patents Online FPO. www.freepatentsonline.com, (accessed on August 3, 2022).

56. Lens. www.lens.org/lens, (accessed on August 3, 2022).

57. Prior Smart. www.priorsmart.com, (accessed on August 3, 2022).

58. BIOSIS Previews clarivate.com/webofsciencegroup/solutions/webofscience-biosis-previews/, (accessed on August 2, 2022).

59. Embase www.elsevier.com/solutions/embase, (accessed on August 1, 2022).

60. MEDLINE www.nlm.nih.gov/medline/medline_overview.html, (accessed on August 8, 2022).

61. Inspec www.theiet.org/resources/inspec, (accessed on August 1, 2022).

62. Scopus www.elsevier.com/solutions/scopus, (accessed on August 1, 2022).

63. Scifinder www.cas.org/products/scifinder, (accessed on August 2, 2022).

64. PMC Pubmed Central. www.ncbi.nlm.nih.gov/pmc, (accessed on August 2, 2022).

65. Science Direct. www.sciencedirect.com, (accessed on August 2, 2022).

66. New Drug Application (NDA). www.fda.gov/Drugs/DevelopmentApprovalProcess/HowDrugsareD evelopedandApproved/ApprovalApplications/NewDrugApplicationNDA, (accessed on August 3, 2022).

67. European public assessment reports. www.ema.europa. EU/ema/index.jsp?curl=pages/medicines/ landing/epar_search.jsp&mid=WC0b01ac058001d125, (accessed on August 3, 2022).

68. Adis Insight. adisinsight.springer.com, (accessed on August 3, 2022).

69. Scopus vs Web of Science. hlwiki.slais.ubc.ca/index.php/Scopus_vs._Web_of_Science, (accessed on August 3, 2022).

70. Web of Science. clarivate.com/webofsciencegroup/solutions/web-of-science/, (accessed on August 3, 2022).

71. Reaxys: The Quickest Path from Q to A. www.elsevier.com/solutions/reaxys, (accessed on August 3, 2022).

72. STN–The choice of patent experts. www.cas.org/products/stn, (accessed on August 3, 2022).

73. iScienceSearch. isciencesearch.com/iss/default.aspx, (accessed on August 3, 2022).

74. Google Scholar. scholar.google.com, (accessed on August 3, 2022).

75. Europe PMC. europepmc.org, (accessed on August 3, 2022).

76. Access to Research for Development and Innovation ARDI.www.wipo.int/ardi/en, (accessed on August 3, 2022).

77. M. Vazquez, M. Krallinger, F. Leitner, A. Valencia, Text mining for drugs and chemical compounds: methods, tools and applications, Molecular Informatics 30(6-7) (2011) 506–519.

78. N. Kang, Using natural language processing to improve biomedical concept normalization and relation mining, 2013.

79. O. Kononova, T. He, H. Huo, A. Trewartha, E.A. Olivetti, G. Ceder, Opportunities and challenges of text mining in materials research, Science 24(3) (2021) 102155.

80. M.C. Swain, J.M. Cole, ChemDataExtractor: A Toolkit for Automated Extraction of Chemical Information from the Scientific Literature, Journal of Chemical Information and Modeling 56(10) (2016) 1894–1904.

81. A. Constantin, S. Pettifer, A. Voronkov, PDFX: fully-automated PDF-to-XML conversion of scientific literature, Proceedings of the 2013 ACM symposium on Document engineering, 2013, pp. 177–180.

82. M.-T. Luong, T.D. Nguyen, M.-Y. Kan, Logical structure recovery in scholarly articles with rich document features, Multimedia Storage and Retrieval Innovations for Digital Library Systems, IGI Global2012, pp. 270–292.

83. D. Tkaczyk, P. Szostek, M. Fedoryszak, P.J. Dendek, Ł. Bolikowski, CARMINE: automatic extraction of structured metadata from scientific literature, International Journal on Document Analysis and Recognition (IJDAR) 18(4) (2015) 317–335.

84. M. Mahdavi, R. Zanibbi, H. Mouchere, C. Viard-Gaudin, U. Garain, ICDAR 2019 CROHME+ TFD: Competition on recognition of handwritten mathematical expressions and typeset formula detection, 2019 International Conference on Document Analysis and Recognition (ICDAR), IEEE, 2019, pp. 1533–1538.

85. H. Mouchère, R. Zanibbi, U. Garain, C. Viard-Gaudin, Advancing the state of the art for handwritten math recognition: the CROHME competitions, 2011–2014, International Journal on Document Analysis and Recognition (IJDAR) 19(2) (2016) 173–189.

86. E. Stamatatos, N. Fakotakis, G. Kokkinakis, Automatic extraction of rules for sentence boundary disambiguation, Proceedings of the Workshop on Machine Learning in Human Language Technology, Citeseer, 1999, pp. 88–92.

87. S. Bird, E. Klein, E. Loper, Natural language processing with Python: analyzing text with the natural language toolkit, " O'Reilly Media, Inc."2009.

88. M. Hannibal, M. Johnson, An improved non-monotonic transition system for dependency parsing, Proceedings of the 2015 conference on empirical methods in natural language processing, 2015, pp. 1373–1378.

89. D. Jessop, S. Adams, S. Willighagen, E. Hawizy, P. Murray-Rust, P, OSCAR4: A flexible architecture for chemical text-mining J. Cheminf 3 (2011) 41.

90. L. Hawizy, D.M. Jessop, N. Adams, P. Murray-Rust, ChemicalTagger: A tool for semantic text-mining in chemistry, Journal of cheminformatics 3(1) (2011) 1--3.

91. Y. Tsuruoka, Y. Tateishi, J.-D. Kim, T. Ohta, J. McNaught, S. Ananiadou, J.i. Tsujii, Developing a robust part-of-speech tagger for biomedical text, Panhellenic conference on informatics, Springer, 2005, pp. 382–392.

92. A. Bies, J. Mott, C. Warner, English news text treebank: Penn treebank revised, (2015).

93. Y. Tateisi, J.i. Tsujii, Part-of-Speech Annotation of Biology Research Abstracts, LREC, 2004.

94. T. Rocktäschel, M. Weidlich, U. Leser, ChemSpot: a hybrid system for chemical named entity recognition, Bioinformatics 28(12) (2012) 1633–1640.

95. D.M. Jessop, S.E. Adams, E.L. Willighagen, L. Hawizy, P. Murray-Rust, OSCAR4: a flexible architecture for chemical text-mining, Journal of cheminformatics 3(1) (2011) 1–12.

96. M. Krallinger, F. Leitner, O. Rabal, M. Vazquez, J. Oyarzabal, A. Valencia, CHEMDNER: The drugs and chemical names extraction challenge, Journal of cheminformatics 7(1) (2015) 1–11.

97. K.M. Hettne, R.H. Stierum, M.J. Schuemie, P.J. Hendriksen, B.J. Schijvenaars, E.M.v. Mulligen, J. Kleinjans, J.A. Kors, A dictionary to identify small molecules and drugs in free text, Bioinformatics 25(22) (2009) 2983–2991.

98. D.M. Lowe, R.A. Sayle, LeadMine: a grammar and dictionary driven approach to entity recognition, Journal of cheminformatics 7(1) (2015) 1–9.

99. L. Hawizy, D.M. Jessop, N. Adams, P. Murray-Rust, ChemicalTagger: A tool for semantic text-mining in chemistry, Journal of Cheminformatics 3(1) (2011) 17.

100. J. Polanski, Chemoinformatics: from chemical art to chemistry in silico, (2019).

101. T. Engel, Basic overview of chemoinformatics, Journal of chemical information and modeling 46(6) (2006) 2267–2277.

102. W.L. Williams, L. Zeng, T. Gensch, M.S. Sigman, A.G. Doyle, E.V. Anslyn, The Evolution of Data-Driven Modeling in Organic Chemistry, ACS Central Science 7(10) (2021) 1622–1637.

103. J.B.O. Mitchell, Machine learning methods in chemoinformatics, WIREs Computational Molecular Science 4(5) (2014) 468–48.

104. T. Engel, J. Gasteiger, Chemoinformatics: basic concepts and methods, John Wiley & Sons2018.

105. N. Schneider, N. Fechner, G.A. Landrum, N. Stiefel, Chemical topic modeling: Exploring molecular data sets using a common text-mining approach, Journal of chemical information and modeling 57(8) (2017) 1816–1831.

106. J. Degen, C. Wegscheid-Gerlach, A. Zaliani, M. Rarey, On the Art of Compiling and Using'Drug-Like'Chemical Fragment Spaces, ChemMedChem: Chemistry Enabling Drug Discovery 3(10) (2008) 1503–1507.

107. K.M. Jablonka, L. Patiny, B. Smit, Making the collective knowledge of chemistry open and machine actionable, Nature Chemistry 14(4) (2022) 365–376.

108. M. Palmblad, Visual and semantic enrichment of analytical chemistry literature searches by combining text mining and computational chemistry, Analytical chemistry 91(7) (2019) 4312–4316.

109. M. Valletta, 2nd Workshop on Building and Evaluating Resources for Biomedical Text Mining Tuesday, 18 th March 2010.

110. J. Hakenberg, S. Schmeier, A. Kowald, E. Klipp, U. Leser, Finding kinetic parameters using text mining, Omics: a journal of integrative biology 8(2) (2004) 131-152.

111. I. Rojas, M. Golebiewski, R. Kania, O. Krebs, S. Mir, A. Weidemann, U. Wittig, Storing and annotating of kinetic data, In silico biology 7(2 Supplement) (2007) 37–44.

112. S. Heinen, B. Thielen, D. Schomburg, KID-an algorithm for fast and efficient text mining used to automatically generate a database containing kinetic information of enzymes, BMC bioinformatics 11(1) (2010) 1–9.

113. D. Jiao, D.J. Wild, Extraction of CYP chemical interactions from biomedical literature using natural language processing methods, Journal of chemical information and modeling 49(2) (2009) 263–269.

114. C. Feng, F. Yamashita, M. Hashida, Automated extraction of information from the literature on chemical-CYP3A4 interactions, Journal of chemical information and modeling 47(6) (2007) 2449–2455.

115. D. Fourches, J.C. Barnes, N.C. Day, P. Bradley, J.Z. Reed, A. Tropsha, Cheminformatics analysis of assertions mined from literature that describe drug-induced liver injury in different species, Chemical research in toxicology 23(1) (2010) 171–183.

116. F.A. Ilias, O., Lansdall-Welfare, T. et al.(2013). Research methods in the age of digital journalism: Massive-scale automated analysis of news-content: Topics, style and gender, Digital Journalism 1(1) 102–116.

117. O. Kononova, H. Huo, T. He, Z. Rong, T. Bottari, W. Sun, V. Tshitoyan, G. Ceder, Text-mined dataset of inorganic materials synthesis recipes, Scientific Data 6(1) (2019) 1–11.

10 CNN Use and Performance in Virtual Screening

Bhargav Chandegra, Prajesh Prajapati, Jigna Prajapati

1 INTRODUCTION

Chemistry is always driven by the search for novel chemical compounds and structural frameworks with applications in various fields of chemistry like sensing, therapeutic, and research [1]. While developing these structures typically requires the meticulous, scientific, and precise synthesis of numerous potential structures or the screening of desired molecules, most of these attempts may be classified as accidental discoveries that were not made with time and money considerations or a logical strategy in mind [2]. This was owed to the fact that cutting-edge technologies were not being introduced to get control of the problem. These procedures resulted in raised costs often conflated with the uncertainty in the direction of the desired result [3]. In recent years, there has been an inclination towards adopting scientific strategies because this not only helps in ruling out probabilities but also reduces the underlying unreliability. As a result, in silico methods have significantly increased in popularity and have become an essential component of both academic and industry research, guiding the design and discovery of chemical compounds [4]. Among the few terms used recently to describe the relatively new and rapidly expanding field of molecule discovery are bioinformatics and cheminformatics [5]. Computational methods that describe the degree of interaction between different ligands and targets along with attractive 3D visualization are becoming increasingly important for the identification of potential lead compounds from databases [6–8]. In silico methods such as computational methods and the strategies to develop new molecules mainly rely on the exploration of increasingly evolving chemical spaces. New medications like antivirals [9], antibiotics [10], catalysts[11], battery materials [12], and other molecules with customized features need to be discovered and developed, which calls for a paradigm shift to research in unexplored regions of the enormous chemical universe.

This was one step towards easing the task of filtering out the molecular image data sets by in silico studies. The visual identification of patterns enabled by deep learning brought a tremendous advantage over conventional methods in order to deal with massive amounts of data. It also allows recognition and classification of the molecular data according to various parameters and potential [13, 14]. This visual application of deep learning enables the prioritizing of different compounds from data libraries as and when required. It is known that the generation of any therapeutic chemical entity requires a thorough understanding of the metabolic pathways. Therefore, any rational approach to a given condition must prioritize assessing many metabolic pathways and selecting a viable biological target [15]. The primary objective of the molecular design process is to identify and explore effective ligands, particularly for target proteins. The availability of several robotic and computerized screening techniques has led to the testing and screening of numerous ligands based on their potential to interact with specific biological receptors for a particular action of need [16]. The escalation of biomedical data involves the design of novel methodologies for screening molecules for drug discovery and other synthesis processes. The industries are adopting artificial intelligence

DOI: 10.1201/9781003353768-10

and machine learning in particular with big data. In current times, visual screening is becoming an important factor in drug discovery and prioritizing potential active compounds. The screening takes place according to criteria such as structure-based [17] and ligand-based methodologies [18]. The computerized classification of image data sets has been made easier by the development of artificial intelligence and its popular segment deep learning. It is growing and developing swiftly. The merits of convolutional neural networks (CNNs) over traditional methods of classification include reduced costs and speeding up of image interpretation. They can attain enhanced accuracy and curb various challenges concerned with visual recognition.

2   VIRTUAL SCREENING

Traditionally trial and error has been the most preferable strategy in the field of sciences like pharmaceutical, chemical, or biological, but this method has major limitations (e.g., it is time consuming and costly). High-throughput screening was necessary to overcome these challenges problem [19], and thus novel approaches have been developed. While it is true that computations do not actually replace experimental research, it has become abundantly evident that an effective synergy between the two approaches is crucial for assisting aspiring experimentalists in the rational synthesis and screening of drugs [20]. It has been established that integrating random screening and rational design leads to a major improvement in the molecular discovery process and results in the identification of ligands. More and more reliable quantum-mechanical properties for millions of molecules are accessible in regulated datasets. Virtual screening is among the in silico methods of screening that filter the chemicals, drugs, or biological molecules by their structure or image. Virtual screening has become a crucial tool in our endeavor to discover new chemically similar molecules [21]. In screening databases for the lead compounds, a broad variety of equivalent and contradictory methodological techniques are available. Thus, over the past three decades, many design processes have been rationalized and sped up by the use of computational methods [22]. Virtual screening (VS) is a standard and rational approach for prioritizing compounds from (very) numerous compound libraries that have a strong probability of binding to an interest target [23]. A number of techniques and digital content are using target and ligand-based virtual screening and are expanding rapidly. VS techniques can quickly and cheaply scan lots of available molecules like ZINC [24] or MolPORT [25], and prioritize which ones should be evaluated, synthesized internally, or bought from outside vendors. Likewise, novel approaches to creating data libraries by computer have increased and improved the chemical and molecular fields, such as the Enamine REAL [26], which includes about 17 billion established compounds with approximately 2 billion drug-related compounds, making VS more feasible.

VS techniques can reduce the search area to a few hundred compounds with the desired attributes for further study; however, they are not always efficient in identifying the most active molecule. Currently, VS is an essential factor in the molecule design process. It is typically implemented as a hierarchical workflow, incorporating various techniques to prioritize potentially active substances (either sequentially or concurrently).

There are three broad classes of virtual screening procedures:

- Structure-based docking approaches with high resolution [27], [28]
- QSAR methods (Ligand) [29], [30]
- Structure-based and ligand-based threading techniques [31]
- Inverse virtual screening (Docking) [32]

Physical principles in high-resolution image interpretation are used to observe structures for docking. These approaches have the advantage of assisting in the development of new active binders, and currently cover about one-third of all human proteins but at a higher computational cost and with less accuracy than ligand-based approaches. Despite being more accurate and efficient than docking methods, when it comes to the case of ligand-based techniques, ligands that bind with the desired target protein must be identified in advance for ligand-based methods. Such established collections of ligands are not available for most human protein targets. Threading/structure-based methods have similar productivity and accuracy as both approaches and these methods address the drawbacks of structure as well as ligand-based docking approaches For example, structure-base require high-resolution and ligand-base required advance knowledge of binder [23].

The two main categories of VS techniques are as follows: 1. Structure-based techniques (focusing on the similarity of the drug and receptor-binding affinity) and 2. ligand-based techniques, which rely on the affinity between new identified compounds and the chemicals that are already familiar.

2.1 STRUCTURE-BASED METHODS

As the name suggests, if there is a need for a structure-based approach then it is clear that structural information is mandatory including information about molecular as well as the receptor structural aspects. Also the structure image should be in 3D so that every bonds, the rotations and active sites can be identified [33]. Information such as the binding site and the protein is also needed. Molecular docking is a popular technique in this method, and is used to predict multiple poses of binding in the receptor structure. This also helps in estimation of binding affinity [34]. An advantage of protein–ligand docking is that it has the capacity to enrich the probably active compounds over inactive compounds [35]. However, there are some shortcomings in positioning poses individually. Throughout the method, many number of ligand poses can be created on the basis of target structure, sometimes even thousands [36]. The generated poses are ranked by scoring function. The scoring functions are of the following types [37]:

Physics-based methods – These methods are based on force field mechanics. These include Van der Waal forces, electrostatic forces, etc. [38].

Empirical methods – These methods work on weighed energy terms, which includes rotatable bonds, or sometimes the surface area that is accessible [39].

Experience and understanding methods – These methods are solely based on the statistical data of analysis that has detected atom pair potentials [40].

Other methods – The most recent methods are based on the deep learning area of artificial intelligence. These methods try to establish the connection between different molecules, receptors, and ligands in terms of interaction to predict binding affinity, QSAR relationships for screening, etc.

2.2 LIGAND-BASED METHODS

Ligand-based technologies include comparatively mature technologies such as QSAR modeling [41], ligand-based pharmacophores [42], and molecular similarity search [43]. In contrast to structure-based counterparts, these methods only need information about ligands. This type can be improved by sequence-based protein data, and is called proteochemometric modelling. This modelling merges the information about ligand and target together in a single model. This can then predict the molecule activity bioassay [43,44]. These modelling techniques depend on the target they attach to as well as ligand affinity. Compared to ligand-based techniques, these provide an added benefit. There has been tremendous growth during the past few years in the effective and popular applications that employ a range of methods, such as pharmacophore-based search, spanning similarity analysis, scoring, fast docking, theoretical graph approaches as well as various machine learning tools, etc.

3 ARTIFICIAL INTELLIGENCE-BASED VIRTUAL SCREENING

3.1 MOLECULAR FORMATS

To promote computer-aided molecular design by the aid of artificial intelligence and specifically machine learning, first and foremost we need to understand the structural characteristics of the molecules such as structure of the compound, receptor characteristics and data representation of the compound, compounds in cheminformatics, etc. [46]. Cheminformatics is an expanding science that explores how computers are used to collect, process, and interpret chemical data. It also focus on various aspects of design and screening like structural representation of molecule, defining the characteristics of a compound, gathering and storing the large chemical data with necessary characteristics, characterization and prediction of various probabilities to modify structure, and in silico (computerized) compound screening and design. The discipline has progressed for more than 30 years [47]. Thus, a plethora of similarity-based compound representations have been proposed for generation of libraries and to process the data along with these advancements. The chemical compounds are easy to represent in graphs, which are also known as molecular graphs [48], and are mainly interpreted with nodes or vertices that suggest the atoms and edges or links with information about the bonds. These molecular graphs have the ability to differentiate various structural isomers, but the major limitation of these graphs is that they cannot represent the molecules with 3D structural conformations. Therefore, to include 2D and 3D conformations various chemical file formats are used including MDL (Molfile), Protein Data Bank (PDB), and Structure Data Format (SDF). In this context, PDB only deals with preserving the structural data, whereas in the case of SDF, there are more functions and advantages such as the ability to record the descriptors and many more chemical properties, so it is more preferable than PDB for molecular screening analysis. This all files had a major disadvantage that was less data storage capacity so many researches carried out and many more chemical liner notations were introduced. Thus, in the 1980s Weininger et al. developed a format called SMILES, an abbreviation for "simplified molecular-input line-entry system." This format was developed to be utilized to convert the 3D structure of a molecule into a series of symbols that computer programs can easily understand; this compound storing method can also store structural data in ASCII format whereas other methods storing functionality was different like some methods records the compounds codes directly [49]. Many other formats have been made including "SYBYL" Line Notation (SLN) [50], "Wiswesser line notation (WLN)" [51], and "ROSDAL" [52]. There are no distinct strings for encoding chemical compounds, especially for large and structurally complicated molecules, despite the fact that they are memory-efficient, so one solution can be used if chemical structures can be consistently classified and gathered using the Morgan algorithm [53]. The loss of coordinate information is another drawback of these other formats, which makes it necessary to use structure generation tools like PRODRG to anticipate native molecule structure [54].

3.2 CHEMICAL REPRESENTATION

In cheminformatics and chemical biology, target molecule or carrier properties like chemical, structural, or physical are stored using machine learning attributes. These data are then referred or represented during synthesis of novel chemical compounds or molecules; this phenomenon is called chemical representation [55]. These features and the data it increases the process by predicting of more essential chemical compounds and allow for the extraction of a compound's key characteristics and offer the potential for the development of predictors that may group novel structures with related characteristics. All dimensions can be used to represent substances like 0D, 1D, 2D, and 3D in which 1D is the simpler one because it contains the information about molecular mass and atom number, which are no more complex than 2D and 3D for compound classification. Thus, for 2D and 3D chemical representations a machine learning approach and cheminformatics analysis is carried out [56],[57].

4 ARTIFICIAL NEURAL NETWORKS

The design and the motivation behind artificial neural networks (ANNs) was the human brain, which has biological neurons situated in multi-layered linked node analogs [58]. Therefore, it was observed that how our brain interpret things. After observation and brainstorming by scientists, they concluded with the question that can we make such brain artificially? As result, the first ANN was developed in the 1950s by Bernard Widrow of Stanford University; it was called "Perceptron." Although the discovery was revolutionary it had some limitations. For example, it only consisted one layer for the input; therefore, at that time multi-layered was in demand of research [59]. The major limitation of the perceptron that it was a single layered so, in case of XOR logical relationship it was not able to solve it correctly because XOR logical relationships need at least two inputs [60]. This limitation was countered by introducing the multi-layered perceptron by adding other hidden layers and a system mainly based on back-propagation. This technique is used to modify a neural network's methods and relies on the loss (or margin of error) recorded during a preceding cycle (iteration) [61]. Other deep learning strategies and models were developed as fundamental elements of the ANN for the processing of multi-dimensional and unorganized data (Figure 10.1). These types of reports were then processed by other novel mechanisms like natural language processing (NLP) and also by computerized in silico statistical strategies also known as machine learning and artificial intelligence (AIML). Numerous studies found that DNN was more accurate than many conventional machine learning methods at predicting drug toxicity, solubility, biological activity, and ADMET parameters [61,62]. Dealing with the high-dimensional data is challenging but over time lots of different deep learning frameworks have been built with various feature extraction and dimension reduction algorithms. The CNN is a popular deep learning system for the purpose of image analysis and is made of various layers such as convolution filter layers, fully connected multi-layer perceptrons, and convolution pooling layers [64]. Both convolutional layer and next to the convolutional layer that is pooling layer identify small patterns in the data that are already been present in the visual information that are used for input dimension relating to the fully connected layers. Over time greater expansion in applications and features and expanding the tool, for example, this tool now can interpret the image of the protein which is presented as 3D image. To interpret this type of structure and image 3D-CNN models are used. Using these tools not only the 3D structure protein can be interpreted but also other optical chemical structures in images and other documents [65].

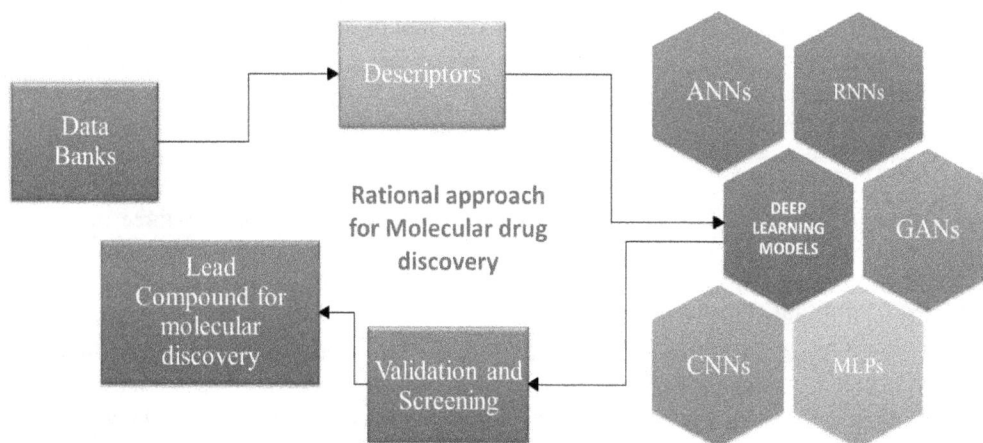

FIGURE 10.1 This representation shows various deep learning models and rational approach for design a molecule. Deep learning models may be used as virtual screening tools.

5   CHALLENGES ADDRESSED BY NEURAL NETWORKS

Artificial intelligence has become the focus of attention in scientific as well as non-scientific communities. Numerous researchers have analyzed the mechanisms around deep learning technology during the last few decades. Deep learning technology is considered a systemic approach, and is usually centered on automating cognitive functions in a similar fashion as a human being would do. The first challenge that CNN addresses is image classification [66]. This is the case because higher sample recognition is one of the domains where humans have always preferred non-rational methods over AI algorithms. The ever-increasing burden of increased bulk of datasets in cheminformatics [67], bioinformatics [68], forensic sciences [69], pharmaceutical [70], life sciences, and other relevant domains [71] require constant development of novel techniques approaches in the analytical dimension. Artificial intelligence and its segments such as machine learning is constantly explored. The domain of neural networks is increasingly used and adopted because of their systemic mechanisms in the chemical industry that work on the synthesis of new molecules, particularly with respect to bulky data. The findings of the same in the international conference was arranged by the "Big Data in Chemistry" [13]. Machine learning technologies can enhance big data analysis, particularly when applied to high-throughput screening (HTS) campaigns. For example, Rodrguez Pérez et al. investigated the comparison of structure-based and protein–ligand interaction fingerprinting (IFPs) and the prediction of ligand-binding mechanisms for protein kinases [14].

6   CONVOLUTIONAL NEURAL NETWORKS

Convolutional neural networks, or more commonly CNNs, are a specific kind of neural network with a well-defined, grid-like topology that is used for data processing such as time-series data, which is meant to be visualized to a 1D grid collecting samples at periodic intervals, and image data, to be used to visualize as a 2D grid composed of pixels [72]. Convolutional networks have proved to be promising in practical executions. The term "convolutional neural network" is implied specifically to be a kind of a network that makes use of convolution mathematical operations [73]. A specific kind of linear processing is convolution. Convolutional networks are simply neural networks with one or more of these layers using convolution rather than standard matrix multiplication. For the purpose of image recognition, individual unit in a feed-forward neural network is correlated to the individual pixels that form the respective image [74]. Although there happens to be no link between nodes of a layer, and there is possibility loss of dimensional settings of an image, the CNN is designed to resolve this problem. CNNs process different patches of an image to their respective nodes, rather than pixels. Such patches are enabled to extricate specific characteristics, which make up the convolutional filters [75].

6.1   Components of a CNN

There are numerous layers that make up the whole CNN. A CNN comprises three main layers (Figure 10.2).

6.1.1   Convolutional Layers

A convolutional layer can be considered as the basic building block or the structural unit of the CNN. It is mainly composed of a particular set of filters (also called kernels), and it must be trained in order to learn the necessary parameters. The filter size is usually more undersized than the actual size of the picture. Then the activation map is formed because every filter is convolved with the picture. For convolution, the filter is passed right across the height and width of the image. Calculating the matrix multiplication between each filtration system and the input is fundamental. This is done at each specific position [76].

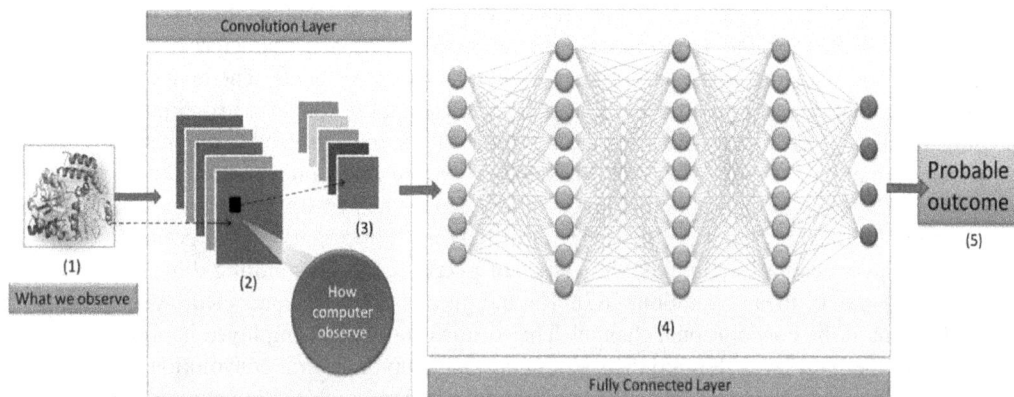

FIGURE 10.2 This representation shows the mechanism of the CNN by giving an input of a protein structure, various processing layers in which (2) convolutional filter, (3) pooling layer (4) fully connected layer. By these layers, probable outcome may observe.

6.1.2 Fully Connected Multilayer Perceptron

These layers are the final layers of the network, which is also known as a feed-forward network. The output is from the final pooling layers or the convolutional layer, which is then flattened in order to be fed to the fully connected layer. As soon as it is passed through the fully connected layers, the final layer then utilizes the activation function. It is then used for prediction of the probabilities of the input [77].

6.1.3 Pooling Layers

The max pooling layer in CNNs serves the purpose of extraction of patterns after identifying the recurring nature of the same from data of the image and then subsequently fitting into the input dimensions of the layers that are entirely connected [78]. This layer's main job in the network is to make it easier to minimize the number of parameters. The parameters of the response tensor are hence decreased. This further facilitates in:

- Helping in reduction of over fitting
- Extraction features
- Reduction in multiplication
- Aiding greater competence

There are mainly two kinds of pooling layers:

a) Max Pooling
b) Average Pooling

As far as Max Pooling is concerned, an example of which is shown in the, the maximum value is picked from each position, this value is then positioned in the suitable site of the output matrix with the help of a kernel of size n*n. where the estimate of each of the results is estimated for every node and placed in the right location of the transformation matrix in the case of average pooling using a kernel of size n*n, which is shifted across the matrix [74-76]. This ability of CNNs to extract image data is utilized to carry out the analysis of protein structures in 3D approaches. Protein structures are considered as 3D images to be analyzed by neural networks. The multiple layers that outline the CNN can transform various inputs. This is achieved by the use of convolutional filters [81]. This is what makes CNNs different from other deep learning models. Contrary to other deep learning technologies, CNN is capable of extracting certain smaller portioned characteristics features from images [82]. These characteristic features are known as receptive fields. The convolutional filters are passed through at every layer of the model. The entire data is convolved by filters, resulting in new feature maps of the input. After the passing of the filters, the extracted features undergo a non-linear transformation [78–80]. After that, the same procedure keeps on repeating for the rest of the convolutional layers. CNNs use pooling layers to broaden the perceptron of the convolutional layers [86]. The pooling layers also serve the purpose of making it invariant to local deformation of the input. These layers can also cumulate the values of pixels of the neighboring dimensions. This is carried out by max or mean operations [87]. The merging of the fully connected layers takes place at the end point of the convolutional channel. The softmax function is employed in order to feed the activations in the final layer [83–85]. The CNNs are made up of several convolutional filters and a greater number of convolutional as well as pooling layers. This enables the network to incorporate the image data extracted by the various filters at multiple different levels [91]. Convolutions offer multiple functions such as image prediction, processing, blurring and detecting edges [92], [93]. The convolutional filter is moved and superimposed over the initial image at individual point and the outcome is traced in a pattern to a matrix on the basis of detection of the desired characteristic by the convolutional filter. The CNN filters are devised to bring out certain characteristics from the original images such as curves and edges to mark out to a feature map [94], [95]. CNNs use more filters to trace further feature maps. This can go on for numerous new layers and hence the level of predictability becomes remarkably high. The finished classification and prediction of the image is based on these characteristic features. This whole process is referred as deep learning. CNNs with multiple hidden layers with more intricate structures and more features than other machine learning technologies. These algorithms have made extraordinary progress in visual identification [96]. The CNN scoring functions are considered promising for the purpose of predicting affinity and selection of image poses. These serve applications in structure-based drug design pipelines [97], [98]. While dealing with protein–ligand complexes, protein–ligand scoring is a crucial step. Identifying the suitable binding pose becomes an important requirement to carry effectively out the process of virtual screening. The prediction of the binding affinity is also an important application of machine learning technologies [99]. Many of these hurdles are overcome by the CNN approaches. Machine learning models are designed to utilize the increasing bulk data from data libraries. There are several issues that need to be fixed such as certain neural networks are hard to interpret and the algorithm becomes difficult to train. There are difficulties with both the quantity and quality of something like the input data that must be used to train the algorithms. It is crucial to fine-tune the settings in order to boost the effectiveness of the virtual screening, subsequently maximizing the performance and the probability of the desired outcome.

7   APPLICATIONS OF CNN IN VIRTUAL SCREENING

7.1   PREDICTION OF BINDING AFFINITY

Some frequently used approaches are under SBVS (structure-based virtual screening) is known as docking. Docking is employed to validate the extent of binding of small molecules to the target structure in a "lock and key" pattern [100]. Throughout the docking process various probably ligand conformers, also known as docking poses, are reiteratively tested at the binding site in order to detect an acceptable ligand pose giving the best binding affinity [101]. The binding affinity of an individual pose is generated by the scoring function (SF) of the docking program. The SF is therefore critically important for docking programs in SBVS [102]. These can be determined as estimation of strength of the docked pose of a ligand binding to the target. Scoring functions are commonly used for fast assessment of protein–ligand interactions, so building an effective and robust SF is a step in the direction of accelerating the VS procedure [103]. Detailed 3D representations of a protein–ligand interaction are used as input by CNN scoring algorithms. The key aspects of protein–ligand interactions that correspond to binding are automatically learned by a CNN scoring function to distinguish between known binders and non-binders as well as between correct and wrong binding poses [75-77].

Here are a few virtual screening uses of current CNN models.

7.2 PROTEIN–LIGAND-BINDING CNN MODELS

Prediction of binding affinity has always been a mammoth task because protein and ligand binding is like lock and key; only the correct ligand can bind with the specific protein to show the desired effect. Since the advent of computational chemistry in the field of scientific research, a lot has been done to overcome limitations and achieve this goal. CNNs are designed to address challenges. Various CNN models have been developed, scrutinized, applied, and validated for this purpose. In this context, Jiménez et al. developed a novel approach with fast machine learning technology for the estimation of probable affinity of binding of the ligands towards the protein by the application of cutting-edge 3D CNN technology. A comparison of the aforementioned method with various other approaches to machine learning and scoring methods was conducted by using multiple different data sets like PDBbind. As a result a test-set showed that the Pearson's parameter of the correlation coefficient between the both was 0.82 and another parameter RMSE showed 1.27 pK units. However, the accuracy remained significantly dependent on the particular protein type used. Each prediction was validated within a second. KDEEP is already a desirable scoring function for existing computational chemistry workflows due to its high speed, economical performance, and simplicity of usage. When a computerized molecular design approach was introduced it enabled identification of a desired molecule from numerous molecules. The ability of the previous selecting functions to predict binding affinity was insufficient [106], so for effective and successful rational drug design, a correct prediction of the combinational binding affinity between protein and ligand is necessary. As a result, numerous approaches for predicting binding affinity have been created. In the course of the last few years, deep learning technology has become a popular tool to carry out the task much more efficiently. Kwon et al. showed a selective and sensitive neural network model employed to predict the protein–ligand complex and structure's binding affinity. The binding affinity of a protein is predicted by this particular model using a collection of networks that are trained individually. They are made up of several 3D CNN channels. The dataset used for the training was PDBbind, which examined and tested 3772 protein–ligand complexes. According to the result the experimental data and the anticipated binding affinities had a Pearson correlation of 0.827, which was greater than that of the most recent scoring functions for binding affinity predictions. The similarity between the protein and ligand was shown to be the most significant structural factor for determining binding affinity [107]. Wang et al. developed a novel CNN model, called SE-OnionNet. Using two squeeze-and-excitation (SE) modules, they made it possible to computationally estimate the binding interactions of a protein–ligand pair. A protein–drug molecular complex's 3D structure was used to generate a feature map using the OnionNet. In order to enhance the non-linear expression of this proposed network and the overall functioning and final performance of the model, the SE component was introduced between two layers, the second and third. To improve the performance of the model, three other optimizers – Stochastic Gradient Descent, Adam, and Adagrad – were also used. The majority of biomolecules, especially protein, were used as training examples, and the benchmark was the comparison of scoring function evaluations. According to experimental findings, the model outperformed OnionNet, Pafnucy, and AutoDock Vina. As an example, they used the macrophage migration inhibitor factor (PDB ID: 6cbg) and discovered that the model was accurate and the outcomes of the predictions robust [108]. The smaller chemical compounds and their binding affinity to the receptor proteins is a very important factor for drug repositioning and drug discovery. Wang et al. developed a novel method that includes deep learning technology, namely CSConv2d, for prediction of interfaces between protein–ligand. Also, this method used 2D structural representation enhanced by DEEPScreen model. Additionally, a spatial attention technique known as feature abstraction was included. Data experiments that use the ChEMBLv23 datasets revealed that CSConv2d outperforms certain state-of-the-art DTIs (drug–target interactions) prediction approaches, including DeepConv-DTI, CPI-Prediction, CPI-Prediction+CS, DeepGS, and DeepGS+CS in predicting protein–ligand-binding affinity. The docking data of a protein (PDB ID: 5ceo), ligand (Chemical ID: 5OD), and many kinase inhibitors were used to systematically evaluate the robustness [109]. The foundation of computational design of a drug is formed by the calculation of affinity of binding, which is the binding affinity between the two biological equivalent chemical compounds. These equivalent compounds happen to be ligand and protein. The prediction of the strength of binding of the protein and the ligand should be reasonably accurate. This is crucial for the entire process of drug discovery. Li et al. developed a framework driven by datasets called DeepAtom. The data architecture of 3D Convolutional Neural Network (3D–CNN), DeepAtom could successfully extract the binding interface between the relevant atoms in an automated manner. These interactional patterns came from the complicated voxelized structure. This compact model was intended to significantly improve the capability of model description in contrast to other CNN-based methods. Despite the low amount of data available for model training, the performance was not affected. Using authentication tests on the autonomous Astex Diverse Set and the PDBbind v.2016 standard, they demonstrated that the DeepAtom strategy, which relies less on feature engineering, outperformed other cutting-edge scoring techniques. They created a model benchmark dataset as well to improve the performance. The results achieved by using the novel dataset for training DeepAtom models were highly promising. This model has the potential to be adopted in computational drug development procedures, and has potential applications in virtual screening [110].

7.3 DRUG–TARGET-BINDING PREDICTION

The process of discovering new drug–target interactions (DTIs) is crucial. Binary classification, used by the majority of computational techniques created for predicting DTIs, seeks to establish whether a drug–target (DT) pair interacts or not [111]. Predicting the binding affinity, which describes the strength of the interaction between a DT pair, is more significant but also more difficult. Such drugs only are not applicable if the binding affinity is not clear or low. Thus, it is required to search for the methods for calculating DT-binding affinities, which are therefore particularly useful. In DT-related databases, the amount of newly collected accessible affinity data has increased, making it possible to predict binding affinities using cutting-edge deep learning approaches [112]. In this context, Shim et al. proposed DT-binding affinities predicted using a correlation model that used a 2D CNN to calculate the outer products of two similarity matrices for the medications and targeting unit vectors. According to the validation outcomes on various accessible datasets, the proposed model was a highly effective technique for DT-binding affinity prediction and may be extremely useful in the drug development process [113]. Hakime et al. recently improvised a model that can predict the binding affinities. This model uses the sequence information of targets as well as drugs. For the DT-binding affinity predictions mainly 2D and 3D structured interpretation models of CNNs are used. CNNs are combined with modeling of protein classifications and complex 1D signs in a new revolutionary strategy to achieve this objective (CNNs). According to the findings, the proposed deep learning-based model, which employs 1D representation of both targets and drugs, and is an effective technique for predicting DT-binding affinity. A medicine and a target were created in high-level representations by the model in one of their main benchmark datasets using the CNN model, which outperformed the algorithm developed by KronRLS and SimBoost, which in itself is a cutting-edge technique for DT-binding affinity prediction (Table 10.1) [114]. Another model developed by Shao et al. is a novel approach for DTI prediction called DTIGCCN. The model works on the basis of a graph convolutional network (GCN), which is spectral-based. The spectral graph was created for

TABLE 10.1
CNN models with the features and datasets used

DEEP LEARNING TOOLS	TYPES	DATASETS	FEATURES	REFERENCE
AttentionDTA	Model based on 1-DCNN	DAVIS & KIBA	Use of attention mechanism for prediction of binding affinity, better representational ability. Better performance in terms of AUPR, MSE, CI.	[118]
DeepDTA	Works on the basis of sequence information of Drug & Target	KIBA	Used for predicting drug–target affinity.	[114]
OnionNet	Based on rotation-free element-pair specific interaction.	PDBind database	Accuracy of prediction is highly sensitive.	[119]
CSatDTA	Convolution-based mechanism for self-attention to a molecule	DAVIS & KIBA	Higher efficiency for DTA (Drug– Target Affinity) prediction.	[117]
DeepCPI	Automated unsupervised representational learning	ChEMBL	The low-dimensional features are better extracted and learned.	[120]
WideDTA	Utilization of textual sequence information	DAVIS & KIBA	Better prediction of Drug–Target Binding affinity.	[121]
DeepPocket	3D CNN Network, classification & segmentation model	scPDB v.2017	Ligand-binding site prediction & segmentation from protein structures.	[122]
PADME DeepAffinity	Molecular Graph Convolution (MGC)	DTI, ToxCast	Prediction of interaction patterns between compounds & protein. It is enabled to fix & resolve cold target problems.	[123]
K_{Deep}	3D CNN Network	Multiple diverse data sets	Prediction can be concluded in a fraction of second. Attractive scoring function for computational chemistry	[106]
SE-OnionNet	Squeeze & Excitation modules for computational prediction	CASF benchmark	Capable of extracting feature maps from 3D structure of protein, and drug molecular complex by using SE module for further improvement of performance of the model.	[108]
AutoDock vina	Molecular docking	PDBind database	Capable of carrying out multiple docking such as site-specific, blind docking, protein–ligand, etc.	[124]
Pafnucy	3D Convolution network to produce feature map of representation	Astex diverse datasets	Better scoring functions, allows model to "learn" to extract features.	[125]
DeepScreen	Based on 2D representations for DTI prediction	ChEMBL	Suited for large-scale DTI predictions.	[126]
DeepGS	Advanced embedding technique to encode amino acids	Multiple	Extraction of local chemical context from amino acids & molecular structure from drugs.	[127]
DeepAtom	Data-driven framework, extracts binding related Interaction patterns from voxelised structures	Astex diverse datasets	Promising results for computational drug development procedures & molecular docking.	[128]

extraction of the features. The extracted features of both the target the drug profiles were used. Then, a CNN network was employed for extraction purposes. The latent interaction features between the drug and the target were extracted. In the end, all the extracted features were combined together, then fed together into a classifier. The classifier carried out the prediction. The DTIGCCN model has its own merits. The extracted features are more sophisticated. Another aspect also entirely the prediction process of the drug and target is their correlation, This model has better performance than the conventional DTI methods [115]. Torng et al. developed a graph-convolutional (Graph–CNN) framework. They employed a selection of common drug and protein-binding sites to examine fixed-size models of protein pockets; also it was first developed an unsupervised graph-auto encoder. Two Graph–CNNs were trained for the purpose of extraction of features automatically right from 2D ligand graphs, or from pocket graphs, correspondingly, driven by binding classification labels. It was shown that the Graph–CNN architecture can efficiently capture protein–ligand-binding interactions without relying on target–ligand complexes, and that graph-auto encoders can learn representations of fixed-size intended for protein pockets ranging in their sizes. Visualization capacities of the key pocket residues as well as that of the ligand atoms, which contribute to the classification decisions, confirmed the ability of the networks to detect the interface residues, along with the ligand atoms, both of which would be present within the pockets and ligands, respectively [8]. Lingling et al. developed a type of method based on GAN networks (generative adversarial networks). These networks are semi-supervised GANs. This model could predict the binding affinity. This approach is divided into two parts. One half comprises two GANs to carry out the task of feature extraction and the other half is a regression network to execute prediction. This model is enabled to learn proteins and drug features of the data, regardless of it being labeled or unlabeled. This is owed to the semi-supervised mechanism. The performance of this method was evaluated using multiple datasets. By using experimental results, this method was able to achieve effective performance. An added advantage was the utilization of freely available unlabeled data. In particular, when only a small amount of labeled data is available in such tasks, the results show promise for increasing efficient performance in numerous areas that require applications of biomedical extractions, such as drug–target and protein–protein interactions. This is the first, one of a kind semi-supervised GAN approach for binding affinity prediction purposes [116]. Another model for drug–target affinity is is expected that will use a convolution model combined with self-attention (CSatDTA). This model involves the application of convolution-based mechanisms for self-attention to the molecule of a drug and the target sequences. This mechanism will predict drug–target affinity (DTA) in a highly effective manner. There were some limitations with other convolution methods that this model successfully overcame in this regard. Some challenges also include the limited specificity of a region of information over which a particular CNN model is designed to perform. Also, comprehensive details are sometimes excluded. For capturing of long-range interactions, self-attention is a recent advanced technique. This technique is primarily used in tasks such as sequence modeling drawing, but a comparison between the sequence-based approaches concluded that CSatDTA could outperform the previous approaches [117].

8 CHALLENGES AND FUTURE OUTLOOK

The abundance of information in data banks in the digital world makes deep learning an immensely popular tool. It holds dominance in the realm of medical sciences, pharmaceutical development, cheminformatics, bioinformatics, etc. The challenges that lie ahead are the estimation of the required quantity as well as the quality of data to be fed as input for the various models of technologies of deep learning, to work in the direction of the prediction of the desired result. Generic prediction as well as accurate prediction can be required, depending on the demand of the situation. The prediction models and their prediction requirement also vary in complexity of an image interpretation. Furthermore, increased data bulk in deep learning does not necessarily mean better output. The quality of the data should be unambiguous and error-free to achieve the desired output. Scientists are not always sure if a particular deep learning model will work sufficiently well in the physical world. Another challenge is something referred to as "black box." This creates a situation where the identified features are not distinguishable enough for recognition. Although there are some models that work towards resolving the black box issue, some filters are abstract and make the interpretation challenging. Nevertheless, selective models such as heatmaps to highlight certain characteristics in the image often require human interpretation. A lot of other challenges exist to increase precision and predictability and decrease complexity of the models. There are requirements of fine tuning and rejecting unrelated perceptron connections to name some.

9 CONCLUSION

Artificial intelligence will inevitably become an indispensable part of future technologies. The predictive models will prove to be promising methodologies in visual interpretation in the domains of drug discovery, molecular design, and diagnostics. Deep learning technologies will continue to grow for the purpose of computer-aided detection and information retrieval. The main goal behind deep learning virtual screening is interpretation, representation. The hope is to augment the extent of classification and feature extraction to overcome the hurdles of large data representation.

REFERENCES

1. A. S. Reddy, S. P. Pati, P. P. Kumar, H. N. Pradeep, and G. N. Sastry, "Virtual screening in drug discovery --a computational perspective.," *Curr. Protein Pept. Sci.*, vol. 8, no. 4, pp. 329–51, 2007, [Online]. Available: www.ncbi.nlm.nih.gov/pubmed/17696867.
2. S. F. L. da Silva Rocha, C. G. Olanda, H. H. Fokoue, and C. M. R. Sant'Anna, "Virtual Screening Techniques in Drug Discovery: Review and Recent Applications," *Curr. Top. Med. Chem.*, vol. 19, no. 19, pp. 1751–1767, Oct. 2019, doi: 10.2174/1568026619666190816101948.
3. H. Dowden and J. Munro, "Trends in clinical success rates and therapeutic focus," *Nat. Rev. Drug Discov.*, vol. 18, no. 7, pp. 495–496, Jul. 2019, doi: 10.1038/d41573-019-00074-z.
4. G. Schneider, "Automating drug discovery," *Nat. Rev. Drug Discov.*, vol. 17, no. 2, pp. 97–113, Feb. 2018, doi: 10.1038/nrd.2017.232.
5. GUHA RAJARSHI & BENDER ANDREAS, *APPROACHES IN COMPUTATIONAL APPROACHES IN Edited by*. 2012.
6. X. Lin, X. Li, and X. Lin, "A Review on Applications of Computational Methods in Drug Screening and Design," *Molecules*, vol. 25, no. 6, p. 1375, Mar. 2020, doi: 10.3390/molecules25061375.
7. W. Torng and R. B. Altman, "3D deep convolutional neural networks for amino acid environment similarity analysis," *BMC Bioinformatics*, vol. 18, no. 1, p. 302, Dec. 2017, doi: 10.1186/s12859-017-1702-0.
8. W. Torng and R. B. Altman, "Graph Convolutional Neural Networks for Predicting Drug-Target Interactions," *J. Chem. Inf. Model.*, vol. 59, no. 10, pp. 4131–4149, Oct. 2019, doi: 10.1021/acs.jcim.9b00628.
9. S. Majumdar *et al.*, "Deep Learning-Based Potential Ligand Prediction Framework for COVID-19 with Drug–Target Interaction Model," *Cognit. Comput.*, Feb. 2021, doi: 10.1007/s12559-021-09840-x.
10. X. Tan, Y. Liang, Y. Ye, Z. Liu, J. Meng, and F. Li, "Explainable Deep Learning-Assisted Fluorescence Discrimination for Aminoglycoside Antibiotic Identification," *Anal. Chem.*, vol. 94, no. 2, pp. 829–836, Jan. 2022, doi: 10.1021/acs.analchem.1c03508.
11. V. N. Shinde, N. Bhuvanesh, A. Kumar, and H. Joshi, "Design and Syntheses of Palladium Complexes of NNN/CNN Pincer Ligands: Catalyst for Cross Dehydrogenative Coupling Reaction of Heteroarenes," *Organometallics*, vol. 39, no. 2, pp. 324–333, 2020, doi: 10.1021/acs.organomet.9b00695.
12. A. Yan, T. Sokolinski, W. Lane, J. Tan, K. Ferris, and E. M. Ryan, "Applying transfer learning with convolutional neural networks to identify novel electrolytes for metal air batteries," *Comput. Theor. Chem.*, vol. 1205, p. 113443, Nov. 2021, doi: 10.1016/j.comptc.2021.113443.

13. I. V. Tetko and O. Engkvist, "From Big Data to Artificial Intelligence: chemoinformatics meets new challenges," *J. Cheminform.*, vol. 12, no. 1, pp. 12–14, 2020, doi: 10.1186/s13321-020-00475-y.

14. R. Rodríguez-Pérez, F. Miljković, and J. Bajorath, "Assessing the information content of structural and protein–ligand interaction representations for the classification of kinase inhibitor binding modes via machine learning and active learning," *J. Cheminform.*, vol. 12, no. 1, p. 36, Dec. 2020, doi: 10.1186/s13321-020-00434-7.

15. N. Brown, "The History of Artificial Intelligence and Chemistry," *Artif. Intell. Drug Discov.*, vol. 75, p. 7, 2020.

16. M. K. Warmuth, J. Liao, G. Rätsch, M. Mathieson, S. Putta, and C. Lemmen, "Active Learning with Support Vector Machines in the Drug Discovery Process," *J. Chem. Inf. Comput. Sci.*, vol. 43, no. 2, pp. 667–673, Mar. 2003, doi: 10.1021/ci025620t.

17. Q. Li and S. Shah, "Structure-Based Virtual Screening," 2017, pp. 111–124.

18. A.-J. Banegas-Luna, J. P. Cerón-Carrasco, and H. Pérez-Sánchez, "A review of ligand-based virtual screening web tools and screening algorithms in large molecular databases in the age of big data," *Future Med. Chem.*, vol. 10, no. 22, pp. 2641–2658, Nov. 2018, doi: 10.4155/fmc-2018-0076.

19. R. Lahana, "How many leads from HTS?," *Drug Discov. Today*, vol. 4, no. 10, pp. 447–448, Oct. 1999, doi: 10.1016/S1359-6446(99)01393-8.

20. S. P. Leelananda and S. Lindert, "Computational methods in drug discovery," *Beilstein J. Org. Chem.*, vol. 12, pp. 2694–2718, Dec. 2016, doi: 10.3762/bjoc.12.267.

21. J. Lyu *et al.*, "Ultra-large library docking for discovering new chemotypes," *Nature*, vol. 566, no. 7743, pp. 224–229, Feb. 2019, doi: 10.1038/s41586-019-0917-9.

22. W. P. Walters, M. T. Stahl, and M. A. Murcko, "Virtual screening—an overview," *Drug Discov. Today*, vol. 3, no. 4, pp. 160–178, Apr. 1998, doi: 10.1016/S1359-6446(97)01163-X.

23. H. Zhou, H. Cao, and J. Skolnick, "FINDSITE comb2.0: A New Approach for Virtual Ligand Screening of Proteins and Virtual Target Screening of Biomolecules," *J. Chem. Inf. Model.*, vol. 58, no. 11, pp. 2343–2354, Nov. 2018, doi: 10.1021/acs.jcim.8b00309.

24. T. Sterling and J. J. Irwin, "ZINC 15–Ligand Discovery for Everyone," *J. Chem. Inf. Model.*, vol. 55, no. 11, pp. 2324–2337, Nov. 2015, doi: 10.1021/acs.jcim.5b00559.

25. "MOLPORT." www.molport.com/.

26. "Enamine REAL." enamine.net/library-synthesis/real-compounds.

27. M. McGibbon, S. Money-Kyrle, V. Blay, and D. R. Houston, "SCORCH: Improving structure-based virtual screening with machine learning classifiers, data augmentation, and uncertainty estimation," *J. Adv. Res.*, Jul. 2022, doi: 10.1016/j.jare.2022.07.001.

28. J. Ricci-Lopez, S. A. Aguila, M. K. Gilson, and C. A. Brizuela, "Improving Structure-Based Virtual Screening with Ensemble Docking and Machine Learning," *J. Chem. Inf. Model.*, vol. 61, no. 11, pp. 5362–5376, Nov. 2021, doi: 10.1021/acs.jcim.1c00511.

29. A. Bustamam *et al.*, "Artificial intelligence paradigm for ligand-based virtual screening on the drug discovery of type 2 diabetes mellitus," *J. Big Data*, vol. 8, no. 1, p. 74, Dec. 2021, doi: 10.1186/s40537-021-00465-3.

30. Z. Jiang, J. Xu, A. Yan, and L. Wang, "A comprehensive comparative assessment of 3D molecular similarity tools in ligand-based virtual screening," *Brief. Bioinform.*, vol. 22, no. 6, Nov. 2021, doi: 10.1093/bib/bbab231.

31. F. Shahid *et al.*, "Identification of Potential HCV Inhibitors Based on the Interaction of Epigallocatechin-3-Gallate with Viral Envelope Proteins," *Molecules*, vol. 26, no. 5, p. 1257, Feb. 2021, doi: 10.3390/molecules26051257.

32. X. Xu, M. Huang, and X. Zou, "Docking-based inverse virtual screening: methods, applications, and challenges," *Biophys. Reports*, vol. 4, no. 1, pp. 1–16, Feb. 2018, doi: 10.1007/s41048-017-0045-8.

33. Z. Wang *et al.*, "Combined strategies in structure-based virtual screening," *Phys. Chem. Chem. Phys.*, vol. 22, no. 6, pp. 3149–3159, 2020, doi: 10.1039/C9CP06303J.

34. V. B. Sulimov, D. C. Kutov, and A. V. Sulimov, "Advances in Docking," *Curr. Med. Chem.*, vol. 26, no. 42, pp. 7555–7580, Jan. 2020, doi: 10.2174/0929867325666180904115000.

35. A. Fischer, M. Smieško, M. Sellner, and M. A. Lill, "Decision Making in Structure-Based Drug Discovery: Visual Inspection of Docking Results," *J. Med. Chem.*, vol. 64, no. 5, pp. 2489–2500, Mar. 2021, doi: 10.1021/acs.jmedchem.0c02227.

36. G. Klebe, "Virtual ligand screening: strategies, perspectives and limitations," *Drug Discov. Today*, vol. 11, no. 13–14, pp. 580–594, Jul. 2006, doi: 10.1016/j.drudis.2006.05.012.

37. J. Li, A. Fu, and L. Zhang, "An Overview of Scoring Functions Used for Protein–Ligand Interactions in Molecular Docking," *Interdiscip. Sci. Comput. Life Sci.*, vol. 11, no. 2, pp. 320–328, Jun. 2019, doi: 10.1007/s12539-019-00327-w.

38. M. A. Azam and S. Jupudi, "Structure-based virtual screening to identify inhibitors against Staphylococcus aureus MurD enzyme," *Struct. Chem.*, vol. 30, no. 6, pp. 2123–2133, Dec. 2019, doi: 10.1007/s11224-019-01330-z.

39. K. K. Reddy, S. K. Singh, S. K. Tripathi, C. Selvaraj, and V. Suryanarayanan, "Shape and pharmacophore-based virtual screening to identify potential cytochrome P450 sterol 14α-demethylase inhibitors," *J. Recept. Signal Transduct.*, vol. 33, no. 4, pp. 234–243, Aug. 2013, doi: 10.3109/10799893.2013.789912.

40. C. Shen, J. Ding, Z. Wang, D. Cao, X. Ding, and T. Hou, "From machine learning to deep learning: Advances in scoring functions for protein–ligand docking," *WIREs Comput. Mol. Sci.*, vol. 10, no. 1, Jan. 2020, doi: 10.1002/wcms.1429.

41. B. J. Neves, R. C. Braga, C. C. Melo-Filho, J. T. Moreira-Filho, E. N. Muratov, and C. H. Andrade, "QSAR-Based Virtual Screening: Advances and Applications in Drug Discovery," *Front. Pharmacol.*, vol. 9, Nov. 2018, doi: 10.3389/fphar.2018.01275.

42. S. Pal *et al.*, "Ligand-based Pharmacophore Modeling, Virtual Screening and Molecular Docking Studies for Discovery of Potential Topoisomerase I Inhibitors," *Comput. Struct. Biotechnol. J.*, vol. 17, pp. 291–310, 2019, doi: 10.1016/j.csbj.2019.02.006.

43. Y. O. Adeshina, E. J. Deeds, and J. Karanicolas, "Machine learning classification can reduce false positives in structure-based virtual screening," *Proc. Natl. Acad. Sci.*, vol. 117, no. 31, pp. 18477–18488, Aug. 2020, doi: 10.1073/pnas.2000585117.

44. A. Tropsha, "Best Practices for QSAR Model Development, Validation, and Exploitation," *Mol. Inform.*, vol. 29, no. 6–7, pp. 476–488, Jul. 2010, doi: 10.1002/minf.201000061.

45. D. Sydow *et al.*, "Advances and Challenges in Computational Target Prediction," *J. Chem. Inf. Model.*, vol. 59, no. 5, pp. 1728–1742, May 2019, doi: 10.1021/acs.jcim.8b00832.

46. J. Gasteiger, "Chemoinformatics: a new field with a long tradition," *Anal. Bioanal. Chem.*, vol. 384, no. 1, pp. 57–64, Jan. 2006, doi: 10.1007/s00216-005-0065-y.

47. H. Chen, T. Kogej, and O. Engkvist, "Cheminformatics in Drug Discovery, an Industrial Perspective," *Mol. Inform.*, vol. 37, no. 9–10, p. 1800041, Sep. 2018, doi: 10.1002/minf.201800041.

48. R. Kojima, S. Ishida, M. Ohta, H. Iwata, T. Honma, and Y. Okuno, "kGCN: a graph-based deep learning framework for chemical structures," *J. Cheminform.*, vol. 12, no. 1, p. 32, Dec. 2020, doi: 10.1186/s13321-020-00435-6.

49. D. Weininger, "SMILES, a Chemical Language and Information System: 1: Introduction to Methodology and Encoding Rules," *J. Chem. Inf. Comput. Sci.*, vol. 28, no. 1, pp. 31–36, 1988, doi: 10.1021/ci00057a005.

50. J. J. Vollmer, "Wiswesser line notation: an introduction," *J. Chem. Educ.*, vol. 60, no. 3, p. 192, 1983.

51. H.-G. Rohbeck, "Representation of structure description arranged linearly," in *Software Development in Chemistry 5*, Springer, 1991, pp. 49–58.

52. S. Ash, M. A. Cline, R. W. Homer, T. Hurst, and G. B. Smith, "SYBYL line notation (SLN): A versatile language for chemical structure representation," *J. Chem. Inf. Comput. Sci.*, vol. 37, no. 1, pp. 71–79, 1997.

53. N. M. O'Boyle, "Towards a Universal SMILES representation–A standard method to generate canonical SMILES based on the InChI," *J. Cheminform.*, vol. 4, no. 1, p. 22, Dec. 2012, doi: 10.1186/1758-2946-4-22.

54. A. W. Schüttelkopf and D. M. F. van Aalten, "PRODRG: a tool for high-throughput crystallography of protein–ligand complexes," *Acta Crystallogr. Sect. D Biol. Crystallogr.*, vol. 60, no. 8, pp. 1355–1363, Aug. 2004, doi: 10.1107/S0907444904011679.

55. Haggarty, "Mapping Chemical Space Using Molecular Descriptors and Chemical Genetics: Deacetylase Inhibitors," *Comb. Chem. High Throughput Screen.*, vol. 7, no. 7, 2004, doi: 10.2174/1386207043328319.

56. V. J. Sykora and D. E. Leahy, "Chemical Descriptors Library (CDL): A Generic, Open Source Software Library for Chemical Informatics," *J. Chem. Inf. Model.*, vol. 48, no. 10, pp. 1931–1942, Oct. 2008, doi: 10.1021/ci800135h.

57. J. H. Nettles, J. L. Jenkins, A. Bender, Z. Deng, J. W. Davies, and M. Glick, "Bridging Chemical and Biological Space: 'Target Fishing' Using 2D and 3D Molecular Descriptors," *J. Med. Chem.*, vol. 49, no. 23, pp. 6802–6810, Nov. 2006, doi: 10.1021/jm060902w.

58. A. K. Jain, Jianchang Mao, and K. M. Mohiuddin, "Artificial neural networks: a tutorial," *Computer (Long. Beach. Calif).*, vol. 29, no. 3, pp. 31–44, Mar. 1996, doi: 10.1109/2.485891.

59. B. Widrow and M. A. Lehr, "30 years of adaptive neural networks: perceptron, Madaline, and backpropagation," *Proc. IEEE*, vol. 78, no. 9, pp. 1415–1442, 1990, doi: 10.1109/5.58323.

60. M. L. Minsky and S. Papert, "Perceptrons: an Introduction to Computational Geometry, Cambridge, Mass." London, 1969.

61. D. E. Rumelhart, G. E. Hinton, and R. J. Williams, "Learning representations by back-propagating errors," *Nature*, vol. 323, no. 6088, pp. 533–536, Oct. 1986, doi: 10.1038/323533a0.

62. A. Korotcov, V. Tkachenko, D. P. Russo, and S. Ekins, "Comparison of Deep Learning With Multiple Machine Learning Methods and Metrics Using Diverse Drug Discovery Data Sets," *Mol. Pharm.*, vol. 14, no. 12, pp. 4462–4475, Dec. 2017, doi: 10.1021/acs.molpharmaceut.7b00578.

63. T. M. Whitehead, B. W. J. Irwin, P. Hunt, M. D. Segall, and G. J. Conduit, "Imputation of assay bio-activity data using deep learning," *J. Chem. Inf. Model.*, vol. 59, no. 3, pp. 1197–1204, 2019.

64. A. Ajit, K. Acharya, and A. Samanta, "A Review of Convolutional Neural Networks," in *2020 International Conference on Emerging Trends in Information Technology and Engineering (ic-ETITE)*, Feb. 2020, pp. 1–5, doi: 10.1109/ic-ETITE47903.2020.049.

65. F. Musazade, N. Jamalova, and J. Hasanov, "Review of techniques and models used in optical chemical structure recognition in images and scanned documents," *J. Cheminform.*, vol. 14, no. 1, p. 61, Sep. 2022, doi: 10.1186/s13321-022-00642-3.

66. Y. Cheng, D. Wang, P. Zhou, and T. Zhang, "Model Compression and Acceleration for Deep Neural Networks: The Principles, Progress, and Challenges," *IEEE Signal Process. Mag.*, vol. 35, no. 1, pp. 126–136, Jan. 2018, doi: 10.1109/MSP.2017.2765695.

67. K. Rajan, H. O. Brinkhaus, A. Zielesny, and C. Steinbeck, "A review of optical chemical structure recognition tools," *J. Cheminform.*, vol. 12, no. 1, p. 60, Dec. 2020, doi: 10.1186/s13321-020-00465-0.

68. V. Jhalia and T. Swarnkar, "A Critical Review on the Application of Artificial Neural Network in Bioinformatics," in *Data Analytics in Bioinformatics*, Wiley, 2021, pp. 51–76.

69. S. Alkaabi, S. Yussof, H. Al-Khateeb, G. Ahmadi-Assalemi, and G. Epiphaniou, "Deep Convolutional Neural Networks for Forensic Age Estimation: A Review," 2020, pp. 375–395.

70. S. A. Damiati, "Digital Pharmaceutical Sciences," *AAPS PharmSciTech*, vol. 21, no. 6, p. 206, Aug. 2020, doi: 10.1208/s12249-020-01747-4.

71. J. Liu, J. Li, H. Wang, and J. Yan, "Application of deep learning in genomics," *Sci. China Life Sci.*, vol. 63, no. 12, pp. 1860–1878, Dec. 2020, doi: 10.1007/s11427-020-1804-5.

72. R. Yamashita, M. Nishio, R. K. G. Do, and K. Togashi, "Convolutional neural networks: an overview and application in radiology," *Insights Imaging*, vol. 9, no. 4, pp. 611–629, Aug. 2018, doi: 10.1007/s13244-018-0639-9.

73. D. R. Sarvamangala and R. V. Kulkarni, "Convolutional neural networks in medical image understanding: a survey," *Evol. Intell.*, vol. 15, no. 1, pp. 1–22, Mar. 2022, doi: 10.1007/s12065-020-00540-3.

74. L. Alzubaidi *et al.*, "Review of deep learning: concepts, CNN architectures, challenges, applications, future directions," *J. Big Data*, vol. 8, no. 1, p. 53, Dec. 2021, doi: 10.1186/s40537-021-00444-8.

75. J. Cai, L. Lu, Z. Zhang, F. Xing, L. Yang, and Q. Yin, "Pancreas Segmentation in MRI Using Graph-Based Decision Fusion on Convolutional Neural Networks," 2016, pp. 442–450.

76. S. Albawi, T. A. Mohammed, and S. Al-Zawi, "Understanding of a convolutional neural network," in *2017 International Conference on Engineering and Technology (ICET)*, Aug. 2017, pp. 1–6, doi: 10.1109/ICEngTechnol.2017.8308186.

77. S. Li, H. Jiang, and W. Pang, "Joint multiple fully connected convolutional neural network with extreme learning machine for hepatocellular carcinoma nuclei grading," *Comput. Biol. Med.*, vol. 84, pp. 156–167, May 2017, doi: 10.1016/j.compbiomed.2017.03.017.

78. M. Sun, Z. Song, X. Jiang, J. Pan, and Y. Pang, "Learning Pooling for Convolutional Neural Network," *Neurocomputing*, vol. 224, pp. 96–104, Feb. 2017, doi: 10.1016/j.neucom.2016.10.049.

79. S. Zhao, T. Zhang, M. Hu, W. Chang, and F. You, "AP-BERT: enhanced pre-trained model through average pooling," *Appl. Intell.*, Mar. 2022, doi: 10.1007/s10489-022-03190-3.

80. Z. Li, S. Wang, R. Fan, G. Cao, Y. Zhang, and T. Guo, "Teeth category classification via seven-layer deep convolutional neural network with max pooling and global average pooling," *Int. J. Imaging Syst. Technol.*, vol. 29, no. 4, pp. 577–583, Dec. 2019, doi: 10.1002/ima.22337.

81. S. Park and C. Seok, "GalaxyWater-CNN: Prediction of Water Positions on the Protein Structure by a 3D-Convolutional Neural Network," *J. Chem. Inf. Model.*, vol. 62, no. 13, pp. 3157–3168, Jul. 2022, doi: 10.1021/acs.jcim.2c00306.

82. J. Cheng, Y. Liu, and Y. Ma, "Protein secondary structure prediction based on integration of CNN and LSTM model," *J. Vis. Commun. Image Represent.*, vol. 71, p. 102844, Aug. 2020, doi: 10.1016/j.jvcir.2020.102844.

83. C. W. Coley, R. Barzilay, W. H. Green, T. S. Jaakkola, and K. F. Jensen, "Convolutional Embedding of Attributed Molecular Graphs for Physical Property Prediction," *J. Chem. Inf. Model.*, vol. 57, no. 8, pp. 1757–1772, Aug. 2017, doi: 10.1021/acs.jcim.6b00601.

84. J. Wang, N. Wen, C. Wang, L. Zhao, and L. Cheng, "ELECTRA-DTA: a new compound-protein binding affinity prediction model based on the contextualized sequence encoding," *J. Cheminform.*, vol. 14, no. 1, p. 14, Dec. 2022, doi: 10.1186/s13321-022-00591-x.

85. P. Karpov, G. Godin, and I. V. Tetko, "Transformer-CNN: Swiss knife for QSAR modeling and interpretation," *J. Cheminform.*, vol. 12, no. 1, pp. 1–12, 2020, doi: 10.1186/s13321-020-00423-w.

86. Y. Xu, J. Wang, M. Guang, C. Yan, and C. Jiang, "Multistructure Graph Classification Method With Attention-Based Pooling," *IEEE Trans. Comput. Soc. Syst.*, pp. 1–12, 2022, doi: 10.1109/TCSS.2022.3169219.

87. Z. Yang *et al.*, "Learning to Predict Crystal Plasticity at the Nanoscale: Deep Residual Networks and Size Effects in Uniaxial Compression Discrete Dislocation Simulations," *Sci. Rep.*, vol. 10, no. 1, p. 8262, Dec. 2020, doi: 10.1038/s41598-020-65157-z.

88. F. Chen, S. Pan, J. Jiang, H. Huo, and G. Long, "DAGCN: Dual Attention Graph Convolutional Networks," in *2019 International Joint Conference on Neural Networks (IJCNN)*, Jul. 2019, pp. 1–8, doi: 10.1109/IJCNN.2019.8851698.

89. K. Balaji, K. Lavanya, and A. G. Mary, "Machine learning algorithm for clustering of heart disease and chemoinformatics datasets," *Comput. Chem. Eng.*, vol. 143, p. 107068, Dec. 2020, doi: 10.1016/j.compchemeng.2020.107068.

90. D. B. Kell, S. Samanta, and N. Swainston, "Deep learning and generative methods in cheminformatics and chemical biology: navigating small molecule space intelligently," *Biochem. J.*, vol. 477, no. 23, pp. 4559–4580, Dec. 2020, doi: 10.1042/BCJ20200781.

91. A. Khan, A. Sohail, U. Zahoora, and A. S. Qureshi, "A survey of the recent architectures of deep convolutional neural networks," *Artif. Intell. Rev.*, vol. 53, no. 8, pp. 5455–5516, Dec. 2020, doi: 10.1007/s10462-020-09825-6.

92. D. Zhou, W. Dong, and X. Shen, "Image zooming using directional cubic convolution interpolation," *IET Image Process.*, vol. 6, no. 6, pp. 627–634, Aug. 2012, doi: 10.1049/iet-ipr.2011.0534.

93. D. Kuzminykh *et al.*, "3D Molecular Representations Based on the Wave Transform for Convolutional Neural Networks," *Mol. Pharm.*, vol. 15, no. 10, pp. 4378–4385, Oct. 2018, doi: 10.1021/acs.molpharmaceut.7b01134.

94. J. Chen, S. Zheng, H. Zhao, and Y. Yang, "Structure-aware protein solubility prediction from sequence through graph convolutional network and predicted contact map," *J. Cheminform.*, vol. 13, no. 1, p. 7, Dec. 2021, doi: 10.1186/s13321-021-00488-1.

95. Y.-C. Lo, S. E. Rensi, W. Torng, and R. B. Altman, "Machine learning in chemoinformatics and drug discovery," *Drug Discov. Today*, vol. 23, no. 8, pp. 1538–1546, Aug. 2018, doi: 10.1016/j.drudis.2018.05.010.

96. F. Yang, J. Liu, Q. Zhang, Z. Yang, and X. Zhang, "CNN-based two-branch multi-scale feature extraction network for retrosynthesis prediction," *BMC Bioinformatics*, vol. 23, no. 1, p. 362, Sep. 2022, doi: 10.1186/s12859-022-04904-7.

97. J. Sunseri, J. E. King, P. G. Francoeur, and D. R. Koes, "Convolutional neural network scoring and minimization in the D3R 2017 community challenge," *J. Comput. Aided. Mol. Des.*, vol. 33, no. 1, pp. 19–34, Jan. 2019, doi: 10.1007/s10822-018-0133-y.

98. J. Hochuli, A. Helbling, T. Skaist, M. Ragoza, and D. R. Koes, "Visualizing convolutional neural network protein-ligand scoring," *J. Mol. Graph. Model.*, vol. 84, pp. 96–108, 2018, doi: 10.1016/j.jmgm.2018.06.005.

99. S. Kumar and M. Kim, "SMPLIP-Score: predicting ligand binding affinity from simple and interpretable on-the-fly interaction fingerprint pattern descriptors," *J. Cheminform.*, vol. 13, no. 1, p. 28, Dec. 2021, doi: 10.1186/s13321-021-00507-1.

100. N. Yasuo and M. Sekijima, "Improved Method of Structure-Based Virtual Screening via Interaction-Energy-Based Learning," *J. Chem. Inf. Model.*, vol. 59, no. 3, pp. 1050–1061, Mar. 2019, doi: 10.1021/acs.jcim.8b00673.

101. X.-Y. Meng, H.-X. Zhang, M. Mezei, and M. Cui, "Molecular Docking: A Powerful Approach for Structure-Based Drug Discovery," *Curr. Comput. Aided-Drug Des.*, vol. 7, no. 2, pp. 146–157, Jun. 2011, doi: 10.2174/157340911795677602.

102. L. Ferreira, R. dos Santos, G. Oliva, and A. Andricopulo, "Molecular Docking and Structure-Based Drug Design Strategies," *Molecules*, vol. 20, no. 7, pp. 13384–13421, Jul. 2015, doi: 10.3390/molecules200713384.

103. M. D. Disney, I. Yildirim, and J. L. Childs-Disney, "Methods to enable the design of bioactive small molecules targeting RNA," *Org. Biomol. Chem.*, vol. 12, no. 7, pp. 1029–1039, 2014, doi: 10.1039/C3OB42023J.

104. H. M. Ashtawy and N. R. Mahapatra, "Task-Specific Scoring Functions for Predicting Ligand Binding Poses and Affinity and for Screening Enrichment," *J. Chem. Inf. Model.*, vol. 58, no. 1, pp. 119–133, Jan. 2018, doi: 10.1021/acs.jcim.7b00309.

105. R. Meli, G. M. Morris, and P. C. Biggin, "Scoring Functions for Protein-Ligand Binding Affinity Prediction Using Structure-based Deep Learning: A Review," *Front. Bioinforma.*, vol. 2, Jun. 2022, doi: 10.3389/fbinf.2022.885983.

106. J. Jiménez, M. Škalič, G. Martínez-Rosell, and G. De Fabritiis, "K DEEP: Protein–Ligand Absolute Binding Affinity Prediction via 3D-Convolutional Neural Networks," *J. Chem. Inf. Model.*, vol. 58, no. 2, pp. 287–296, Feb. 2018, doi: 10.1021/acs.jcim.7b00650.

107. Y. Kwon, W.-H. Shin, J. Ko, and J. Lee, "AK-Score: Accurate Protein-Ligand Binding Affinity Prediction Using an Ensemble of 3D-Convolutional Neural Networks," *Int. J. Mol. Sci.*, vol. 21, no. 22, p. 8424, Nov. 2020, doi: 10.3390/ijms21228424.

108. S. Wang *et al.*, "SE-OnionNet: A Convolution Neural Network for Protein–Ligand Binding Affinity Prediction," *Front. Genet.*, vol. 11, Feb. 2021, doi: 10.3389/fgene.2020.607824.

109. X. Wang, D. Liu, J. Zhu, A. Rodriguez-Paton, and T. Song, "CSConv2d: A 2-D Structural Convolution Neural Network with a Channel and Spatial Attention Mechanism for Protein-Ligand Binding Affinity Prediction," *Biomolecules*, vol. 11, no. 5, p. 643, Apr. 2021, doi: 10.3390/biom11050643.

110. M. A. Rezaei, Y. Li, D. Wu, X. Li, and C. Li, "Deep Learning in Drug Design: Protein-Ligand Binding Affinity Prediction," *IEEE/ACM Trans. Comput. Biol. Bioinforma.*, vol. 19, no. 1, pp. 407–417, Jan. 2022, doi: 10.1109/TCBB.2020.3046945.

111. T. I. Oprea and J. Mestres, "Drug Repurposing: Far Beyond New Targets for Old Drugs," *AAPS J.*, vol. 14, no. 4, pp. 759–763, Dec. 2012, doi: 10.1208/s12248-012-9390-1.

112. Y. Yamanishi, M. Araki, A. Gutteridge, W. Honda, and M. Kanehisa, "Prediction of drug-target interaction networks from the integration of chemical and genomic spaces," *Bioinformatics*, vol. 24, no. 13, pp. i232–i240, Jul. 2008, doi: 10.1093/bioinformatics/btn162.

113. J. Shim, Z.-Y. Hong, I. Sohn, and C. Hwang, "Prediction of drug–target binding affinity using similarity-based convolutional neural network," *Sci. Rep.*, vol. 11, no. 1, p. 4416, Dec. 2021, doi: 10.1038/s41598-021-83679-y.

114. H. Öztürk, A. Özgür, and E. Ozkirimli, "DeepDTA: deep drug–target binding affinity prediction," *Bioinformatics*, vol. 34, no. 17, pp. i821–i829, Sep. 2018, doi: 10.1093/bioinformatics/bty593.

115. K. Shao, Z. Zhang, S. He, and X. Bo, "DTIGCCN: Prediction of drug-target interactions based on GCN and CNN," in *2020 IEEE 32nd International Conference on Tools with Artificial Intelligence (ICTAI)*, Nov. 2020, pp. 337–342, doi: 10.1109/ICTAI50040.2020.00060.

116. L. Zhao, J. Wang, L. Pang, Y. Liu, and J. Zhang, "GANsDTA: Predicting Drug-Target Binding Affinity Using GANs," *Front. Genet.*, vol. 10, Jan. 2020, doi: 10.3389/fgene.2019.01243.

117. A. Ghimire, H. Tayara, Z. Xuan, and K. T. Chong, "CSatDTA: Prediction of Drug–Target Binding Affinity Using Convolution Model with Self-Attention," *Int. J. Mol. Sci.*, vol. 23, no. 15, p. 8453, Jul. 2022, doi: 10.3390/ijms23158453.

118. Q. Zhao, F. Xiao, M. Yang, Y. Li, and J. Wang, "AttentionDTA: prediction of drug–target binding affinity using attention model," in *2019 IEEE International Conference on Bioinformatics and Biomedicine (BIBM)*, Nov. 2019, pp. 64–69, doi: 10.1109/BIBM47256.2019.8983125.

119. L. Zheng, J. Fan, and Y. Mu, "OnionNet: a Multiple-Layer Intermolecular-Contact-Based Convolutional Neural Network for Protein–Ligand Binding Affinity Prediction," *ACS Omega*, vol. 4, no. 14, pp. 15956–15965, Oct. 2019, doi: 10.1021/acsomega.9b01997.

120. F. Wan *et al.*, "DeepCPI: A Deep Learning-based Framework for Large-scale in silico Drug Screening," *Genomics. Proteomics Bioinformatics*, vol. 17, no. 5, pp. 478–495, Oct. 2019, doi: 10.1016/j.gpb.2019.04.003.

121. H. Öztürk, E. Ozkirimli, and A. Özgür, "WideDTA: prediction of drug-target binding affinity," *arXiv Prepr. arXiv1902.04166*, 2019.

122. R. Aggarwal, A. Gupta, V. Chelur, C. V. Jawahar, and U. D. Priyakumar, "DeepPocket: Ligand Binding Site Detection and Segmentation using 3D Convolutional Neural Networks," *J. Chem. Inf. Model.*, p. acs.jcim.1c00799, Aug. 2021, doi: 10.1021/acs.jcim.1c00799.

123. Q. Feng, E. Dueva, A. Cherkasov, and M. Ester, "Padme: A deep learning-based framework for drug-target interaction prediction," *arXiv Prepr. arXiv1807.09741*, 2018.

124. S. Forli, R. Huey, M. E. Pique, M. F. Sanner, D. S. Goodsell, and A. J. Olson, "Computational protein–ligand docking and virtual drug screening with the AutoDock suite," *Nat. Protoc.*, vol. 11, no. 5, pp. 905–919, May 2016, doi: 10.1038/nprot.2016.051.

125. M. M. Stepniewska-Dziubinska, P. Zielenkiewicz, and P. Siedlecki, "Development and evaluation of a deep learning model for protein–ligand binding affinity prediction," *Bioinformatics*, vol. 34, no. 21, pp. 3666–3674, Nov. 2018, doi: 10.1093/bioinformatics/bty374.

126. A. S. Rifaioglu, E. Nalbat, V. Atalay, M. J. Martin, R. Cetin-Atalay, and T. Doğan, "DEEPScreen: high performance drug-target interaction prediction with convolutional neural networks using 2-D structural compound representations," *Chem. Sci.*, vol. 11, no. 9, pp. 2531–2557, 2020, doi: 10.1039/c9sc03414e.

127. X. Lin, "Deepgs: Deep representation learning of graphs and sequences for drug-target binding affinity prediction," *arXiv Prepr. arXiv2003.13902*, 2020.

128. Y. Li, M. A. Rezaei, C. Li, and X. Li, "DeepAtom: A Framework for Protein-Ligand Binding Affinity Prediction," in *2019 IEEE International Conference on Bioinformatics and Biomedicine (BIBM)*, Nov. 2019, pp. 303–310, doi: 10.1109/BIBM47256.2019.8982964.

11 Machine Learning in Improving Force Fields of Molecular Dynamics

Virupaksha A. Bastikar, Alpana V. Bastikar, Varun Talati, Bhupendra Gopalbhai Prajapati

All branches of science are being transformed by machine learning (ML). The implementation of existing ML techniques has already had a significant impact on the difficult and time-consuming computations involved in molecular simulations. Prediction of quantum-mechanical energies and forces, the extraction of free energy surfaces and kinetics, coarse-grained molecular dynamics, generation of molecular equilibrium structures, and compute thermodynamics using deep neural networks are promising applications of ML for molecular simulation.

Deep neural networks that allow for the example-driven design of arbitrary complicated functions and their derivatives have been developed resulting in increased potential of ML. In order to include quick yet accurate potential energy functions in MD simulations, DNNs are thus a very promising option. This is especially true after training on extensive databases obtained from more expensive methods using larger datasets. The ability of neural network potentials to learn many-body interactions is a trait that is particularly intriguing (Figure 11.1). For instance, the SchNet design uses continuous filter convolutions on a graph neural network to learn a collection of characteristics and forecast the forces and energy of the system. To forecast the energies of tiny molecules from their atomistic representations, SchNet was first applied in quantum chemistry. SchNet has the advantage of being universally applicable to different molecular systems. Recently, this has been expanded to include learning average potential of mean force, the parametrization of which presents difficulties. Indeed, so-called coarse-graining (CG) techniques have addressed molecular modeling on a more granular scale in the past, but it is particularly intriguing when combined with DNNs.

MOLECULAR DYNAMICS

All things are comprised of atoms, according to Richard Feynman, who won the 1965 Nobel Prize in Physics, and all living things do can be explained by the movement of atoms. Over the last few decades, biophysics has spent a lot of time trying to understand the nature of these atomic movements. These motions are based on chemical bonds, which are constructed from electron clouds, which are both waves and particles. Without a doubt, understanding these bizarre chemical motions is crucial for the development of new drugs. In favour of binding models that consider conformational changes as well as random variations in structure, the idea that receptors might bind a tiny chemical without experiencing conformational rearrangements has been disproved. Molecular dynamics is a computer simulation technique for analyzing the physical motions of atoms and molecules to elucidate their conformational states, energies, and stability.

DOI: 10.1201/9781003353768-11

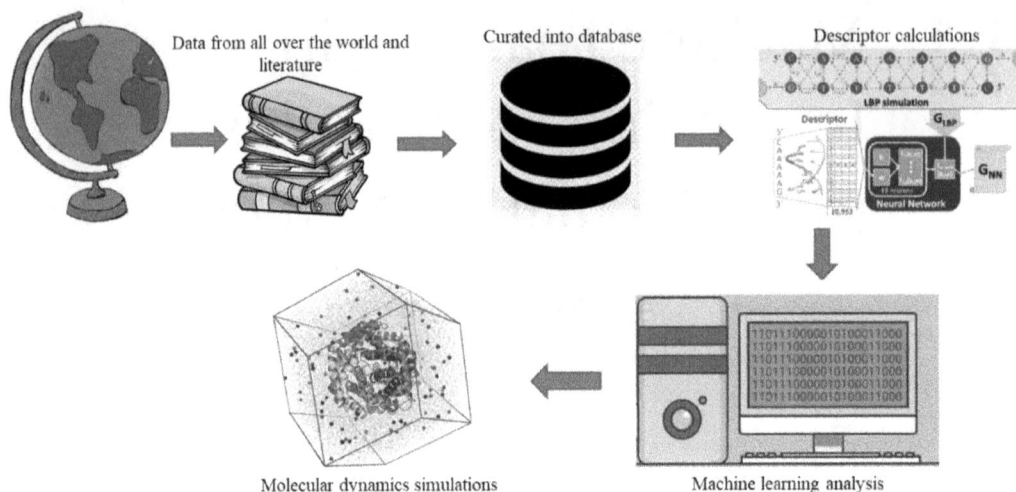

FIGURE 11.1   Molecular Dynamics using Machine Learning.

Molecular dynamics is a method for simulating behaviors of biomolecules and their conformational changes. The atoms and molecules are allowed to interact for a fixed period of time, giving a view of the dynamic "evolution" of the system (Ciccotti et al., 2022). The most popular approach entails numerically resolving Newton's equations of motion for a system of interacting particles. Forces between the particles and their potential energies are typically computed using interatomic potentials or molecular mechanics force fields. By using this technique, the paths of atoms and molecules may be determined. The main domains where the method is applied include chemical physics, materials science, and biology.

It is impossible to estimate the properties of such complex systems analytically since molecular systems often contain a large number of particles; MD simulation gets around this issue by using numerical techniques. Long MD simulations, on the other hand, are theoretically unsound, leading to cumulative mistakes in numerical integration that can be reduced but not totally avoided with the right choice of techniques and settings.

One MD simulation's evolution can be used to ascertain the macroscopic thermodynamic characteristics of a system if it obeys the ergodic hypothesis: An ergodic system's time averages match micro-canonical ensemble averages. Molecular dynamics has also been termed "statistical mechanics by numbers" and "Laplace's vision of Newtonian mechanics" of predicting the future by animating nature's forces and allowing insight into molecular motion on an atomic scale (Zheng et al., 2019).

While crystallographic studies such as these convincingly demonstrate the important role of protein flexibility in ligand binding, the cost and extensive amount of work required to generate it has led many to seek computational techniques that can predict protein motion. Unfortunately, even the fastest supercomputers typically struggle to do the calculations needed to explain the bizarre quantum mechanical motions and chemical reactions of huge molecular systems. Molecular dynamics simulations, first developed in the late 1970s, attempt to overcome this limitation by using simple approximations based on Newtonian physics to simulate atomic motion, thus reducing computational complexity. First, a computer model of the molecular system is constructed from nuclear magnetic resonance (NMR), crystallographic, or homology modeling data. In short, forces arising from interactions between bound and unbound atoms contribute. Chemical bonds and atomic angles are modeled using simple virtual springs, and dihedral angles (i.e., rotations about a bond) are modeled using a sine function that approximates the energy differences between eclipsed and

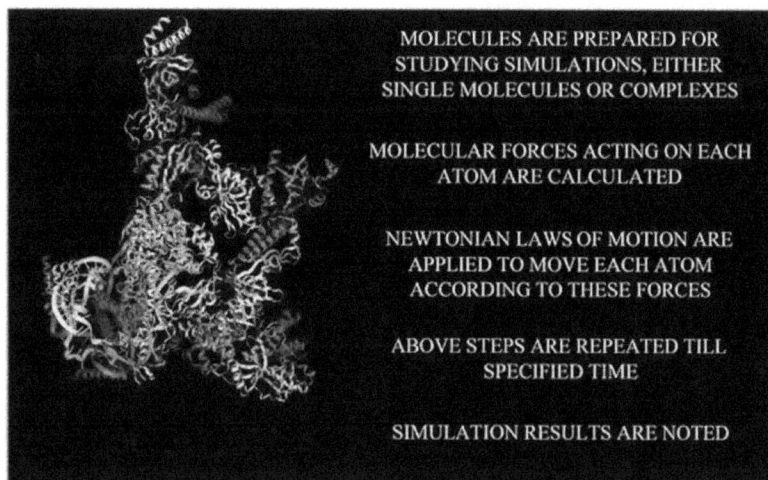

FIGURE 11.2 Steps for Molecular Simulations.

staggered conformations. Unbound forces arise due to van der Waals interactions, which are modeled using the Lennard-Jones potential, and charged (electrostatic) interactions, which are modeled using Coulomb's law (Figure 11.2).

WHAT ARE THESE FORCES?

A set of parameters are defined for types of atoms, bonds, non-bonded interactions, out-of-plane interactions, and other terms in force fields. Values for atomic mass, charge, and Lennard–Jones parameters for atoms types along with equilibrium values for bond angles, dihedral angles, and bond length are included in a typical parameter set.

Functional Form:

The total energy can be written as

$$E_{Total} = E_{Bonded} + E_{Non\text{-}bonded}, \text{ wherein}$$

$$E_{Bonded} = E_{Bond} + E_{Angle} + E_{Dihedral}$$

$$E_{Non\text{-}bonded} = E_{Electrostatic} + E_{van\,der\,Waals}$$

For bonded terms, bond and angle are typically modeled using quadratic energy functions that do not permit breaking of bonds. However, the functional form of dihedral angles depends on the force fields. The non-bonded terms are computationally intensive. Lennard–Jones potential is used for computing van der Waals term and Coulomb's law for electrostatic term. These can be scaled by a constant factor to take into account electronic polarizability.

Electrostatic interactions are represented by Coulomb's energy, which uses atomic charges for covalent, polar covalent, and ionic bonds. Atomic charges are important for the simulation of interaction energy, reactivity, and geometry and have significant contributions towards the potential energy, particularly for polar and ionic compounds. On the basis of experimental information for electron deformation densities, internal dipole moments, and an extended born model, reproducible atomic charges have been developed for force fields. A consistent depiction of chemical bonding and around 100 times higher accuracy in computed structures and energies, coupled with physical interpretation of other force field characteristics, are made possible by uncertainties of less than 10%, or 0.1e.

The simplest approach is using Hooke's law formula, which is

$$E_{Bond} = kij(l_{ij} - l_{0, ij})^2/2, \text{ where}$$

k_{ij} = force constant
l_{ij} = bond length
$l_{0, ij}$ = bond length between atoms i and j when other terms of force field are
     set to 0, often referred as equilibrium bond length.

The force constant k_{ij} describes the vibration frequencies in MD simulations and its values can be estimated using experimental IR spectrum, high-level quantum mechanical calculations, and Raman spectrum.

Dihedral or torsional terms are also included in cases of molecules containing more than four atoms in a row. Torsional motions are important for the determination of the local structure of a macromolecule and the relative stability of various molecular conformations. They are less stiff than bond stretching motions and are required for ensuring the proper degree of rigidity of the molecule and to mimic the significant conformational changes caused by rotations about bonds.

PARAMETERIZATION

For simulations with maximum accuracy and transferability, force field parameterizations follow a well-defined protocol. The steps involved are:

1. Retrieval of X-ray crystal structure or chemical formula.
2. Defining atom types.
3. Obtaining atomic charges.
4. Assigning of bonded and Lennard-Jones parameters.
5. In relation to reference data, computational testing of density and geometry.
6. Similar to step v, computational tests of energetic properties.
7. Secondary validation and refinement.

A crucial part is played by the chemical interpretation of the parameters and reliable experimental reference data. For the simulations of biological macromolecules, the parameters were often obtained from observations of small organic molecules since they are easily available for experimental studies and quantum calculations. However, due to multiple issues divergent force field parameters have been described for biological macromolecules.

With the development of molecular mechanics, the first force fields emerged in the 1960s with the aim of predicting molecular structures, vibrational spectra, and enthalpies of isolated molecules (Gavezzotti, 2006). The force fields were initially developed for studying hydrocarbons, but have been extended for dealing with various organic or functionalized molecules. Some of the earlier force fields have continued to develop and are used today (e.g., MM potentials by Alligner's group).

Highly relevant force fields have been invented to deal with complex systems, such as universal force fields (UFF) (Rappe et al., 1992) and Dreiding (Mayo et al., 1990), which includes the parameters for all periodic table elements. Other force fields are CHARMM (MacKerell et al., 1998);

GROMOS (Oostenbrink et al., 2004) and AMBER (Cornell et al., 1995), which are implemented in the simulation of biomolecules; OPLS (Jorgensen et al., 1996) and COMPASS (Sun, 1998) are used for condensed matter; CFF (consistent force field) (Maple et al., 1994) is adapted for a wide range of organic compounds; and MMFF (Merck molecular force field) (Halgren, 1996) for various molecules. Additionally, many of these force fields are constantly changing, and there are various versions available, such as CHARMM19, CHARMM22, and CHARMM27. With further development polarizable force fields such as PIPF (Polarizable intermolecular potential function) (Gao et al., 1995), AMOEBA (Ponder et al., 2010), and DRF90 (Swart and van Duijnen, 2006) have emerged. AMBER (Wang et al., 2006), CHARMM (Lamoureux and Roux, 2003; Patel and Brooks III, 2004), GROMOS (Geerke and van Gunsteren, 2007), and OPLS (Jorgensen et al., 2007) have also been expanded to include polarization.

MOLECULAR SIMULATION PROBLEMS THAT CAN BE ADDRESSED BY MACHINE LEARNING

1 POTENTIAL ENERGY SURFACES

Modern atomistic modelling in chemistry, biology, and materials science is based on MD and Markov chain Monte Carlo simulations with classical force fields in the Born–Oppenheimer approximation. These techniques perform importance sampling, which is the long-term selection of states from the equilibrium distribution of the molecular system. Terms that depend on the thermodynamic restrictions are present in the reduced potential (e.g., fixed temperature, pressure). However, the accuracy of these simulations' predictions depends on the potential energy surface that underlies them (PES). Therefore, a precise description of the global PES is needed for predictive simulations of the characteristics and functionalities of molecular systems, which can be accomplished by machine learning.

2 FREE ENERGY SURFACES

The collective coordinates are frequently defined in molecular simulations to describe the system's slowest processes. E (free energy) can be mapped (or encoded) in a manner that is generally quite nonlinear. Machine learning techniques have been employed for this, including neural networks and kernel regression. Additionally, machine learning has been applied in conjunction with improved sampling techniques to learn the free energy surface.

3 COARSE GRAINING

An appealing alternative to atomistic models that are exceedingly expensive to run is the use of coarse-grained models of large molecular systems, such as proteins. The definition of a mapping is the first step in designing a coarse-grained model for a system with N atoms into a reduced form. In this situation, the mapping is typically a linear function, defined as a subset or a linear combination of sets of atoms in the original system, in such a way that one can retain some information about the shape of the molecule. For instance, a coarse-grained mapping in protein systems can be defined by swapping out every atom in a residue for a bead that is centred on the appropriate C atom. The best coarse-graining mapping for a given system is not currently defined by a general theory. A few machine learning techniques can be used to enhance the mapping E while maintaining specific characteristics of the original system. Several methods exist to define the model energy function once the coarse-graining mapping has been provided, either to match experimental observables or to replicate particular atomistic system features.

4   KINETICS

The slow component of dynamics is called kinetics. There is a probability distribution of discovering the molecule in configuration for any trajectory originating from a configuration at time t because of the stochastic components in the MD integrator. Learning molecular kinetics from a set of supplied trajectory data is a well-established method. This often takes two phases in order to produce a low-dimensional model of the molecular dynamics that is simple to read and analyse: Using the encoder E, one can (a) locate a low-dimensional latent space representation of the collective variable, and (b) learn a dynamical propagator in that space. Learning the embedding E is more challenging than learning the dynamical propogator because when E is optimised in latent space by maximum likelihood or least regression error, E collapses to simple, uninteresting solutions. By adopting a variational strategy and machine learning techniques, this issue can be avoided.

5   SAMPLING AND THERMODYNAMICS

The time increments involved in MD are on the order of one femtosecond (10^{-15} s), whereas configuration changes that control molecular activity are frequently sporadic and can take anywhere between 10^{-3} and 10^3 s. Even when the forces and possibilities of single-protein folding and unfolding are immediately assessed, replicating these processes via direct MD simulation on a supercomputer may take years to millennia. To get around this sampling issue, machine learning techniques can be used to generate statistically independent samples all at once or even learn how to generate equilibrium samples more quickly.

INCLUSION OF PHYSICS INTO MACHINE LEARNING FOR MOLECULAR DYNAMICS AND SIMULATIONS

1   DATA AUGMENTATION

By artificially producing new training data and applying the known invariances to it, a technique known as data augmentation allows us to memorise invariances. The machine learning model is strengthened by data augmentation, which also aids in roughly predicting the invariances. Despite the fact that hardwiring some invariances into the machine learning model is conceptually challenging or prohibitively expensive, it is a significant machine learning technique nonetheless.

2   BUILDING PHYSICAL CONSTRAINTS INTO MACHINE LEARNING MODEL

Directly incorporating physical restrictions into the machine learning model is the more precise, statistically effective, and elegant way to do it. In order to make the learning problem effective, it is necessary to take into account two connected factors. We start with equivariances: The invariances and symmetries in the machine learning model should match those in the modelled physics issue. The second is parameter sharing: Any invariance or symmetry should be mirrored by distributing model parameters among various network nodes.

3   INVARIANCE AND EQUIVARIANCE

We can lower the problem's dimensionality if we can permanently incorporate the physical symmetries into the machine learning structure. Energy and force are one-and six-dimensional functions that are specified for each point in a six-dimensional configuration space in the case of the O2 molecule. When utilising the invariances, however, we only need to learn a one-dimensional energy function over a one-dimensional space in order to compute the entire force field. We now only learn in the variety of physically relevant solutions, making the learning problem much easier to solve. Equivariance is a crucial idea associated with invariance. If a function transforms in the same way as its argument, it is said to be equivariant. By designating atoms to nodes in a bond graph in classical MD force fields, bound interactions are defined, and exchanging individual atom pairs no longer results in energy preservation. For coarse graining and creating samples from the equilibrium density, it is crucial to remember that the energy is still invariant to the interchange of identical molecules, such as solvent.

The mapping of all training and test data to a reference permutation is a straightforward substitute for including permutation invariance into the machine learning function. Bipartite graph matching techniques are effective for accomplishing this.

4 PARAMETER SHARING AND CONVOLUTIONS

The fact that convolutional layers are equivariant makes it easier to detect an object regardless of where it is, but parameter sharing is what gives convolutional neural networks their true efficiency. All neurons in a traditional dense network are interconnected and have separate properties that are stored in a weight matrix. High-dimensional data become problematic as a result of this. With convolutions, the number of independent parameters is drastically reduced. Prior to introducing bias and nonlinearities, a convolutional layer with a filter computing on a one-dimensional signal.

MACHINE LEARNING FRAMEWORKS FOR MOLECULAR SIMULATIONS

1 DEEP POTENTIAL NET, BEHLER-PARRINELLO, AND ANI

One of the first machine learning applications in the molecular sciences was Behler-Parrinello networks. They integrate all of the pertinent physical symmetries and parameter sharing for this problem with the goal of learning and forecasting potential energy surfaces from QM data. Deep Potential net, a comparable method, treats each atom in a local coordinate frame without rotational and translational degrees of freedom (Figure 11.3A). In the ANI network, for instance, the Behler-Parrinello approach has been improved by being expanded to more complex functions involving two neighbours. ANI has been trained using density functional theory and coupled-cluster data across a broad chemical space, whereas Behler-Parrinello networks have primarily been utilised to predict the same molecular system in order to execute MD simulations that are unaffordable via straight ab initio QM MD.

2 DEEP TENSOR NEURAL NETWORK, SCHNET, AND CONTINUOUS CONVOLUTIONS

The family of deep tensor neural networks (DTNNs), with its most recent addition SchNet, was one of the first deep learning architectures to learn to represent molecules or materials. DTNN and its extension SchNet learn a multiscale representation of the attributes of molecules or materials from vast data sets, whereas chemical compounds are compared in terms of prespecified kernel functions in kernel-based learning approaches. The word-to-vec technique to language processing, which involves learning and encoding in a parameter vector the function of a word within its grammatical or semantic context, served as the model for DTNNs. DTNNs learn the representation that is pertinent for the purpose of forecasting these variables as they are end-to-end taught to do so, such as potential energies (Figure 11.3B). Deep convolutional neural networks are used by SchNet. Traditionally, discrete convolution filters have been used in convolutional neural networks because they were created for computer vision using pixelated pictures. However, because QM parameters like the energy are so sensitive to slight position changes, like the stretching of a covalent bond, the particle positions of molecules cannot be discretized on a grid. Due to this, SchNet developed continuous convolutions, which are represented by neural networks that generate filters and translate rototranslationally invariant interatomic distances into convolution filter values. In order to imitate

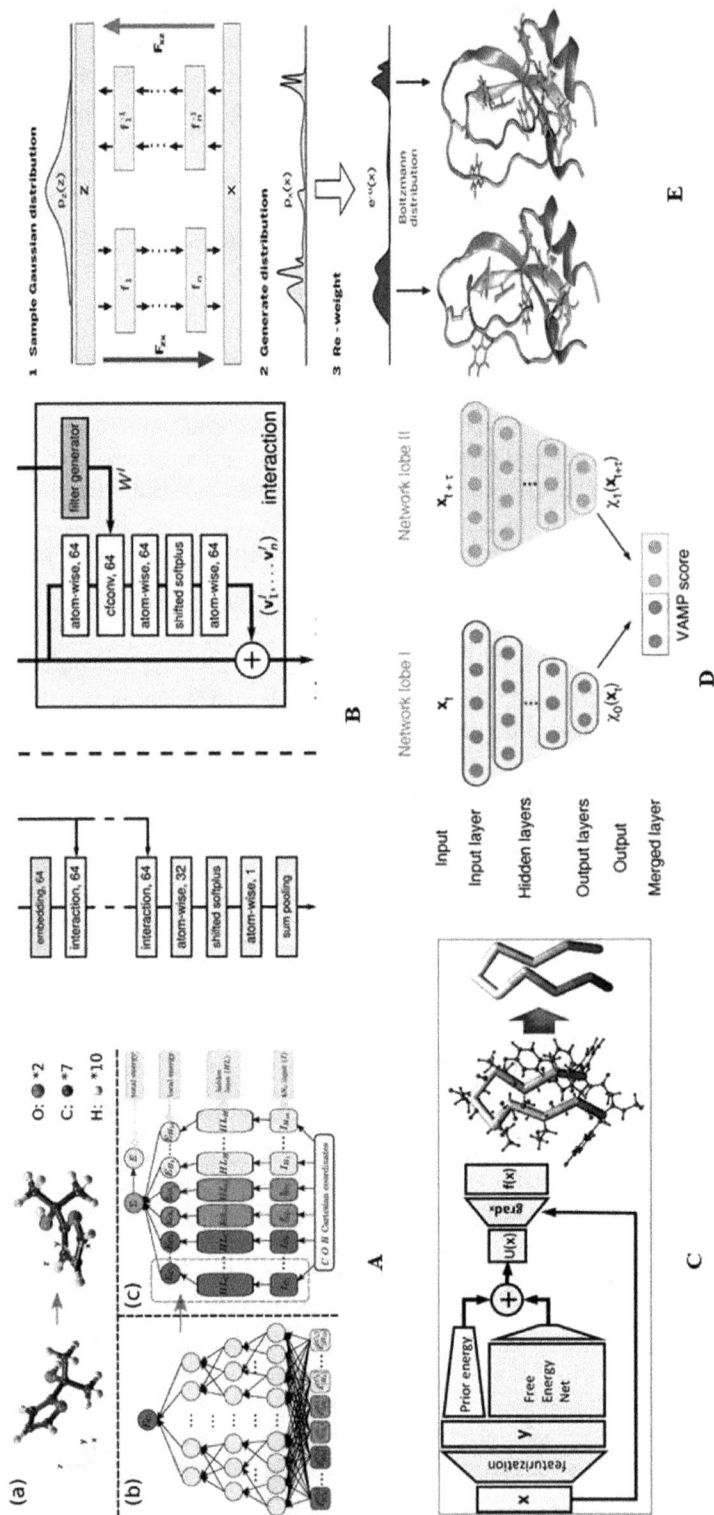

FIGURE 11.3   A–Deep Potential Net; B–Deep Tensor Neural Network; C–Course grain Network; D–VAMPnet; E–Boltzmann Generators.

MD, DTNNs and SchNet have both attained extremely competitive prediction quality throughout chemical compound space and across configuration space. The DTNNs are becoming more and more well-liked research tools due to their high prediction accuracy, scalability to enormous data sets, and capacity to derive fresh chemical insights using their learned representation.

3   COARSE GRAINING CGNETS

For molecular systems, machine learning has been utilised to define coarse-grained models. For specific systems, the force-matching loss has been minimised using both deep neural networks and kernel approaches for given coarse-graining mappings. It has been demonstrated in both situations that including physical restrictions is essential to the model's performance. Regions of the configurational space that are physically prohibited, such as configurations with broken covalent bonds or overlapping atoms, are not sampled and are not included in the training data, which are obtained by atomistic molecular dynamic simulations. Without further restrictions, the machine is unable to make predictions that are far from the training data, and as a result, it is unable to forecast with sufficient accuracy that the energy should diverge as it approaches physically banned regions. As the first layer converts the Cartesian coordinates into internal coordinates like distances and angles, CGnet predicts a rototranslationally invariant energy, similar to Behler-Parrinello networks and SchNet (Figure 11.3C). Additionally, because CGnet computes the gradient of the total free energy with respect to the input configuration self-consistently, it predicts a conservative and rotation-equivariant force field. By minimising the force-matching loss of this prediction, the network is trained.

4   KINETICS VAMPNETS

The difficult and error-prone process of building Markov state models by (a) looking for the best features, (b) merging them into a low-dimensional representation, (c) clustering, (d) estimating the transition matrix, and (e) coarse graining was replaced by the introduction of VAMPnets. Instead, VAMPnets employ a single end-to-end learning strategy in which a deep neural network takes the role of each of these processes. (Figure 11.3D) This is conceivable because loss functions that are suitable for simultaneously training the embedding and the propagator are available using the variational technique for conformational dynamics and VAMP principles.

5   SAMPLING/THERMODYNAMICS BOLTZMANN GENERATORS

In order to understand how to sample equilibrium distributions, Boltzmann generators were developed. In contrast to conventional generative learning, a Boltzmann generator is trained to effectively sample utilising inputs such as dimensionless energy rather than attempting to learn the probability density from data (Figure 11.3E) In a Boltzmann generator, there are two components:

1. A trained generative model that suggests samples from a probability distribution that enables us to assess.
2. A reweighting method that creates unbiased samples by using predictions from a probability distribution.

CONCLUSION

We have recently gained insight into the inner workings of deep learning algorithms thanks to the rising popularity of explainable AI techniques. In this way, it is now possible to deduce how a deep model resolves a given issue. These inspection techniques may offer scientific insights into the mechanisms that give rise to the anticipated physicochemical quantity when combined with networks that learn a representation, like DTNN/SchNet and VAMPnets, and thereby stimulate the creation of novel theories.

REFERENCES

1. Ciccotti, G., Dellago, C., Ferrario, M., Hernández, E.R., Tuckerman, M.E., 2022. Molecular simulations: past, present, and future (a Topical Issue in EPJB). Eur. Phys. J. B 95, 3. doi.org/10.1140/epjb/s10051-021-00249-x
2. Cornell, W.D., Cieplak, P., Bayly, C.I., Gould, I.R., Merz, K.M., Ferguson, D.M., Spellmeyer, D.C., Fox, T., Caldwell, J.W., Kollman, P.A., 1995. A Second Generation Force Field for the Simulation of Proteins, Nucleic Acids, and Organic Molecules. J. Am. Chem. Soc. 117, 5179–5197. doi.org/10.1021/ja00124a002
3. Gao, J., Habibollazadeh, D., Shao, L., 1995. A Polarizable Intermolecular Potential Function for Simulation of Liquid Alcohols. J. Phys. Chem. 99, 16460–16467. doi.org/10.1021/j100044a039
4. Gavezzotti, A., 2006. Molecular Aggregation: Structure analysis and molecular simulation of crystals and liquids. OUP/International Union of Crystallography.
5. Geerke, D.P., van Gunsteren, W.F., 2007. On the Calculation of Atomic Forces in Classical Simulation Using the Charge-on-Spring Method To Explicitly Treat Electronic Polarization. J. Chem. Theory Comput. 3, 2128–2137. doi.org/10.1021/ct700164k
6. Halgren, T.A., 1996. Merck molecular force field. I. Basis, form, scope, parameterization, and performance of MMFF94. J. Comput. Chem. 17, 490–519. doi.org/10.1002/(SICI)1096-987X(199604)17:5/6<490::AID-JCC1>3.0.CO;2-P
7. Jorgensen, W.L., Jensen, K.P., Alexandrova, A.N., 2007. Polarization Effects for Hydrogen-Bonded Complexes of Substituted Phenols with Water and Chloride Ion. J. Chem. Theory Comput. 3, 1987–1992. doi.org/10.1021/ct7001754
8. Jorgensen, W.L., Maxwell, D.S., Tirado-Rives, J., 1996. Development and Testing of the OPLS All-Atom Force Field on Conformational Energetics and Properties of Organic Liquids. J. Am. Chem. Soc. 118, 11225–11236. doi.org/10.1021/ja9621760
9. Lamoureux, G., Roux, B., 2003. Modeling induced polarization with classical Drude oscillators: Theory and molecular dynamics simulation algorithm. J. Chem. Phys. 119, 3025–3039. doi.org/10.1063/1.1589749
10. MacKerell, A.D.Jr., Bashford, D., Bellott, M., Dunbrack, R.L.Jr., Evanseck, J.D., Field, M.J., Fischer, S., Gao, J., Guo, H., Ha, S., Joseph-McCarthy, D., Kuchnir, L., Kuczera, K., Lau, F.T.K., Mattos, C., Michnick, S., Ngo, T., Nguyen, D.T., Prodhom, B., Reiher, W.E., Roux, B., Schlenkrich, M., Smith, J.C., Stote, R., Straub, J., Watanabe, M., Wiórkiewicz-Kuczera, J., Yin, D., Karplus, M., 1998. All-Atom Empirical Potential for Molecular Modeling and Dynamics Studies of Proteins. J. Phys. Chem. B 102, 3586–3616. doi.org/10.1021/jp973084f
11. Maple, J.R., Hwang, M.-J., Stockfisch, T.P., Dinur, U., Waldman, M., Ewig, C.S., Hagler, A.T., 1994. Derivation of class II force fields. I. Methodology and quantum force field for the alkyl functional group and alkane molecules. J. Comput. Chem. 15, 162–182. doi.org/10.1002/jcc.540150207
12. Mayo, S.L., Olafson, B.D., Goddard, W.A., 1990. DREIDING: a generic force field for molecular simulations. J. Phys. Chem. 94, 8897–8909. doi.org/10.1021/j100389a010
13. Oostenbrink, C., Villa, A., Mark, A.E., Van Gunsteren, W.F., 2004. A biomolecular force field based on the free enthalpy of hydration and solvation: The GROMOS force-field parameter sets 53A5 and 53A6. J. Comput. Chem. 25, 1656–1676. doi.org/10.1002/jcc.20090
14. Patel, S., Brooks III, C.L., 2004. CHARMM fluctuating charge force field for proteins: I parameterization and application to bulk organic liquid simulations. J. Comput. Chem. 25, 1–16. doi.org/10.1002/jcc.10355
15. Ponder, J.W., Wu, C., Ren, P., Pande, V.S., Chodera, J.D., Schnieders, M.J., Haque, I., Mobley, D.L., Lambrecht, D.S., DiStasio, R.A.Jr., Head-Gordon, M., Clark, G.N.I., Johnson, M.E., Head-Gordon, T., 2010. Current Status of the AMOEBA Polarizable Force Field. J. Phys. Chem. B 114, 2549–2564. doi.org/10.1021/jp910674d

16. Rappe, A.K., Casewit, C.J., Colwell, K.S., Goddard, W.A.I., Skiff, W.M., 1992. UFF, a full periodic table force field for molecular mechanics and molecular dynamics simulations. J. Am. Chem. Soc. 114, 10024–10035. doi.org/10.1021/ja00051a040

17. Sun, H., 1998. COMPASS: An ab Initio Force-Field Optimized for Condensed-Phase ApplicationsOverview with Details on Alkane and Benzene Compounds. J. Phys. Chem. B 102, 7338–7364. doi.org/10.1021/jp980939v

18. Swart, M., van Duijnen, P.Th., 2006. DRF90: a polarizable force field. Mol. Simul. 32, 471–484. doi.org/10.1080/08927020600631270

19. Wang, Z.-X., Zhang, W., Wu, C., Lei, H., Cieplak, P., Duan, Y., 2006. Strike a balance: Optimization of backbone torsion parameters of AMBER polarizable force field for simulations of proteins and peptides. J. Comput. Chem. 27, 781–790. doi.org/10.1002/jcc.20386

20. Zheng, L., Alhossary, A.A., Kwoh, C.-K., Mu, Y., 2019. Molecular Dynamics and Simulation, in: Ranganathan, S., Gribskov, M., Nakai, K., Schönbach, C. (Eds.), Encyclopedia of Bioinformatics and Computational Biology. Academic Press, Oxford, pp. 550–566. doi.org/10.1016/B978-0-12-809633-8.20284-7

12 Defining the Role of Chemical and Biological Data and Applying Algorithms

Priya Patel, Kevinkumar Garala, Mihir Raval

1 INTRODUCTION

The development of new drugs is essential for better, more long-lasting healthcare. Its success is closely related to developments in the fields of chemistry and biology research that eventually provide novel chemical compounds for as yet undiscovered but disease-relevant therapeutic targets and signalling pathways [1]. To enable selections on target identification, lead identification, structural modification, and candidate selection, scientists in large pharmaceutical companies often use a variety of data sources and technologies. A substantial percentage of this data is either generated internally or licenced from business providers [2]. For example, having access to vast collections of screening data, patent records, and clinical candidate records can be utilised to find chemical tools or leads for an important target or to evaluate competitive position. Large-scale measurements of toxicity, absorption, distribution, metabolism, and excretion (ADME), as well as toxicity, facilitate the development of predictive models that can be used to prioritise compounds, choose the best candidates for further development, and try to reduce the probability of any detrimental consequences. Academic researchers, in contrast, have usually had to rely on a much smaller number of public domain resources as well as data scattered throughout the literature. Large chemical and pharmacological datasets have always been difficult to access, in part because of concerns about probable intellectual property loss arising from disclosing compound structures.

Large-scale open data for drug discovery, however, has been significantly more accessible in recent years. In particular, there has been a huge increase in the quantity and size of screening databases. the creation of programmes like the NIH Molecular Libraries, high-throughput screening capacities, and availability of primary data [3]. Such data are available in open sources like Pubmed and Pubchem. Furthermore, existing activity databases like BindingDB [4], IUPHARDB [5], and PDSP Ki [6] will be supplemented by the recent transfer of the ChEMBL database from the private sector into the public domain [7]. By establishing guidelines for the deposit of screening data into open sources and helping in the uniformity of the reporting of such data, publishers can also contribute to the process of making data more accessible. There are already a rising number of huge chemical structure databases that provide access to hundreds of thousands of compounds for applications like virtual screening [8] (e.g., PubChem, Zinc [9], and GDB-13 [10]), in addition to screening and bioactivity information. Other publicly accessible databases with information relevant to drug research are also being created, such as DrugBank [11] and DailyMed [12]. ClinicalTrials.gov [13] offers data on clinical-stage experimental pharmaceuticals, DSSTox and TOXNET [14-16] compile toxicity data from a variety of public sources, and they all provide information about approved drugs. Public data is becoming more accessible at the same time that the pharmaceutical sector is taking steps to save costs, like increasing outsourcing and participating in pre-competitive activities. The European Innovative Medicines Initiative [17] and the Pistoia

DOI: 10.1201/9781003353768-12

Alliance [18], both non-profit organisations formed by pharmaceutical firms, research institutions, and technology vendors with the goal of bridging shared precompetitive needs, are both working to advance the development and integration of tools and databases in the public domain. Public–private partnerships are becoming more common, and pharmaceutical corporations are beginning to disclose some of their own formerly proprietary data, as seen in the Structural Genomics Consortium's chemical probes effort [19]. For instance, GlaxoSmithKline recently declared that it would release to the public a sizable dataset of 13,500 compounds having antimalarial activity [20]. It is anticipated that additional businesses will follow this example. Academic, non-profit, and industrial drug discovery are expected to be greatly affected by the availability of public large-scale datasets. Groups will first be given access to the data they require for certain initiatives, such as quick identification of high-quality tool molecules to support target verification or disease model characterization. The datasets will promote the development of new tools and predictive algorithms in the public domain, which will be beneficial to the largest community possible. When we consider the huge diversity of bioinformatics tools and techniques created for functional studies of proteins in light of the exponential growth in the deposition of sequence and structural data since the early 1990s, we see a similarity to this. Similar funding and research surges in computational chemical biology and chemoinformatics could help fill numerous gaps in drug discovery and design. For instance, data mining could be used to identify potential lead optimization approaches and strategies or to uncover essential characteristics and guidelines associated with successful medications. It is possible to predict chemical activity from structure by deriving panels of quantitative structure–activity relationship or classification models from large bio-activity datasets. Through the optimization of alternative activities, such predictions can help reveal the molecular targets of phenotypic tests, forecast or explain pharmacological side effects, and identify prospective therapeutic repurposing prospects. Application of structure-based virtual screening techniques, such as docking and pharmacophore- or molecular similarity-based methods, can also be used to find novel leads. However, as with all predictive techniques, the accuracy and scope of the generated models are largely determined by the quality and relevance of the training data.

For instance, HTS results are often unsaturated and frequently have a high false-positive rate. In the published literature, dose–response studies do not usually sufficiently reflect negative findings. It is common for chemical structures to be pictured or called wrongly. As datasets become more widely accessible, the focus will shift away from data quantity and toward quality, indexing, and organizing. In fact, a number of analyses that evaluate the efficacy of public screening libraries and identify promiscuous or reactive compounds that may be to blame for many false-positive results have already been published [21, 22]; these analyses also examine the accuracy of compound structures in various repositories [23]. A significant improvement in this one area alone will have a significant impact on the rate of discovery of chemical probes that are actually helpful as a basis for the creation of novel and secure treatments. As these public-domain assets spread more quickly, ensuring quality and compatibility will become increasingly difficult. We are beginning to witness an increase in the availability of open-source tools for chemical data processing and analysis to go along with the growth of open data and related research activity. For instance, toolkits and workflow tools like CDK/Taverna [24], Bioclipse [25], RDKit [26], KNIME [27], and OpenBabel [28] are becoming more and more popular, allowing scientists to access the growing number of resources available and facilitating data mining efforts without needing to invest in expensive commercial software – this is similar to projects like BioPerl for the community of bioinformatics research. Parallel attempts are being made to promote connectivity through the development of standards (such as the adoption of the InChI representation for chemical structures) [29] and to better integrate various chemical and drug-discovery data sources [30]. To ensure that the data are as useful as possible, further effort in this area will be necessary. But for collaborative and academic drug development initiatives to be truly successful, scientists will need access to the entire set of resources and information that are available to those in business. There is a substantial amount of information about chemical structure, synthesis, and pharmacology that is only found in patent documents. Despite being easily accessible online, many records are not well-structured enough for extensive searching and analysis. Here the use of machine learning and artificial intelligence comes into the picture.

Academic institutions and the pharmaceutical business have amassed a sizable amount of data over the years to the point that it is now impossible for humans to process all of the information therein. Indeed, the creation of enhanced methods to extract knowledge and effectively advance discovery programmes is warranted by the generalised non-linearity of data correlations and a perception of human inefficiency at integrating data from more than four variables at once [1].

The ever-increasing volume of chemical and biological data, as documented in both publicly available and internally developed datasets, can now be correctly segmented and exposed to correlation and regression analysis due to recent improvements in machine learning/artificial intelligence (ML/AI) algorithms, computing capabilities, and storage capacity. Furthermore, the availability to and construction of tailored ML/AI for a wide range of real-life applications is made possible by falling hardware costs and widespread support for open source tools [31, 32].

Numerous applications of ML/AI have been systematically validated and in-depth reviews have been conducted, including retrosynthetic design [33], de novo design [34], chemical outcome forecasting [35], and drug–target deconvolution [36]. While the later contributions frequently concentrate on the ML/AI model designs and methodologies for standardising expert knowledge, less attention has been paid to data source quality, a crucial step in the model building process.

One could argue that ML/AI for drug discovery still faces a constraint in the collection of data despite the fact that it is expanding. This is particularly true for novel ML/AI applications when annotated data is not always available or when a deep learning algorithm is used and in charge of feature engineering. Another crucial, time-consuming, and underestimated activity by the less informed population is data filtering and quality assurance.

However, using exempt datasets might significantly affect the synchronisation of information and, as a result, the usefulness and quality of the resulting models. Here we highlight the excellent (high quality and comprehensive), the problematic (mid quality and scant), and the ugly (low quality) chemical and biological data for information extraction in ML/AI, taking potential problems into consideration.

2 DATA MINING AND ITS APPLICATION IN HEALTHCARE

Numerous organisations have made substantial and frequent use of data mining. Data mining is growing in popularity, if not need, in the healthcare industry. All parties involved in the healthcare industry can tremendously benefit from data mining techniques. Data mining, for instance, can assist physicians in identifying effective treatments and best practises, healthcare insurers in identifying fraud and abuse, healthcare organisations in making customer relationship management decisions, and patients in receiving better and more reasonably priced healthcare services. Traditional approaches cannot process and analyse the enormous amounts of complicated and voluminous data created by healthcare transactions. The technology and methodology for converting these mountains of data into information that can be used for decision-making are provided by data mining [37].

2.1 HISTORY OF DATABASES AND DATA MINING

Data mining systems have existed since the 1960s and before. Here, data mining focuses solely on file processing. The database management system stage began in the early to mid-1970s. Tools for data modeling and query processing are used in this OLTP. There are three main categories of database management systems that need to be worked. The first one is called advanced database systems, which was first developed in the mid-1980s to the present and works with data models and application-oriented processes. Data warehousing and data mining, which have been in use since the late 1980s, make up the second category. The third category is devoted to web-based database systems, which have been around since the 1990s and include web mining and XML-based database systems. The new generation of the integrated information system, which combines these three main categories, was launched in 2000 [38].

2.2 Work of Data Mining

The use of computing technologies in healthcare is nothing new. The foundation of medical data mining is the computational capacity of cloud solutions and the self-learning capabilities of AI algorithms. In addition, actual data sets to train the model to spot patterns and draw conclusions are needed [39].

The following stages will be followed by the data mining process once all the necessary elements have been put in place:

Acquisition/selection: A target set is constructed using the original data at this stage.
Preprocessing: Data is formatted and standardised.
Mining: The process of actually finding patterns and knowledge is called mining.
Interpretation: Gaining knowledge from the patterns discovered.

2.3 Data Mining in Healthcare

In the healthcare sector, data mining is becoming increasingly common, if not essential. Data mining technologies can be very advantageous for all stakeholders involved in the healthcare sector.

- **Brain tumor segmentation**
  Six researchers have finished their study on using K-means clustering and deep learning to categorise brain tumors. MRI scans were used to create the initial data sets, which were then fed into a data mining system for preprocessing and algorithmic analysis. After passing the data down a pipeline of several statistical classifiers and geometric identification models, the system was able to differentiate between benign and malignant tumors. The resulting average accuracy turned out to be 95.62%, much higher than expected or achieved previously in similar experiments. Scientists added generated data to the MRI scans to improve the model's training. In order to enhance the volume of data for deep learning algorithms, the team took the original photos and applied cropping, flipping, distortion, and noise. Deep learning algorithms require vast amounts of labeled data for training. The end result was a system that has a remarkable accuracy of 98.3% for classifying brain cancers.

- **Identifying and preventing fraud**
  A group of Italian researchers examined fraud-related behaviour trends in 183 hospitals around Lombardia.
  The procedure was divided into two steps:
  The team found batches of hospitals with similar heart failure treatment protocols using K-means clustering. The second stage was supervised by human auditors, who aided the algorithmic model by cross-validating outliers based on fraud-related behaviour, thereby streamlining the process of discovering outliers – abnormal behaviour patterns. The researchers was able to identify two hospitals whose patterns suggested potential fraud. Since the data mining algorithms were just being tested, nothing more was done [39].

- **Treatment effectiveness**
  Applications for data mining can be created to assess the efficacy of medical interventions. By comparing and contrasting the causes, symptoms, and treatment options, data mining can provide an analysis of the most successful course of action.

- **Healthcare management**

  Applications for data mining can be created to more accurately identify and monitor high-risk patients and chronic disease states, plan effective interventions, and lower the number of hospital admissions and claims to help control healthcare costs. Massive amounts of data and statistics were analysed using data mining to look for patterns that would point to a bioterrorist strike.

- **Customer relationship management**

  Customer relationship management is a key strategy for controlling interactions between businesses and their clients, primarily banks and merchants. It is equally crucial in the healthcare industry. Call centres, doctors' offices, billing departments, inpatient facilities, and ambulatory care facilities are just a few places where customers may contact businesses.

- **Pharmaceutical industry**

  Pharmaceutical companies are using technology to manage their inventories and create new goods and services. For a firm's competitive position and corporate decision-making, a thorough grasp of the knowledge concealed in pharma data is essential [40,41].

- **Matching specialists and patients**

  It may not always be possible for patients with rare diseases to get the care they need. A recent study found that by employing data mining techniques, medical professionals could improve a doctor's ability to diagnose these patients. It can also help patients find medical professionals familiar with their disease. This reduces the chance of error, saves time, decreases expenses for the patient and the provider, and raises the standard of care.

- **Increased diagnostic precision**

  Clinicians in the healthcare industry may identify patients more quickly and with greater certainty with the help of data mining. A qualified doctor must still make the final decision, but software with AI capabilities can quickly analyse a lot of data. Blood tests, X-rays, and MRI images can all be quickly processed and categorised to help with the early diagnosis of malignancies and other problems. Speed and accuracy of interpretation are crucial in the treatment of complex disorders with equivocal symptoms.

2.4   EXPANDING DATA MINING IN HEALTHCARE

The three-system approach is the best method for expanding data mining beyond the bounds of academic study. The best method to make a real-world improvement with any healthcare analytics endeavour is to implement all three systems. Regrettably, only a small number of healthcare institutions use all three of these platforms.

The analytics system:

  The analytics system includes the tools and knowledge needed to gather data, analyse it, and standardise metrics. The system's core is built on the aggregation of clinical, patient satisfaction, financial, and other data into an enterprise data warehouse.

The content system:

  The content system includes a knowledge work standardisation component. It integrates care delivery with best practises supported by evidence. Every year, substantial advances in clinical best practise are made by scientists, but as was already noted, it sometimes takes a while for these advancements to be used in actual clinical settings. Organizations can swiftly implement the newest medical standards thanks to a robust content system.

TABLE 12.1
Primary healthcare with analysis

S.No	Types of disease	Data mining tool	Technique	Algorithm	Traditional method
1.	Tuberculosis	WEKA	Naïve Bayes Classifier	KNN	Probability Statistics
2.	Dengue	SPSS Modeler		C5.0	Statistics
3.	Kidney Dialysis	RST	Classification	Decision Making	Statistics

The deployment system:

Driving change management over new hierarchical structures is part of the deployment system. Implementing group structures that facilitate the consistent, enterprise-wide adoption of best practises is particularly important. To encourage the adoption of best practises throughout a company, a true hierarchical transformation is necessary.

2.5 Results of Comparative Analysis of Various Diseases in Healthcare

Numerous experts have provided detailed comparisons of data mining applications in the healthcare industry. Data mining technologies are mostly used to forecast outcomes from information gathered on healthcare issues. In order to anticipate the precision level in various healthcare situations, several data mining technologies are used. The medical issues on the provided list have been looked at and assessed.

Table 12.1 provides examples of the primary healthcare issues, particularly on the disease side and analysis outcomes. These illnesses are widespread issues in people. The traditional methods of statistical applications are also provided and contrasted in order to examine the impact of data mining applications for diagnosing certain conditions.

2.6 Advantages of Data Mining in Healthcare

The workflow of healthcare organisations is made simpler and more automated by a data framework. Healthcare organisations can save time and effort in decision-making by integrating data mining into their data frameworks. The best informational support and expertise are provided to healthcare professionals through predictive models. The goal of predictive data mining in medicine is to develop a predictive model that is understandable, yields trustworthy predictions, and aids physicians in improving their processes for diagnosing patients and formulating treatment plans. When there is uncertainty regarding the relationship between different subsystems and when conventional analysis techniques are ineffective, as is frequently the case with nonlinear associations, biomedical signal processing communicated by internal guidelines and reactions to improve the condition is crucial a part of data mining [42,43].

2.7 Future of Data Mining in Healthcare

Providers can further benefit from the advantages of healthcare data mining as it becomes more widely used and the underlying technology develops. In the near future, advancements are predicted to emerge that will allow for:

- More effective revenue cycle management for healthcare institutions
- Improved management of uncommon disorders
- Greater chances of survival for cancer patients

- Significant enhancements in patient care quality, notably for disadvantaged groups
- National preventative measures against infectious diseases

The healthcare industry is already undergoing a revolution thanks to data mining, and this change is unstoppable. Machine learning algorithms are being used to aggregate and process more data. The sector is becoming more adaptable and resilient thanks to real-time analytics able to withstand any storm that may come its way [39].

3 BIOINFORMATICS AND DATABASES BUILDING AND SKILLS BY APPLYING ALGORITHMS

The use of computer science and technology in the study of biology is known as bioinformatics. Data about biological processes is stored there, examined, and interpreted. Anyone working in the field of biology needs to be proficient in bioinformatics since it enables the effective management of massive volumes of data.

3.1 DATA MINING

The practise of collecting information from huge data sets in order to uncover patterns, trends, or other important information is known as data mining. Bioinformatics, a discipline that use computer science and statistics to analyse and understand data pertaining to biology, is heavily reliant on data mining. Applications for bioinformatics include gene sequencing, drug development, and disease research.

Data mining can be used to locate new therapeutic targets, identify genes linked to certain disorders, and forecast how medications will behave in the body. Infectious illness outbreak trends, the transmission of infectious diseases, and the success of public health initiatives may all be tracked via data mining.

3.2 ALGORITHMS

Algorithms are a set of procedures that can be used to address issues. In order to examine data and forecast how proteins and genes will behave, bioinformatics algorithms are utilised. For instance, depending on a protein's amino acid sequence, algorithms can be used to predict how the protein will fold. In bioinformatics, algorithms are crucial because they can expedite the process of data analysis and prediction. It is crucial to remember that algorithms are only as good as the data they are provided. Predictions produced based on incomplete or faulty data will also be erroneous [44].

Statistics show that every 18 months, the volume of biological data roughly doubles. Only 606 sequences totaling 680,000 nucleotide bases were present in GenBank's initial database of nucleic acid sequences in 1982 [45]. In its database as of February 2013, there are currently 150 billion nucleotide bases and 162 million biological sequences. An essential area of bioinformatics research is how to steer biological research by extracting insights from these vast amounts of data.

In order to extract useful information from complicated biological data, it is first important to tackle the issue of storing and managing enormous amounts of data on the assumption that the data accurately reflect the underlying meaning of biology. An essential strategy for developing artificial intelligence is machine learning. It has been extensively employed in the field of bioinformatics and can handle the automatic learning of machines without explicit programming [46–48].

Machine Learning Algorithm

We have seen revolutionary advancements in biomedical research, biotechnology, and enormous expansion of biomedical data during the last few decades. The issue now is how to extract relevant information from the growing body of biomedical data. On the one hand, bioinformatics has emerged as a challenging new discipline as a result of the quick growth of biotechnology and biological data analysis techniques. On the other hand, a huge number of efficient and well-scalable algorithms have been created as a result of the ongoing development of biological data mining technologies. It is important to focus on and conduct research on how to combine machine learning and bioinformatics to effectively interpret biomedical data. We should examine how to employ data mining for efficient biomedical data analysis in particular, and we should lay out some possible research topics that could encourage the creation of more potent biological machine learning algorithms.

3.3 BASIC DATA MINING PROCESS

The field of data mining blends traditional statistical methods with computer science algorithms to extract knowledge from massive amounts of data for application in science, computation, or industry. We provide an exhaustive description of the data mining process from six perspectives, as shown in Figure 12.1.

1. Cleaning up data. Data sets frequently contain missing data and inconsistent data because of the rise in heterogeneous data. The task of extracting information is severely hampered by poor data quality. Therefore, the first stage in data mining is to remove any inconsistent or missing data.
2. Integration of data. If the study's data come from many sources, they must be uniformly aggregated.
3. The selection of data. It is important to accurately choose pertinent information based on the study's subject matter.
4. Data transformation. Additional qualities or functionalities that are helpful for the data mining process should be integrated in addition to transforming or combining data into a form that is suitable for mining.
5. Data analysis. The best model for the situation should be chosen and any necessary adjustments made.
6. Mode assessment. After learning from the data, the right metrics to assess the model should be chosen.

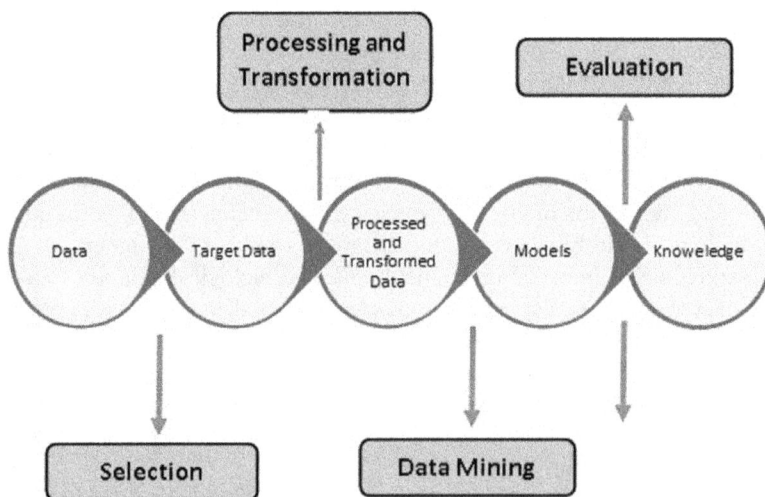

FIGURE 12.1 The steps for data mining process.

The major goal of the data mining phase is to choose one or a combination of these stages correctly and discover an efficient and trustworthy solution to the problem at hand. Machine learning has become a popular tool for bioinformatics analysis in recent years. Data mining has several different machine learning methods that are created independently of one another.

3.4 ASSOCIATION RULE MINING ALGORITHM

Association rule mining, one of the most significant subfields of data mining, may locate relationships and recurring patterns among a collection of elements in a database. It comprises two related issues: (1) Decide on the minimal support criterion and apply it. Get frequent itemsets from the database. (2) Find association rules that fulfil the required constraints for frequent itemsets using the minimum confidence method. In addition to being crucial for corporate data analysis, association rule mining has also found success in a wide range of other industries, including the analysis of virtual shopping baskets and medical data. The Apriori method is a common association rule-based mining technique with applications in protein structure prediction and sequence pattern mining. Apriori is the foundation from which many machine learning techniques in data mining are derived [49]. The fundamental approach to association rule mining makes use of the database's strong relationships are examined using a few metrics. The two measurement techniques that are most frequently utilised are minimum support and minimum confidence. The Apriori algorithm employs a guided approach to discover association rules among database elements.

- **Classification Algorithm**
  One of the most researched machine learning tasks is classification. The basis for classification is the predicted attribute, which predicts the class of the user-specified target attribute. The two most important problems in genomics are genome categorisation and sequence annotation. Numerous methods, such as fuzzy sets, neural networks, evolutionary algorithms, and rough sets, are employed in the mining of biological sequences. Neural networks, decision trees, naïve Bayesian networks, and rule learning with evolutionary algorithms are only a few examples of the numerous generic categorisation models.

Clustering Algorithm
Machine learning's clustering technique may group together sequences that share certain features and investigate the useful information of unidentified sequences from well-known structures and functionalities. As a result, the grouping of biological sequences has a significant impact on bioinformatics research. Clustering differs from classification in that it does not use a predetermined category. Each cluster has a set of shared traits. With the help of cluster analysis, data with similar features are grouped into one category and then subjected to further analysis.

With the advancement of artificial intelligence in recent years, the clustering algorithm has emerged as a hot topic for research in the area of machine learning. Both domestic and international researchers have studied clustering algorithms in greater detail in order to increase the processing capability of vast amounts of data. There are a number of outstanding clustering algorithms that have been developed, including clustering algorithms based on granularity, uncertainty, entropy, and clustering integration algorithms.

In addition to the ones described above, there are many other algorithms. Each algorithm has unique properties and hence can not be used in every circumstance. We can apply and investigate each algorithm more effectively if we are aware of its benefits and drawbacks [50–52].

The processing, searching, and data mining of DNA sequence data are all part of bioinformatics. The creation of methods to store and search DNA sequences [53] has resulted in significant advancements in computer science, particularly in the fields of database theory, machine learning, and string searching algorithms. Even simple algorithms for this problem typically work well in other contexts, such as text editors, but in DNA sequences, the small amount of different letters leads to near-worst case behaviour in these algorithms. Without annotations, which mark the positions of genes and regulatory elements on each chromosome, data sets covering full genomes' worth of DNA sequences, such as those created by the Human Genome Project [54], are challenging to use. Gene discovery algorithms [55] can identify regions of DNA sequence that display the distinctive patterns connected to protein- or RNA-coding genes, allowing researchers to anticipate the presence of specific gene products in an organism even before they have been isolated experimentally.

3.5 Need for Data Mining in Bioinformatics

The complete set of genetic data present in each human cell, or the human genome, has now been identified. Identifying the mechanics underpinning biological processes like growth and ageing, as well as clearly tracking our evolution and its relationships with other species, are all expected to be made possible by understanding these genetic instructions. The sheer amount of information accessible is the main barrier preventing researchers from discovering the information they want. Like the majority of natural scientists, biologists are largely educated to obtain fresh data. Previously, biology lacked the resources to examine vast databanks like the human genome database. Fortunately, the field of computer science has been developing techniques and strategies well adapted to aid biologists in managing and analysing the enormous amounts of data that promise to significantly advance the state of humanity. One such technology is data mining [56].

4 POWER OF ALGORITHMS IN BIOLOGY

For many years, there has been a lengthy and productive connection between computer science and biological science. Numerous computational techniques were influenced by the higher-level design principles of biology, and researchers depend on them to evaluate and adapt massive databases. The identification of physiologically active molecules is a crucial challenge in the disciplines of drug development and drug adaptation for different purposes. The cumbersome wet-lab evaluation of several substances is eliminated by computerised high-throughput screening, although this process can take some time. Computational biology is the study of using computer resources to solve biologically relevant problems.

The term "bioinformatics" is often used to describe the field of computational biology. The terms "computational biology" and "bioinformatics" are closely interlinked, but there appears to be a widespread belief that computational biology refers to exercises primarily focused on building algorithms to address problems with biological relevance and bioinformatics refers to actions primarily focused on building as well as with help of computational techniques to assess accessible biological information.

Many reference genomes of different species have indeed become publically available as a consequence of the endeavour to sequence and compare numerous distinctive genomes of a specific species over the last ten years. Datasets from projects like the 1000 Genome Project, 100K Genome Project, 1001 Arabidopsis Genomes Project, Rice Genome Annotation Project, and the Bird 10,000 Genomes (B10K) Project, for instance, today are publicly available. The very first continuous, full human reference genome, also known as Telomere-to-Telomere (T2T) and genome in a bottle (GIAB) data, support such endeavours. The data obtained for these and additional projects are not restricted to only short read data, since many metadata types, such as long reads, ultra-long reads, optical maps, mass spectrometry data, etc., may now be produced in a high-throughput way [57].

An algorithm is a process used to carry out a computation or solve an issue. In either hardware-based or software-based processes, algorithms function as a detailed sequence of instructions that carry out predetermined operations sequentially. Algorithms are already a part of self-driving cars, smart speakers, and smartphones in modern life. Deep-learning algorithms in biology delve through data in a manner that humans cannot, spotting patterns that can sometimes go undetected. Algorithms are being used by investigators to categorise cell imaging, link genomic data, advance drug development, and perhaps detect correlations amongst various kinds of information such as genetics, imaging, and electronic health information. The process of building algorithms for bio-informatics involves two interrelated processes that must be completed in order to solve challenges with biological significance. The first stage is to propose an intriguing biology question and to build a model of biological fact that allows the computational formulation of the stated topic. The creation of an algorithm that resolves the stated computing issue is the second phase. Understanding a biological fact is necessary for the first stage, while expertise of algorithmic principles is necessary for the latter. Traditional algorithmic methodologies use the resources, most notably time and space, which the created algorithm uses to resolve the issue to evaluate the effectiveness of the algorithm. Nevertheless, as the algorithm's solution comes from a query having biological significance, the biological relevance of the results should also be used to assess the algorithm's quality [58].

The runtime and space assumption of an algorithm as well as the biological relevance of the solutions it generates together make up its quality in terms of solving problems with biological function. The modeling of the biological fact that resulted in the framing of the computational problem that the algorithm solves is what determines these two characteristics of an algorithm's excellence. In order to build an effective algorithm that addresses an issue of biological relevance, it is necessary to engage in a multidisciplinary process that alternates among both modeling biological fact and building the algorithm. This process must continue until a satisfactory equilibrium among the algorithm's running time and space assumption and the biological significance of the responses it generates is reached. As a result, it is indeed important to grade the algorithm by the biological significance of the results it generates. The amount of exchanging between modeling and constructing naturally relies on how directly connected the issue handled by the algorithm is to a specific biological application. The investigators letting programme find molecule traits on its own rather than defining what makes a structure prediction greater or lesser correct. Researchers accomplish this as a result of their discovery that the traditional method of imparting such information can bias an algorithm in support of individual characteristics, prohibiting this from discovering further significant attributes.

The Toronto-based CEO of the Canadian business Deep Genomics, Brendan Frey, focuses on genomic datasets with the intention of diagnosing and managing diseases. At the University of Toronto, Frey's research group created algorithms that were trained on genomic and transcriptome information from healthy human cells. Within that data, these computers created predictive models of RNA-processing events like splicing, transcription, and polyadenylation. The algorithms were able to find mutations and mark such as harmful once they were tried to apply to clinical data [59]. Particularly in the past twenty years as molecular data have progressively been collected, scientists have now been leaning more and more on advanced computational techniques. Investigators today frequently employ computational methods for scanning huge datasets, such as BLAST [60]. Algorithms are frequently used in genome sequencing and assembling to expedite data collection and processing [61–63]. Additionally, computational techniques have been created enabling combining multiple forms of functional datasets and applying those to model regulatory networks and other cellular interconnections [64–66]. In fact, over the past several years, a number of bioinformatics teams have been founded with the goal of creating new computational techniques to help solve the deepest puzzles in bioscience.

Biological systems have served as a source of motivation for computer programmers, particularly when creating optimization algorithms. Neural networks are a type of computational technique that were first developed as part of early work on artificial intelligence in the 1960s. They have been employed in a variety of machine learning algorithms, from image classification to calculating rocket trajectories [67–69]. Some methodologies, such genetic algorithms [70], which have been used extensively for the past two decades, were motivated by general processes in DNA sequence development. The investigation of how self-organization originates through local contacts has also been spurred by the investigation of social insects [71] and particle swarms [72]. Substantial use of these concepts has been made in multi-agent optimization models [73,74]. Other techniques have also used biological ideas to create modern computing frameworks, such as non-negative matrix factorization [75] and population protocols [76,77]. Since every approach that has been developed has resulted in an effective implementation, these merely depended on a cursory (and occasionally incorrect) comprehension of the biological functions they were founded on, and hence techniques typically did not immediately result in new biological discoveries. Particularly in comparison to how new computational techniques have already been created to assist investigators in learning novel biological concepts, the implementation of these techniques to the biological issue (i.e., the biological system on its own) has very seldom been used to inform the development of improved algorithms by computer programmers. As a result, the two directions – using computational methods to investigate biology and depending on biological concepts to develop computational methods – remain substantially distinct. The method is being used by researchers to identify genome alterations and forecast changes in the structure of individual cells [78].

Convolution neural networks (CNNs) accelerate relying on a field-programmable gate matrix is implemented by Chen et al (FPGA) [79]. Although CNNs have many uses in bioinformatics, deploying a CNN architecture on a low-power system is difficult since CNNs need a lot of computational power. Chen's input lessens the impact of such a restriction. To innovate novel time-space tradeoffs, Bannai et al. [80] extended current work on determining the shortest distinct subsequence. They also provided an effective approach for the generalisation of determining the shortest distinctive substrings with k-mismatches.

Researchers and programmers of bioinformatics algorithms might learn from being familiar with the idea of computational complexity because the great majority of bioinformatics problems are computationally hard. Reviewing the key heuristic techniques and their relationship to High-Performance Computing (HPC), we look at what computational complexity is and how it relates to the core issues in bioinformatics.

4.1 DIFFICULTY OF ALGORITHMS IN COMPUTATIONAL BIOLOGY

The effectiveness of an algorithm, also known as its computational complexity, is the amount of resources demanded for its implementation, including the number of hours and space necessary to finish the algorithm's elementary activities (i.e., the extent of the essential memory) [81]. Computational biology's research of algorithmic complexity offers guidelines for creating effective applications for handling, modeling, and interpreting biological information. Numerous polynomial algorithms are used in computational biology [82–87].

A Classic Computational Challenge is Multiple Sequence Alignment

In bioinformatics, among the most significant issues is multiple sequence alignment (MSA). The sequence alignment that maximises the underlying evolutionary or structural links between the sequences is considered to be biologically appropriate. MSA makes use of probabilities for amino acid substitutions that are represented in the popular PAM [88] and BLOSUM [89] databases. During proteomic and genomic studies, such as protein structure determination and phylogenetic construction, MSAs are essential. In gene sequences of nucleic and amino acids, they are utilised to identify repeated areas, giving structural, functional, and evolutionary information [90]. Accurate MSA approaches are now a crucial prerequisite for large-scale sequence comparisons due to the current growth in sequencing database [91], especially in light of the current international efforts for the collection and exchange of massive amounts of genetic information [92].

Complexity and Artificial Intelligence

The difficulty of machine learning algorithms is becoming a more pertinent and crucial topic in this field as deep learning for computational biology problems is being widely adopted [93–95]. In light of the limitless possibilities for selecting the limited subset, it has now been shown that it is impossible to determine if a machine learning algorithm can generate forecasts regarding a particular big information source by selecting a limited number of data points (i.e., learnability) [96]. Notwithstanding deep learning's capacity to analyse vast quantities and a wide range of dataset, applying it to actual concerns, including all who arise in healthcare and medicine, presents a number of computing resources difficulties. Well over 107 pictures and much above 104 classes make up the ImageNet datasets, which is used to train CNNs for image processing applications [97]. There are 110M variables in the Bidirectional Encoder Representations from Transformers (BERT) language architecture [98]. Deep learning is therefore mostly carried out on spread memory using graphics processing units (GPUs) as the primary hardware [99]. Tensor processing units (TPUs), specifically made processors created by Google for deep learning applications and utilised for training the AlphaGo algorithm [100], have gained a lot of notoriety since. Technological advancements are creating the groundwork for machine learning-specific embedded applications, like Edge TPUs for edge computing and the Internet of Things (IoT), which include compact design and reduced power requirements.

Heuristics

Discovering sub-optimal solutions is frequently just one alternative in actual issues since it would be too difficult to resolve the issue completely, regardless of whether an exponential method is accessible. Heuristic algorithms are typically used in the development of methods that sacrifice accuracy, completeness, and optimality towards rapidity. Noteworthy work on heuristic techniques for finding adequate answers along with various computational challenges in bioinformatics have been done. Local search algorithms, greedy algorithms, and probabilistic algorithms are traditional examples of heuristic methods used to address various issues [101]. An illustration of a greedy heuristic for the problem of clustering data points is the frequently utilised agglomerative technique in hierarchical clustering. As opposed to certain other heuristics, estimation algorithms ensure the accuracy of the solution by indicating how far away it is to the ideal option. It is important to note that approximating algorithms are just applicable to optimising issues, i.e., whatever challenge in which the optimal option is determined provided an optimal control constraint, typically defined by a cost or objective function to be maximised or minimised. The approximation ratio (i.e., a representation of how distant a response is from the optimum), can be calculated using this approach. Consequently, other generalized heuristic techniques, like local search algorithms, do not really offer sufficient insight, whereas approximation algorithms ensure the accuracy of the provided suboptimal answer. The Sorting by Reversals algorithm is an illustration of an approximation algorithm in the area of bioinformatics. The lowest group of modifications (such as fragment distortions and translocations) among the genes of two closely related organisms is determined using this technique [102].

Computational Biology Using High-Performance Computing

Identifying viable answers in computational biology depends on developing methods that do not suffer from the exponential explosion characteristic in the comprehensive research methodology. Through the use of supercomputers, which can handle billions of processes per second, enormously parallel processing is guaranteed to be performed in an acceptable duration. HPC is the term used to describe such simultaneous processing on expensive hardware. When handling a variety of issues with enormously big occurrences, HPC tools can be especially useful [103–105].

Researchers face both opportunities and challenges as a consequence of the massive increases in biological databases, which is mostly the result of technology developments including next sequencing. Investigators' capacity to interpret the expanding deluge of genomic, metagenomic (the examination of the architecture and use of complete nucleotide sequences that have been extracted and examined from a large sample of organisms, usually microbes), structural, and interactome (beneficial interconnections among molecules that occur potentially inside of a cells or across the organism's body) databases will determine whether the commitment of discovering the causes of disorders like cancer, obesity, Alzheimer's, autism spectrum disorder, and several more is realised, as well as how best to comprehend biology's fundamental principles. Bioinformatics should keep continuously innovating even while incorporating the greatest concepts from several branches of data science. In addition to having a wide variety of scientific techniques at their disposal, scientists also collect datasets from incredibly complicated phenomena that are ultimately governed by development. Researchers can understand biology in perspective of evolvement by creating algorithms that take advantage of the architecture of biological data [105].

5 CONCLUSION

The development of big data has made biology a data-intensive discipline in the age of genetics and genomics. To meet the data analysis issues in this expanding industry, new computational algorithms and software tools are required. The use of genetic algorithms to enhance data mining methods was demonstrated. A brief introduction to molecular biology and bioinformatics followed. This can be utilised for drug treatment taking into account how the genes react to medications, allowing for the detection of fatal diseases like AIDS and Alzheimer's disease. It can also be used to trace the course of human evolution and to determine the mechanisms behind biological processes like development and ageing. Therefore, data mining is a useful way to address the issue of huge data that academics encounter in their quest to unravel the mysteries of our existence.

REFERENCES

1. Rodrigues, T: The good, the bad, and the ugly in chemical and biological data for machine learning. Drug Discovery Today: Technologies (2019) 32–33; 3–8.
2. Gaulton, A, Overington, JP: Role of open chemical data in aiding drug discovery and design. Future Med. Chem. (2010) 2(6), 903–907.
3. Austin CP, Brady LS, Insel TR, Collins FS. NIH Molecular Libraries Initiative. Science (2004) 306(5699), 1138–1139.
4. Liu T, Lin Y, Wen X, Jorrisen RN, Gilson MK. Binding DB: a web-accessible database of experimentally determined protein-ligand binding affinities. Nucleic Acids Res. (2007), 35(Database issue), D198–201.
5. Harmar AJ, Hills RA, Rosser EM et al. IUPHAR-DB: the IUPHAR database of G protein-coupled receptors and ion channels. Nucleic Acids Res. (2009) 37(Database issue), D680–685.
6. PDSP Database, http://pdsp.med.unc.edu/pdsp.php
7. Wellcome Trust press release www.wellcome.ac.uk/News/Media-office/Press-releases/2010/WTX058 219.htm
8. Villoutreix BO, Renault N, Lagorce D, Sperandio O, Montes M, Miteva MA. Free resources to assist structure-based virtual ligand screening experiments. Curr. Protein Pept. Sci. (2007), 8(4), 381–411.
9. Irwin JJ, Shoichet BK. ZINC – a free database of commercially available compounds for virtual screening. J. Chem. Inf. Model (2005), 45(1), 177–182.
10. Blum LC, Reymond JL. 970 million drug like small molecules for virtual screening in the chemical universe database GDB-13. J. Am. Chem. Soc. (2009), 131(25), 8732–8733.
11. Wishart DS, Knox C, Guo AC et al. DrugBank: a knowledgebase for drugs, drug actions and drug targets. Nucleic Acids Res. (2008), 36(Database issue), D901–D906.
12. DailyMed http://dailymed.nlm.nih.gov/dailymed/about.cfm
13. Clinical Trials homepage, www.clinicaltrials.gov
14. Richard AM. DSSTox website launch: improving public access to databases for building structure-toxicity prediction models. Preclinica 2, (2004), 103–108.

15. Richard AM, Williams CR. Distributed structure-searchable toxicity (DSSTox) public database network: a proposal. *Mutat. Res.* (2002), 499(1), 27–52.

16. Hochstein C, Arnesen S, Goshorn J. Environmental health and toxicology resources of the United States National Library of Medicine. Med. Ref. Serv. Q. (2007), 26(3), 21–45.

17. Hunter AJ. The Innovative Medicines Initiative: a pre-competitive initiative to enhance the biomedical science base of Europe to expedite the development of new medicines for patients. Drug Discov. Today. (2008), 13(9–10), 371–373.

18. Barnes MR, Harland L, Foord SM et al. Lowering industry firewalls: pre-competitive informatics initiatives in drug discovery. Nat. Rev. Drug Discov. (2009), 8(9), 701–708.

19. Edwards AM, Bountra C, Kerr DJ, Willson TM. Open access chemical and clinical probes to support drug discovery. Nat. Chem. Biol. (2009), 5(7), 436–440.

20. GSK announces 'open innovation' strategy to help deliver new and better medicines for people living in the world's poorest countries – press release www.gsk.com/media/ pressreleases/2010/2010_pressrelease_10009. Htm.

21. Feng BY, Simeonov A, Jadhav A et al. A high-throughput screen for aggregation-based inhibition in a large compound library. J. Med. Chem. (2007), 50(10), 2385–2390.

22. Soares KM, Blackmon N, Shun TY et al. Profiling the NIH Small Molecule Repository for compounds that generate H_2O_2 by redox cycling in reducing environments. Assay Drug Dev. Technol., (2010), 8(2); 152–174.

23. Young D, Martin T, Venkatapathy R, Harten P. Are the chemical structures in your QSAR correct? QSAR Comb. Sci. (2008), 27(11–12), 1337–1345.

24. Kuhn T, Willighagen EL, Zielesny A, Steinbeck C. CDK-Taverna: an open workflow environment for cheminformatics. BMC Bioinformatics (2010), 11(1), 159.

25. Spjuth O, Helmus T, Willighagen EL et al. Bioclipse: an open source workbench for chemo-and bioinformatics. BMC Bioinformatics (2007), 8, 59.

26. RDKit: cheminformatics and machine learning software www.rdkit.org/

27. Berthold MR, Cebron N, Dill F et al. KNIME: The Konstanz Information Miner. In: Data Analysis, Machine Learning and Applications. Preisach C, Schmidt-Thieme L (Eds). Springer-Verlag, Berlin, (2008), 319–326.

28. Open Babel: the open source toolbox http://openbabel.org

29. Heller SR, McNaught AD. The IUPAC international chemical identifier (InChI). Chem. Int. (2009), 31(1); 7.

30. Belleau F, Nolin MA, Tourigny N, Rigault P, Morissette J. Bio2RDF: towards a mashup to build bioinformatics knowledge systems. J. Biomed. Inform. (2008), 41(5), 706–716.

31. de Almeida AF, Moreira R, Rodrigues T. Synthetic organic chemistry driven by artificial intelligence. Nat Rev Chem (2019);3:589–604.

32. Mayr A, Klambauer G, Unterthiner T, Steijaert M, Wegner JK, Ceulemans H, et al. Large-scale comparison of machine learning methods for drug target prediction on ChEMBL. Chem Sci (2018);9(24):5441–51.

33. Segler MHS, Preuss M, Waller MP. Planning chemical syntheses with deep neural networks and symbolic AI. Nature (2018);555(7698):604–10.

34. Button A, Merk D, Hiss JA, Schneider G. Automated de novo molecular design by hybrid machine intelligence and rule-driven chemical synthesis. Nat Mach Intell (2019);1:307–15.

35. Coley CW, Jin W, Rogers L, Jamison TF, Jaakkola TS, Green WH, et al. A graph-convolutional neural network model for the prediction of chemical reactivity. Chem Sci (2019);10(2):370–7.

36. Rodrigues T, Werner M, Roth J, da Cruz EHG, Marques MC, Akkapeddi P, et al. Machine intelligence decrypts beta-lapachone as an allosteric 5-lipoxygenase inhibitor. Chem Sci (2018);9(34):6899–903.

37. Koh HC, Tan G. Data mining applications in healthcare. J Healthc Inf Manag. 2005;19(2):64–72.

38. M. Durairaj, V. Ranjani, Data Mining Applications In Healthcare Sector: A Study, International Journal Of Scientific & Technology Research, 2013;10(2):29–35.

39. https://demigos.com/blog-post/data-mining-in-the-healthcare-industry-benefits-examples/ (Accessed on 20 Dec 2022).

40. HianChye Koh and Gerald Tan, Data Mining Applications in Healthcare, Journal of Healthcare Information Management – Vol 19, No 2.

41. Prasanna Desikan, Kuo-Wei Hsu, Jaideep Srivastava. Data Mining For Healthcare Management, 2011SIAM International Conference on Data Mining, April, 2011.

42. www.analyticssteps.com/blogs/6-benefits-using-data-mining-healthcare (Accessed on 20 Dec 2022)

43. www.javatpoint.com/data-mining-in-healthcare (Accessed on 20 Dec 2022)

44. https://climbtheladder.com/bioinformatics-skills/ (Accessed on 20 Dec 2022)

45. Bilofsky H. S., Burks C., Fickett J. W., Goad W. B., Lewitter F. I., Rindone W. P., et al. The GenBank genetic sequence databank. Nucleic Acids Res. 1986; 14: 1–4.

46. Li J., Wong L., Yang Q. Guest editors' introduction: data mining in bioinformatics. IEEE Intell. Syst. 2005; 20: 16–18.

47. Larranaga P., Calvo B., Santana R., Bielza C., Galdiano J., Inza I., et al. Machine learning in bioinformatics. Brief. Bioinform. 2006; 7: 86–112

48. Yang A, Zhang W, Wang J, Yang K, Han Y, Zhang L. Review on the Application of Machine Learning Algorithms in the Sequence Data Mining of DNA. Front Bioeng Biotechnol. 2020;8:1032.

49. Zhang, W., Ma, D., and Yao, W. Medical diagnosis data mining based on improved Apriori algorithm. J. Netw. 2014; 9:1339.

50. Yang A, Zhang W, Wang J, Yang K, Han Y and Zhang L. Review on the Application of Machine Learning Algorithms in the Sequence Data Mining of DNA. Front. Bioeng. Biotechnol. 2020; 8:1032.

51. Watson, M. Illuminating the future of DNA sequencing. Genome Biol. 2014; 14:108.

52. Pearson, W. R., and Lipman, D. J. Improved tools for biological sequence comparison. Proc. Natl. Acad. Sci. U.S.A. 1988; 85: 2444–2448.

53. Bergeron, Bryan. Bioinformatics Computing. New Delhi: Pearson Education, 2003.

54. Luscombe, N.M., Greenbaum, D. and Gerstein, M : "What is Bioinformatics? A Proposed Definition and Overview of the Field." Methods of Information in Medicine. 2001; 40(4): 346–358.

55. Shah, Shital C. and Kusiak, Andrew, "Data Mining and Genetic Algorithm Based Gene Selection", Artificial Intelligence in Medicine 2004; 31(2139): 183–196.

56. https://core.ac.uk/download/pdf/53188632.pdf (Accessed on 22 Dec 2022).

57. C. Boucher, Special Issue: Algorithms in Bioinformatics, Algorithms 2023;16(1); 21.

58. C. N. S. Pedersen, 'Basic Research in Computer Science,' 2000.

59. S. Webb, "Deep learning for biology," Nature, 2018; 554(7693): 555–557.

60. S. F. Altschul, W. Gish, W. Miller, E. W. Myers, and D. J. Lipman, "Basic local alignment search tool," J. Mol. Biol., 1990; 215(3): 403–410.

61. D. Gusfield, "CSE280A Algorithms on Strings, Trees, and Sequences," arXiv, 2017, NY: Cambridge University Press, New York.

62. M. C. Schatz, A. L. Delcher, and S. L. Salzberg, Assembly of large genomes using second-generation sequencing, Genome Res., 2010; 20(9): 1165–1173.

63. C. Trapnell and S. L. Salzberg, "How to map billions of short reads onto genomes," Nat. Biotechnol., 2009; 27(5): 455–457.

64. C. L. Myers, C. Chiriac, and O. G. Troyanskaya, Discovering biological networks from diverse functional genomic data, Methods Mol. Biol., 2009; 563: 157–175.

65. C. Huttenhower et al., Detailing regulatory networks through large scale data integration, Bioinformatics.2009; 25(24): 3267–3274.

66. U. Alon. An Introduction to Systems Biology: Design Principles of Biological Circuits (1st ed.), 2006, Chapman and Hall/CRC. https://doi.org/10.1201/9781420011432

67. Bishop, Christopher M. Neural networks for pattern recognition. Oxford University Press, 1995.

68. Hopfield JJ. Neural networks and physical systems with emergent collective computational abilities. Proc Natl Acad Sci U S A. 1982;79(8): 2554–8.

69. Hebb, Donald Olding. The organization of behavior: A neuropsychological theory. Psychology Press, 2005.

70. Goldberg DE. Genetic algorithms. Pearson Education India; 2013.

71. Dorigo, M., Birattari, M. and Stutzle, T., Ant colony optimization. IEEE computational intelligence magazine,2006; 1(4): 28–39.

72. Kennedy, James, and Russell Eberhart. "Particle swarm optimization." In Proceedings of ICNN'95-international conference on neural networks, 1995; 4: 1942–1948.

73. Deneubourg, J.L., Goss, S., Franks, N., Sendova-Franks, A., Detrain, C. and Chrétien, L., 1991, February. The dynamics of collective sorting robot-like ants and ant-like robots. In From animals to animats: proceedings of the first international conference on simulation of adaptive behavior (pp. 356–365).

74. Ferber J, Weiss G. Multi-agent systems: an introduction to distributed artificial intelligence. Reading: Addison-Wesley; 1999 Feb 25.

75. D. D. Lee and H. S. Seung, Learning the parts of objects by non-negative matrix factorization, Nature, 1999; 401(6755): 788–791.

76. Aspnes J, Ruppert E. An introduction to population protocols. Middleware for Network Eccentric and Mobile Applications. 2009: 97–120.

77. Chazelle, B., 2009, January. Natural algorithms. In Proceedings of the twentieth annual ACM-SIAM symposium on Discrete algorithms (pp. 422–431). Society for Industrial and Applied Mathematics.

78. A. Maxmen, Deep learning sharpens views of cells and genes, Nature, 2018; 553: 9–10.

79. Chen, C.; Li, Z.; Zhang, Y.; Zhang, S.; Hou, J.; Zhang, H. Low-Power FPGA Implementation of Convolution Neural Network Accelerator for Pulse Waveform Classification. Algorithms 2020, 13, 213.

80. Bannai, H.; Gagie, T.; Hoppenworth, G.; Puglisi, S.J.; Russo, L.M. More Time-Space Tradeoffs for Finding a Shortest Unique Substring. Algorithms 2020, 13, 234.

81. Time Complexity: How to measure the efficiency of algorithms - KDnuggets. www.kdnuggets.com/ 2020/06/time-complexity-measure-efficiency-algorithms.html (accessed Jan. 06, 2023).

82. Cooley JW, Tukey JW. An algorithm for the machine calculation of complex Fourier series. Math. Comput. 1965; 19:297–297

83. Dijkstra EW. A note on two problems in connexion with graphs. Numer. Math. 1959; 1:269–271

84. Knuth DE, Morris JH Jr, Pratt VR. Fast Pattern Matching in Strings. SIAM J. Comput. 1977; 6:323–350

85. Saitou N, Nei M. The neighbor-joining method: a new method for reconstructing phylogenetic trees. Mol. Biol. Evol. 1987; 4:406–425

86. Pevzner PA, Tang H, Waterman MS. An Eulerian path approach to DNA fragment assembly. Proc. Natl. Acad. Sci. U. S. A. 2001; 98:9748–9753

87. Durbin R, Eddy SR, Eddy S, et al. Biological Sequence Analysis: Probabilistic Models of Proteins and Nucleic Acids. 1998.

88. Dayhoff M, Schwartz R, Orcutt B. A model of evolutionary change in proteins. Atlas of protein sequence and structure 1978; 5:345–352.

89. Henikoff S, Henikoff JG. Amino acid substitution matrices from protein blocks. Proc. Natl. Acad. Sci. U. S. A. 1992; 89:10915–10919.

90. Bawono P, Dijkstra M, Pirovano W, et al. Multiple Sequence Alignment. Methods Mol. Biol. 2017; 1525:167–189.

91. Stephens ZD, Lee SY, Faghri F, et al. Big Data: Astronomical or Genomical? PLoS Biol. 2015; 13:e1002195.

92. Saunders G, Baudis M, Becker R, et al. Leveraging European infrastructures to access 1 million human genomes by 2022. Nat. Rev. Genet. 2019.

93. Tang B, Pan Z, Yin K, et al. Recent Advances of Deep Learning in Bioinformatics and Computational biology. Front. Genet. 2019; 10:214.

94. Padovani de Souza K, Setubal JC, Ponce de Leon F de Carvalho AC, et al. Machine learning meets genome assembly. Brief. Bioinform. 2019; 20:2116–2129.

95. Baichoo S, Ouzounis CA. Computational complexity of algorithms for sequence comparison, short-read assembly and genome alignment. Biosystems. 2017; 156–157:72–85.

96. Ben-David S, Hrubeš P, Moran S, et al. Learnability can be undecidable. Nature Machine Intelligence 2019; 1:44–48.

97. Krizhevsky A, Sutskever I, Hinton GE. ImageNet classification with deep convolutional neural networks. Commun. ACM 2017; 60:84–90.

98. Devlin J, Chang M-W, Lee K, et al. BERT: Pre-training of Deep Bidirectional Transformers for Language Understanding. arXiv [cs.CL] 2018.

99. Ben-Nun T, Hoefler T. Demystifying Parallel and Distributed Deep Learning: An In-Depth Concurrency Analysis. arXiv [cs.LG] 2018.

100. Silver D, Huang A, Maddison CJ, et al. Mastering the game of Go with deep neural networks and tree search. Nature 2016; 529:484–489.

101. Papadimitriou CH, Steiglitz K. Combinatorial Optimization: Algorithm and Complexity. 1982.

102. Bafna V, Pevzner PA. Genome Rearrangements and Sorting by Reversals. SIAM J. Comput. 1996; 25:272–289.

103. D. Samanta, S. Dutta, M. G. Galety, and S. Pramanik, "A Novel Approach for Web Mining Taxonomy for High-Performance Computing," Lect. Notes Networks Syst., vol. 291, pp. 425–432, 2022, doi: 10.1007/978-981-16-4284-5_37/COVER.

104. H. Hafsi, H. Gharsellaoui, and S. Bouamama, "Genetically-modified Multi-objective Particle Swarm Optimization approach for high-performance computing workflow scheduling," Appl. Soft Comput., vol. 122, p. 108791, Jun. 2022, doi: 10.1016/J.ASOC.2022.108791.

105. G. H. Cervi, C. D. Flores, and C. E. Thompson, "Metagenomic Analysis: A Pathway Toward Efficiency Using High-Performance Computing," Lect. Notes Networks Syst., vol. 236, pp. 555–565, 2022, doi: 10.1007/978-981-16-2380-6_49/COVER.

13 Optimization and Quantification of Uncertainty in Machine Learning

*Shama Mujawar, Atharva Tikhe,
Bhupendra Gopalbhai Prajapati*

1 INTRODUCTION

1.1 AIM AND SCOPE OF CHAPTER

Machine learning extracts models from data in order to predict new results from a similar unknown data. The last decade saw exponential growth in the generation of novel machine learning models and their real-world applications. All the machine learning models have inseparable relation with uncertainty. "Learning" is essentially the ability of a model to generalize from the data in such a way that it is not too overconfident (overfitting) or too under-confident (underfitting). There is a trade-off among these two and the problem of low accuracy of the model can be solved (in part) by quantifying and optimizing the uncertainty that the model suffers from.

Therefore, approaches towards uncertainty in machine learning are of immense importance as machine learning is significantly used in solving practical problems which expect certain safety standards. In the field, uncertainty has long been considered as something closely related with standard normal distributions in probability and predictions based on the distributions. New challenges and hurdles have been identified by experts on this topic and these challenges require new methods of quantifying, optimizing, and hence reducing the total uncertainty in models (1).

In this chapter, we review the current methods of quantification of uncertainty and methods of optimization thereof. The scope is limited to giving an account of the relevant methods of uncertainty quantification (QC) and optimization.

1.2 CHAPTER ORGANIZATION

We organize the chapter as follows. Subsection 1.3 of this section defines uncertainty that comes with machine learning models, and here we briefly describe the two main types of uncertainties and give an account of newer classifications. In Section 2, we mention the Uncertainty Quantification (UQ) and an account of the literature associated with it. Section 3 describes optimization methods and processes currently applied to the uncertainty models in order to efficiently mitigate the problems arising due to the total uncertainty of a model hampering its performance when applied to practical, real-world problems.

1.3 DEFINING UNCERTAINTY AND ITS TYPES

We can define uncertainty as something that is unknown or something lacking exactness. Therefore, it can be understood that there has to be a balance between uncertainty and required precision. Thus,

DOI: 10.1201/9781003353768-13

239

uncertainty is inversely proportional to precision. The complexity of the system or the problem affects precision or exactness in the information required to characterize it. This point was discussed by Booker and Ross (2011). They also gave an account of prominence of the role of Zadeh's fuzzy sets and logic, which has influenced research in this field since 1973 (2). Booker and Ross further mentioned uncertainty assessment and how it involves identification and classification, quantification, and characterization of uncertainties involved in the given problem.

1.3.1   Two Types of Uncertainty in Machine Learning

Traditionally, to handle uncertainty, probability theory was the most widely used tool. However, a probabilistic model is not capable of resolving two different sources of uncertainty as it only captures the knowledge in a probability distribution (1). Understanding the difference is important to reduce the uncertainty as the techniques applied to different forms of uncertainties are different. The two main types of uncertainty are aleatoric and epistemic uncertainty, AU and EU, respectively (4).

When there is variability in the outcome of a model due to internal randomness, the uncertainty is called aleatoric. Hence, aleatoric uncertainty is the statistical randomness within the experiment. Hüllermeier and Waegeman (2021) presented a primary example of AU, an event of coin flipping; here, the data-generating process is stochastic and is inherently irreducible. Even the best model can provide two possible outcomes, heads or tails, but does not provide a definite answer. This is caused by agent's ignorance and does not relate to inherent randomness. This type, however, can be reduced by adding information. Therefore, AU is the irreducible part of the total uncertainty, whereas EU is the reducible part. In some cases, relevance of differentiation between these two types is not needed; for example, when the model has no other option than to predict (e.g., lack of "reject" option), the source of the uncertainty might not be relevant. However, the ultimate prediction/decision can be refused with a rejection option (5, 6). This argument of whether distinction is necessary is covered by Hüllermeier and Waegeman (2021).

1.4   FOUR MAIN SOURCES OF UNCERTAINTY

In this sub-section, we will examine four main sources of uncertainty: noise, parameter uncertainty, uncertainty in model specification, and uncertainty due to extrapolation. This source classification is in reference to work presented by Jalaian *et al.* (2019).

(1) **Noise:** Inherent noise in the data generation process is extremely common in statistical models. Even though a model is trained well and the choices it makes seem correct, a lot of noisy data can preclude the accuracy of predictions. Quantification of uncertainty in such situations is important as the outcome, even though uncertain, may falsely attribute to high model confidence.

(2) **Parameter uncertainty:** Parameters shape the form of a model. For example, with linear regression, the parameters are slope and intercept whereas in NN it is weight matrices and bias terms. The goal of a machine learning model is to estimate the true values of these parameters to use them on unseen data. Considering a finite amount of input data, achieving true values of parameters is not possible and thus uncertainty is introduced.

(3) **Uncertainty in model specification:** Unlike the first two sources of uncertainty, which are statistically quantitative, model specification is not quantitative. The context of the model can be subjective and using such models can lead to inaccurate results with no exact knowledge of uncertainty (8). A solution may be to perform exploratory analysis and model assessment.

(4) **Uncertainty due to extrapolation:** Jalaian *et al.* (2019) explains this source of uncertainty as the uncertainty that occurs "when predictions are made on new data outside the range of the training data." In any model, if we try and extrapolate (i.e., make predictions for points

outside the observed data), the accuracy of the model decreases; this means the model is not confident on its predictions due to lack of enough data. A solution can be a process of flagging the points outside observed data as outliers.

1.5   Classification of Uncertainty Based on Input Domain

The inability of a deep neural network to recognize or extract input domain-based knowledge may give rise to uncertainties based on the input data domain. There are three main classes: in-domain uncertainty, out-of-domain uncertainty, and domain-shift uncertainty.

1.5.1 In-domain uncertainty: In this case, the test data distribution and training data distribution seem as though they are not different and hence the input taken at the test time creates in-domain uncertainty and therefore the deep neural network fails to explain the in-domain sample drawn as it lacks in-domain knowledge. It can be caused by both model uncertainty due to underlying network design or structure and data uncertainty due to complexity of the problem data. To tackle in-domain uncertainty, the training data has to be high quality and the process has to be rigorous (9, 10).

1.5.2 Out-of-domain uncertainty: In contrast to in-domain uncertainty, out-of-domain uncertainty arises from input data that is drawn from unseen data. The distribution of this unseen data is different and separate from the training or seen distribution and therefore it may lose correlation. The DNN fails to explain an out-of-domain sample because it lacks the out of domain knowledge. From the network point of view, without domain-shift, it seems that the training subspace and the prediction task are completely different (10, 11, 12).

1.5.3 Domain-shift uncertainty: Real-world events and the data representing them can create shifts in the training distribution; an input taken from such a distribution causes uncertainty to arise due to the variability of the original data. Therefore, the training data is not sufficiently covered by the network and thus it fails to explain this domain shift because it does not see the input sample during the training phase and hence has low confidence. From the modeler's point of view, not much can be done to reduce the uncertainty as it is almost impossible to capture all the domain-shift uncertainty realizations. A possible way to tackle this situation would be to involve the shift data in the training data set (10, 13, and 14).

2   UNCERTAINTY QUANTIFICATION (UQ)

2.1   Stochastic UQ in Deep Learning Models

Bayesian approximation and ensemble learning are the two most commonly used UQ methods in estimating uncertainty in DNNs. As mentioned in Section 1.3.1, there are two types of uncertainty: aleatoric (AU) (data uncertainty) and epistemic (EU) (model uncertainty). Data uncertainty can be further categorized as: homoscedastic and heteroscedastic (15). Hence, the total uncertainty or the predictive uncertainty (PU) is the sum of AU and EU.

$$PU = AU + EU \tag{1}$$

2.1.1   Monte Carlo (MC) Dropout Method

The Monte Carlo (MC) method can be effective in the approximation of posterior inference. But it is computationally expensive. As a solution, MC dropout was introduced. It uses dropouts in the computation of PU as regularization terms (16). It is effective against overfitting problems in NN and DNNs. It involves randomly dropping some units of the NN to prevent overfitting (co-tuning).

TABLE 13.1
Studies applying MC dropout methods reproduced and updated from a review by Abdar *et al.* (2021b)

Study	Year	Application
(Kendall et al., 2015)	2015	Semantic segmentation
(Leibig et al., 2017)	2017	Diabetic retinopathy
(Choi et al., 2017)	2017	Regression
(Jungo et al., 2018)	2018	Brain tumor segmentation
(Wickstrøm et al., 2020)	2018	Polyps segmentation
(Vandal et al., 2018)	2018	Predict flight delays
(DeVries and Taylor, 2018)	2018	Medical image segmentation
(Tousignant et al., 2019)	2019	MRI images
(Norouzi et al., 2019)	2019	MRI images segmentation
(Roy et al., 2019)	2019	Brain images (MRI) segmentation
(Filos et al., 2019)	2019	Diabetic retinopathy
(Harper and Southern, 2020)	2020	Emotion prediction
(Abdar et al., 2021a)	2021	Medical Image Classification
(Mirikharaji et al., 2021)	2021	Skin Lesion Segmentation
	2015	Semantic segmentation
(Leibig et al., 2017)	2017	Diabetic retinopathy
(Choi et al., 2017)	2017	Regression
(Jungo et al., 2018)	2018	Brain tumor segmentation
(Wickstrøm et al., 2020)	2018	Polyps segmentation
(Vandal et al., 2018)	2018	Predict flight delays
(DeVries and Taylor, 2018)	2018	Medical image segmentation
(Tousignant et al., 2019)	2019	MRI images
(Norouzi et al., 2019)	2019	MRI images segmentation
(Roy et al., 2019)	2019	Brain images (MRI) segmentation
(Filos et al., 2019)	2019	Diabetic retinopathy
(Harper and Southern, 2020)	2020	Emotion prediction
(Abdar et al., 2021a)	2021	Medical Image Classification
(Mirikharaji et al., 2021)	2021	Skin Lesion Segmentation

2.1.2 Markov Chain Monte Carlo (MCMC)

MCMC can be used to approximate inference. This method involves taking a value (d_0), which is randomly drawn from the distribution $q(d_0)$ or $q(d_0|x)$. Stochastic transition applied to d_0 will give the equation:

$$D_t \sim q(D_t|D_{t-1}, x) \tag{2}$$

A D_t is chosen and repeated T number of times, and the outcome (random variable) then converges in the distribution to the pinpointed posterior (17, 18).

2.1.3 Variational Inference (VI)

Variational inference (VI) is a method that approximates probability densities using optimization. It is faster than MCMC sampling 2.1.2. It involves postulating a family of densities and further finding the member of that family that is close to the target (19).

2.1.4 Bayesian Active Learning (BAL)

Active learning methods guide the data selection processes and include fewer data points while improving the model's training. The main task for AL methods is to correctly define the acquisition function (to find the condition under which a sample is most informative for the model). However, currently available methods lack scalability when dealing with high-dimensional data (17, 20).

2.1.5 Bayes By Backprop (BBB)

Blundell *et al.* (2015) introduced a novel and efficient algorithm for learning a probability distribution on the weights of a neural network, called Bayes by Backprop. This process is significant in obtaining better prediction outcomes. They showed that principled version of regularization provides comparable performance to dropout on the MNIST classification data. They also used the learnt uncertainty in weights to improve generalization in non-linear regression problems.

2.1.6 Variational Autoencoders (VAEs)

VAEs are yet another method to model the posterior. An autoencoder has two parts: (a) an encoder and (b) a decoder. The encoder maps high-dimensional samples to a low-dimensional latent variable v, and the decoder reproduces the original samples using v. A probabilistic model $P\theta (x)$ of sample x in a data space with a latent variable v in a latent space can be given as Abdar *et al.* (2021b):

$$p_0(x) = \overset{r}{\underset{v}{}} p_0(x|v)p(v) \tag{3}$$

2.1.7 Laplacian Approximations

Penny *et al.* (2007) describes LA as a method of approximating a posterior distribution with a Gaussian centred at maximum of posteriori (MAP). This is the application of Laplace's method with

$$f(\theta) = p(\lambda/\theta)p(\theta) \tag{4}$$

2.1.8 Ensemble Techniques

Bayesian NNs learn a distribution over weights and are predominantly used to estimate predictive uncertainty. But this comes at the cost of significant modifications to the training process and is generally computationally expensive. Lakshminarayanan *et al.* (2016) proposed an alternative method to the Bayesian NNs, which is simplistic and requires less hyperparmeter tuning yielding high-quality PU estimates.

Determination of quality of PU is challenging because actual uncertainty values cannot be estimated. Therefore, two evaluation measures are used in calibration of the uncertainty and shifting the distribution. Calibration measures the discrepancy between long-run frequencies and subjective forecasts. Domain-shift or distribution-shift is the change in distribution of data between the training phase and testing phase. In ensemble learning, the predictive performance is better as the ensembles combine models to discover more powerful models (17).

Deep Ensemble (DE)

To achieve better outcomes, there has to be little shift in the distribution in the testing datasets and training datasets. DEs provide better model uncertainty estimates when learners encounter OoD data. Jain *et al.* (2019) presented a elementary approach to improve the uncertainty estimates relating to ensembles. This was done by increasing the overall diversity in ensemble predictions across all inputs; this is called Maximize Overall Diversity (MOD). They tested and concluded that

MOD significantly improved the predictive performance for OoD data by applying it to numerous NN ensembles. They showed performance enhancement with no reported in-distribution performance on regression datasets among various other datasets.

Bayesian Deep Ensemble or Deep Ensemble Bayesian

Fersini *et al.* (2014 used an ensemble learning approach to solve the noise sensitivity issue related to language ambiguity that resulted in more accurate predictions of polarity. The proposed ensemble method employed Bayesian model averaging, where the reliability and uncertainty of each single model was considered. Pearce *et al.* (2018) suggested an alteration to the pre-existing approximate Bayesian inference method. They regularized the parameters related to values derived from a distribution that could be set equal to the prior. To obtain uncertainty estimates, one of the most promising frameworks is Deep BAL (DBAL) with MC dropout (17). Generative adversarial networks (GAN) failure, also called model collapse, generates overconfident predictions in DBAL methods. This was described by Pop and Fulop (2018). In order to solve the mode collapse issue and improve MC dropout on DNNs, Pop and Fulop (2018) created DE-BAL. They also presented a novel AL technique to test the expressive power of ensembles in enhancing the DBAL technique, which had reported GAN failures. In another study done by Pearce *et al.* (2018), they proposed a new ensemble of NNs, an approximate Bayesian ensembling approach to regularize the parameters regarding values drawn from a distribution.

2.2   Deterministic UQ Methods in Deep learning

In section 2.1, we covered Bayesian methods and esemble methods of uncertainty estimation. In this section, we cover the *deterministic* methods used for estimating uncertainty in DNNs.

2.2.1   Single Deterministic Methods

A deterministic neural network consists of parameters that are deterministic in nature and therefore generate same outcome at each iteration of a forward pass. The uncertainty on a prediction y' is calculated on a single forward pass within the network. There exist roughly two categories. One where a single network is specifically created, modeled, and trained to quantify uncertainty, where the uncertainty quantification influences the training phase and therefore the predictions of the network. Another method where additional components are used in order to quantify uncertainty, where it is usually applied in the testing phase; hence methods in this category do not affect the network's prediction (10). We will follow the notation given by Gawlikowski *et al.* (2021) in their review for these categories as internal and external uncertainty quantification methods, respectively.

Internal Uncertainty Quantification: These methods do not follow the direct pointwise maximum-a-posteriori estimation and instead predict the parameters of a distribution over the predictions. Gawlikowski *et al.* (2021) further discussed this point for classification tasks and contribution of softmax to the accuracy of uncertainty estimates. They also discuss behaviour of Dirichlet distribution over categorical distributions.

External Uncertainty Quantification: As mentioned previously, using additional components for estimating uncertainty does not affect the models' prediction as they are applied on trained models and hence are separate from the prediction process. Raghu *et al.* (2019) presented "direct uncertainty prediction" and argued that if the prediction and uncertainty estimation tasks are performed by using a single method, the uncertainty estimation is biased by the prediction task. They recommended training two neural networks, one solely for predicting the outcome and another for predicting the uncertainty based on the predictions of the first network.

Advantages of single deterministic methods over other similar methods: When it comes to training and evaluation efficiency, single deterministic methods are more computationally efficient than other similar techniques. This is because only one network is required to be trained. For evaluation, often this method can be applied on pre-trained networks. Depending on the actual approach, only a single forward pass is needed for evaluation. Single deterministic methods are more efficient in prediction numbers than all the other methods listed in this section (Bayes, ensemble, and test-time data augmentation).

Disadvantage of single deterministic neural network: Single deterministic neural network methods can become overly sensitive to the internal training procedure and the training data such that a possibility of overfitting might occur if the methods are used in unbalanced manner (10).

2.2.2   Test Time Data Augmentation

Ayhan and Berens proposed a "general purpose uncertainty estimation method that is independent of training procedures as well as the choice of pre trained architectures." It involves creating multiple test samples from each test sample by using traditional data augmentation techniques and then testing all those samples to obtain a predictive distribution to estimate uncertainty. In their application of this method to medical image processing and analysis, Ayhan and Berens used data augmentation methods like geometric and color transformations at test time. Data augmentation methods are commonly used in medical imaging along with deep learning, so in order to obtain uncertainty quantification, it is easy to apply same pre-used augmentation methods making this method most suited for UQ in medical image processing. One of the biggest advantage of test-time data augmentation is the fact that it does not rely on the underlying network structure or the training process, does not require additional data input, and can be easily used with other libraries. All of these aspects make test-time data augmentation a go-to method for medical image processing and analysis.

In sections 2.1.1 to 2.2.2, we provided a summary of uncertainty quantification methods used in deep learning without extensive theoretical and mathematical treatment. As this section is only a part of a review covering uncertainty quantification along with optimization, readers can find details of each method from references. Also, Abdar *et al*. (2021b) provided an extremely comprehensive review of uncertainty quantification in deep learning models where they have covered theoretical aspects in depth.

Gawlikowski *et al*. (2021) reviewed all the four types uncertainty quantification methods in depth along with classification of uncertainty, relation between uncertainty measures and quality (for classification and regression tasks), etc. The theoretical aspects are not repeated here as the review is recent. Readers are encouraged to find mathematical treatment and extensive theoretical accounts in that review.

3   UNCERTAINTY OPTIMIZATION

As much as it is important to quantify uncertainty, it is important to obtain reliable and accurate quantification of the uncertainty of from deep neural networks. To obtain such a state of quantification, "calibration" of uncertainty is required. A well-calibrated model is accurate when it is certain about the outcomes and shows high uncertainty when it is likely to be inaccurate (30.) Therefore, optimization under uncertainty is a process where optimization is done on models that involve uncertainties at data or model level. It is also called stochastic programming or a stochastic optimization problem (31).

Mathematical programming techniques are predominantly used for making decisions under uncertainty in multidisciplinary models and deep neural network setups in many applications. The three main modeling categories of optimization under uncertainty are: (1) stochastic programming, (2) chance-constrained programming, and (3) robust optimization.

3.1.1 Stochastic Programming

Stochastic programming is a mathematical programming paradigm for decision-making under uncertainty that optimizes the expected objective value across all possible types of uncertainty in the system. The modeling is done by generating probability distributions for randomness in uncertain parameters of the network (32,33).

There are roughly two types of stochastic programs employed in this optimization method based on various time stages. Single-stage stochastic programs have no recourse variables and decisions are made before knowing uncertain parameters. On the contrary, stochastic programming with recourse variables can make corrections once uncertainty is revealed. Ning and You (2019) and Birge (1997) show mathematical formulation of a two-stage stochastic programming problem, which results in first-stage objective value and expectation of second-stage objective value. However, two-stage stochastic programming can be computationally expensive and hence requires decomposition algorithms. Van Slyke and Wets (1969) describe a cutting hyperplane algorithm that is equivalent to a partial decomposition algorithm of the dual program, which is used for L-shaped linear programs that arise in stochastic programs.

3.1.2 Chance Constrained Optimization

It was first introduced by Charnes and Cooper (1959) and is an optimization under uncertainty paradigm that adds constraints with a specified probability in uncertain environments when it comes to optimization of an objective. P. Li described chance constraints that are flexible enough to quantify trade-off between reliability and performance. This topic is extensively covered by Ning and You (2019).

3.1.3 Robust Optimization

Stochastic programming and chance constrained optimization rely on probability distributions of uncertain parameters. Alternatively, robust optimization relies on an uncertainty set that contains all possible uncertainty realizations. This set, therefore, is a key component in a robust optimization program (38, 39). The robust optimization works against the worst-case uncertainty realization, which may be a realization giving rise to the largest constraint violation (34).

As the random optimization generates a policy for decision-making that is robust to the worst possible uncertainty in the uncertainty set, the set selection and generation is important such that it captures all the possible uncertainties involved in the system. Conventionally the sets were chosen by the user guided by a few assumptions, but Tulabandhula and Rudin (2014) proposed a new uncertainty set design to obtain higher guarantees on the robustness of the policy, which is dependent on the uncertainty set. The uncertainty sets must be derived from historical data that result in outcomes that are future proof or robust to the future situations.

Recently, Goerigk and Kurtz (2021) presented a data-driven robust optimization approach where they used unsupervised deep learning methods to discover hidden structures from data giving them non-convex uncertainty sets, which were highly robust. These sets were the supersets such that most of the classical uncertainty classes were special cases of their non-convex uncertainty sets.

Ning and You (2018) introduced a new framework for data-driven stochastic robust optimization (DDSRO) for optimization under uncertainty where they used labeled multi-class uncertainty data points. The optimization of the data-driven uncertainty model was performed using *bi-level optimization structure*. The outer optimization program is a two-stage stochastic program that optimizes the expected objective. The inner problem is an adaptive robust optimization program that guarantees robustness of the solution. They also developed a decomposition-based algorithm to reduce the computational cost that accompanies a multi-level optimization problem due to its low efficiency. Lastly, they applied the data-driven stochastic robust optimization to process network design and planning.

When it comes to data-driven programming paradigms, data-driven chance-constrained programming also emerges where the focus is on satisfaction of the worst-case chance constraint rather than optimization of the worst-case expected objective. Data-driven chance constrained program and data-driven robust optimization might seem similar superficially as both of them use ambiguity sets in their uncertainty models, but their models are completely different. Data-driven chance-constrained programs feature constraints that are subject to uncertainty in a probability distribution. In the case of data-driven robust optimization, the focus is only on optimization of the worst-case expectation of an objective function by adopting a family of probability distributions (34).

Krishnan and Tickoo (2020) at Intel Labs introduced a loss function "accuracy versus uncertainty calibration (AvUC)" with two optimization methods "AvU temperature scaling (AvUTS)" and AvUC to improve uncertainty calibration in deep neural networks. Recently, Sinha *et al.* (2022) used a rough set as a tool for testing and analysis of uncertainty; a rough set can be used to discover structural relationships in the data even if it is imprecise or noisy. A rough set reducts formation technique removes uncertainty in the feature set (for feature subset selection). They introduced uncertainty optimization-based reducts (UOR) by formulating an algorithm based on uncertainty optimization to obtain the mentioned reducts of the feature set. The algorithm performed feature selection effectively. They reported an average accuracy of the reducts found by their UOR algorithm to be 96.66%. The comparison results showed that the UOR method finds feature subsets of minimum sizes with similar classification accuracy compared to existing reduct methods.

Meng *et al.* (2022) presented a reliability-based multidisciplinary design optimization. Multidisciplinary optimization is an approach in which optimization of all interacting disciplines is taken into consideration. There exist inherently uncertain parameters when a system is designed that later affect the system's performance. They assumed that some of the problem (uncertain) parameters are in the form of fuzzy numbers. To suit their application, they also considered problem cost as one of the design disciplines as it is important in engineering. Therefore, the uncertainty associated with this parameter will also be optimized. They used sequential optimization and reliability assessment methods as solutions and used PSO and a genetic algorithm to solve the deterministic problem in each iteration. They also designed an autonomous underwater vehicle as a test subject, which included the problem cost analysis and applied to two solutions that they presented. It was determined that if the improvement in reliability of the system is in the range of 0.5 to 0.85, the result is more cost-effective.

4 CONCLUSION

Mitigation of some (if not all) uncertainties in machine learning is important, because uncertainties have the ability to directly affect decision-making. In this chapter, we first defined uncertainty and how it arises and the classification of uncertainty based on input data domain. We then discussed two main types of uncertainties that are common to all the machine learning approaches. To reiterate, AU is the irreducible uncertainty and arises due to inherent randomness in the model data distribution and another is EU, which occurs due to inadequate knowledge, and can therefore be reduced and is the reducible part in Eq 2.1.

Next, we briefly discussed the stochastic and deterministic methods of uncertainty estimation along with ensemble and deep ensemble techniques. These methods have predominantly been applied to deep neural networks and neural networks in general. This was also observed in the "optimization under uncertainty" section. We hence noticed a trend in the literature that most of the uncertainty quantification and optimization paradigms are created and used by keeping deep learning and neural networks in mind. In this section we described three main mathematical programming paradigms and how they are applied for optimization problems. Robust optimization was described in detail by looking at the trends in the literature. Finally, it boils down to usage of these techniques to reduce the effective uncertainty. As machine learning or specifically deep neural networks, convoluted neural networks, and many more are already being used in the industries such as medical imaging, medical image segmentation problems, physics, planning, system design, diabetic retinopathy, tasks and other diseases to solve and aid major problems the reliability and performance confidence become primary factors on which the models are judged. Uncertainty in these models poses a safety concern and limits the usage of these models. No method can reduce the uncertainty to none but there are efforts that can minimize the uncertainty.

REFERENCES

1. Hu¨llermeier, E. and Waegeman, W. (2021). Aleatoric and epis–temic uncertainty in machine learning: an introduction to concepts and methods. *Mach Learn*, 110(3):457–506.
2. Booker, J. M. and Ross, T. J. (2011). An evolution of uncertainty assessment and quantification. *Scientia Iranica*, 18(3):669–676.
3. Zadeh, L. A. (1973). Outline of a new approach to the analysis of complex systems and decision processes. *IEEE Transactions on Systems, Man, and Cybernetics*, SMC- 3(1):28–44.
4. Hora, S. C. (1996). Aleatory and epistemic uncertainty in probability elicitation with an example from hazardous waste management. *Reliability Engineering & System Safety*, 54(2):217–223.
5. Hendrickx, K., Perini, L., Plas, D. V. d., Meert, W., and Davis, J. (2021). Machine Learning with a Reject Option: A survey. *ArXiv*.
6. Choi, S., Lee, K., Lim, S., and Oh, S. (2017). Uncertainty-aware learning from demonstration using mixture density networks with sampling-free variance modeling.
7. Jalaian, B., Lee, M., and Russell, S. (2019). Uncertain Context: Uncertainty Quantification in Machine Learning. *AI Magazine*, 40(4):40–49. Number: 4.
8. Papernot, N., McDaniel, P., Goodfellow, I., Jha, S., Celik, Z. B., and Swami, A. (2017). Practical Black-Box Attacks against Machine Learning. arXiv:1602.02697 (cs).
9. Ashukha, A., Lyzhov, A., Molchanov, D., and Vetrov, D. (2020). Pitfalls of in-domain uncertainty estimation and ensembling in deep learning.
10. Gawlikowski, J., Tassi, C. R. N., Ali, M., Lee, J., Humt, M., Feng, J., Kruspe, A. M., Triebel, R., Jung, P., Roscher, R., Shahzad, M., Yang, W., Bamler, R., and Zhu, X. (2021). A Survey of Uncertainty in Deep Neural Networks. *ArXiv*, abs/2107.03342.
11. Liang, S., Li, Y., and Srikant, R. (2017). Enhancing the reliability of out-of–distribution image detection in neural networks.
12. Shafaei, A., Schmidt, M., and Little, J. J. (2018). A less biased evaluation of out-of-distribution sample detectors.
13. Ovadia, Y., Fertig, E., Ren, J., Nado, Z., Sculley, D., Nowozin, S., Dillon, J. V., Lakshminarayanan, B., and Snoek, J. (2019). Can you trust your model's uncertainty? evaluating predictive uncertainty under dataset shift.
14. Kendall, A. and Gal, Y. (2017a). What uncertainties do we need in bayesian deep learning for computer vision?
15. Kendall, A. and Gal, Y. (2017b). What Uncertainties Do We Need in Bayesian Deep Learning for Computer Vision? arXiv:1703.04977 (cs).
16. Gal, Y. and Ghahramani, Z. (2016). Dropout as a bayesian approxima–tion: Representing model uncertainty in deep learning. In Balcan, M. F. and Weinberger, K. Q., editors, *Proceedings of The 33rd International Conference on Machine Learning*, volume 48 of *Pro–ceedings of Machine Learning Research*, pages 1050–1059, New York, New York, USA. PMLR.
17. Abdar, M., Pourpanah, F., Hussain, S., Rezazadegan, D., Liu, L., Ghavamzadeh, M., Fieguth, P., Cao, X., Khosravi, A., Acharya, U. R., Makarenkov, V., and Nahavandi, S. (2021b). A review of uncertainty

quantification in deep learning: Techniques, applications and challenges. *Information Fusion*, 76: 243–297.

18. Kupinski, M. A., Hoppin, J. W., Clarkson, E., and Barrett, H. H. (2003). Ideal–observer computation in medical imaging with use of Markov-chain Monte Carlo techniques. *J Opt Soc Am A Opt Image Sci Vis*, 20(3):430–438.

19. Blei, D. M., Kucukelbir, A., and McAuliffe, J. D. (2017). Variational inference: A review for statisticians. *Journal of the American Statistical Association*, 112(518):859–877.

20. Tong, S. (2001). Active learning: theory and applications.

21. Blundell, C., Cornebise, J., Kavukcuoglu, K., and Wierstra, D. (2015). Weight uncertainty in neural networks.

22. Penny, W., Kiebel, S., and Friston, K. (2007). CHAPTER 24–Variational Bayes. In Friston, K., Ashburner, J., Kiebel, S., Nichols, T., and Penny, W., editors, *Statistical Parametric Mapping*, pages 303–312. Academic Press, London.

23. Lakshminarayanan, B., Pritzel, A., and Blundell, C. (2016). Simple and scalable predictive uncertainty estimation using deep ensembles.

24. Jain, S., Liu, G., Mueller, J., and Gifford, D. (2019). Maximizing overall diversity for improved uncertainty estimates in deep ensembles.

25. Fersini, E., Messina, E., and Pozzi, F. A. (2014). Sentiment analysis: Bayesian Ensemble Learning. *Decision Support Systems*, 68 :26–38.

26. Pearce, T., Leibfried, F., Brintrup, A., Zaki, M., and Neely, A. (2018). Uncertainty in neural networks: Approximately bayesian ensembling.

27. Pop, R. and Fulop, P. (2018). Deep ensemble bayesian active learning: Ad–dressing the mode collapse issue in monte carlo dropout via ensembles.

28. Raghu, M., Blumer, K., Sayres, R., Obermeyer, Z., Kleinberg, B., Mullainathan, S., and Kleinberg, J. (2019). Direct uncertainty prediction for medical second opinions. In Chaudhuri, K. and Salakhutdinov, R., editors, *Proceedings of the 36th International Conference on Machine Learning*, volume 97 of *Proceedings of Machine Learning Research*, pages 5281–5290. PMLR.

29. Ayhan, M. S. and Berens, P. Test-time Data Augmentation for Estimation of Heteroscedastic Aleatoric Uncertainty in Deep Neural Networks. page 9.

30. Krishnan, R. and Tickoo, O. (2020). Improving model calibration with accuracy versus uncertainty optimization. In *Advances in Neural Information Processing Systems*, volume 33, pages 18237–18248. Curran Associates, Inc.

31. Diwekar, U. M. (2003). Optimization Under Uncertainty. In *Introduction to Applied Optimization*, pages 145–208. Springer US, Boston, MA.

32. Birge, J. R. (1997). State-of-the-Art-Survey—Stochastic Programming: Computation and Applications. *INFORMS Journal on Computing*, 9(2):111–133. Publisher: INFORMS.

33. Marti K., K. P. *Stochastic Programming*.

34. Ning, C. and You, F. (2019). Optimization under uncertainty in the era of big data and deep learning: When machine learning meets mathematical programming. *Computers & Chemical Engineering*, 125:434–448.

35. Van Slyke, R. M. and Wets, R. (1969). L-Shaped Linear Programs with Applications to Optimal Control and Stochastic Programming. *SIAM J. Appl. Math.*, 17(4):638–663. Publisher: Society for Industrial and Applied Mathematics.

36. Charnes, A. and Cooper, W. W. (1959). Chance-Constrained Program–ming. *Management Science*, 6(1):73–79. Publisher: INFORMS.

37. P. Li, H. Arellano-Garcia, G. W. Chance constrained programming approach to process optimization under uncertainty.–University of Surrey.

38. A. Ben-Tal, L. E. Ghaoui, A. N. (2009). *Robust Optimization*.

39. Bertsimas, D. and Sim, M. (2004). The Price of Robustness. *Operations Research*, 52(1):35–53. Publisher: INFORMS.

40. Tulabandhula, T. and Rudin, C. (2014). Robust optimization using machine learning for uncertainty sets.

41. Goerigk, M. and Kurtz, J. (2021). Data-driven robust optimization using unsupervised deep learning.

42. Ning, C. and You, F. (2018). Data-driven stochastic robust optimization: Gen–eral computational framework and algorithm leveraging machine learning for optimization under uncertainty in the big data era. *Computers & Chemical Engineering*, 111:115–133.

43. Sinha, A. K., Shende, P., and Namdev, N. (2022). Uncertainty optimization based feature subset selection model using rough set and uncertainty theory. *Int. j. inf. tecnol.*, 14(5):2723–2739.

44. Meng, D., Yang, S., He, C., Wang, H., Lv, Z., Guo, Y., and Nie, P. (2022). Multi-disciplinary design optimization of engineering systems under uncertainty: a review. *International Journal of Structural Integrity*.

Index

For Product Safety Concerns and Information please contact our EU
representative GPSR@taylorandfrancis.com
Taylor & Francis Verlag GmbH, Kaufingerstraße 24, 80331 München, Germany